Management in the Built Environment

The aim of this book series is to provide a platform to build and consolidate a rigorous and significant repository of academic, practice and research publications that contribute to further knowledge relating to management in the built environment. Its objectives are to:

1. Disseminate new and contemporary knowledge relating to research and practice in the built environment
2. Promote synergy across different research and practice domains in the built environment and
3. Advance cutting-edge research and best practice in the built environment

The scope of this book series is not limited to "management" issues per se because this then begs the question of what exactly are we managing in the built environment. While the primary focus is on management issues in the building and construction industry, its scope has been extended upstream to the design management phase and downstream to the post-occupancy facilities management phase. Management in the built environment also involves other closely allied disciplines in the areas of economics, environment, legal and technology. Hence, the starting point of this book series lies with project management, extends into construction and ends with facilities management. In between this spectrum, there are also other management-related issues that are allied with or relevant to the built environment. These can include, for example cost management, disaster management, contract management and management of technology.

This book series serves to engage and encourage the generation of new knowledge in these areas and to offer a publishing platform within which different strands of management in the built environment can be positioned to promote synergistic collaboration at their interfaces. This book series also provides a platform for other authors to benchmark their thoughts to identify innovative ideas that they can further build on to further advance cutting-edge research and best practice in the built environment.

If you are interested in submitting a proposal for this series, please kindly contact the Series Editor or the Publishing Editor at Springer:

Low Sui Pheng (bdglowsp@nus.edu.sg) or
Ramesh Premnath (Ramesh.premnath@springer.com)

AbdulLateef Olanrewaju

Sustainable Construction Management: Research and Practice Companion

Integrating Sustainability Target in Construction

Springer

AbdulLateef Olanrewaju
Department of Construction Management
Universiti Tunku Abdul Rahman
Kampar, Malaysia

ISSN 2522-0047 ISSN 2522-0055 (electronic)
Management in the Built Environment
ISBN 978-981-96-4909-9 ISBN 978-981-96-4910-5 (eBook)
https://doi.org/10.1007/978-981-96-4910-5

This Springer imprint is published by the registered company Springer Nature Singapore Pte Ltd.
The registered company address is: 152 Beach Road, #21-01/04 Gateway East, Singapore 189721, Singapore

I dedicate this book to the cherished memory of my late father, a remarkable individual whose fervent passion for education and learning remained steadfast despite the constraints of limited formal schooling. His unwavering commitment to knowledge and his relentless pursuit of self-improvement have left an indelible mark on my life and continue to inspire me profoundly. May Allah (SWT) grant him forgiveness and bestow upon him the eternal blessings of Aljannah Firdaus for his noble efforts and the significant impact he had on my journey.

Additionally, I extend this dedication to the many writers and scholars who are deeply concerned about the challenges posed by current construction and construction management practices. Their dedication to exploring alternative approaches and solutions in construction management reflects a collective commitment to addressing and overcoming the pressing issues within the industry. Their contributions are vital in advancing the field and fostering positive change.

Preface

The origin of this book can be concerned with the performance of construction companies. While working as a quantity surveyor with a contractor, when I embark on a site visit to prepare an interim valuation, the concern about waste and pollution creation is a source of concern. How construction companies also treat their site workers creates tension in me. But at that time, I had fewer options, and more crucially, I did not know anything about sustainable development. That was about 25 years ago, though. It can be argued, though, that many people are not aware of and concerned about the menace of climate change. I became more aware of and concerned about the menace of construction activities as a lecturer. I became familiar with the concept of sustainable development during my postgraduate studies about 20 years ago. The awareness created an instinct in me, which led to my interest in and desire for sustainable construction. During these periods, I researched various aspects of construction sustainability. I edited a book on greening affordable housing and developed a business for sustainable maintenance practices, among other research. This interest also underpinned my participation in the development of the M.Sc. Sustainable Construction Management. The programme that I developed during its development stage led to its eventual accreditation. It is a unique programme in its own right. The students have enrolled in the programme from different backgrounds, including civil engineering, construction management, mechanical engineering, architecture, and project management. Not all of them work in the construction sector. Many are overseas students. This comprised mainly the collection of recent research in this scope, together with my lesson notes, especially for the mentioned postgraduate. Parts of my lesson notes for the undergraduate programme also help me shape part of the contents of this book. Their feedback reinforces the need to have a curriculum in academics that prepares students for the roles and responsibilities involved. I am indeed grateful for feedback from the students during the lectures and tutorials.

So, the idea of sustainable construction management as a focus for a book emerged due to my involvement in teaching, of course, not without the impact of my industrial experience. It does take me a long time to realise the vacuum created at the back of a dedicated book for sustainable construction management, although some articles were published on related topics like sustainable construction, sustainable project management, and green buildings. You will come across most of the related topics in this book. However, many questions are yet to be answered. There are many books on sustainable construction. Sustainable construction is different from sustainable construction management by all means. Even though this book calls for a dedicated professional sustainable construction manager, the roles and functions of construction managers are currently performed by experienced civil engineers, architects, and quantity surveyors. This is argued to be inadequate considering the scope of their competencies. Their competencies are insufficient, fragmented, and biased toward their standard and conventional functions and responsibilities.

However, this book is not about how to design, construct, and operate construction projects. Rather, it is about how to coordinate all the project stakeholders from the strategic stage to the operation and deconstruction stages of the projects, including the design team, construction, facilities management, deconstruction (demolition), contractors, and clients. The sustainable construction managers are moderators between all the parties to the contracts. The competencies that sustainable construction managers bring to the contract determine the performance of the construction projects. A sample specification for a sustainable construction project for USD 1 billion based on the 2018 estimate is attached as an appendix for reference. This book comprises theories, principles, and practical insight into the paradigm of sustainable construction management. Three chapters are based on data from surveys on various aspects of sustainable construction management.

This document is designed to function as a handbook, providing a centralised source of information on specific sustainable construction management strategies, practices, and principles. It compiles pertinent information on sustainable construction management into a single, easily accessible document together with guidelines, procedures, and best practices. Without having to examine several sources, users may easily access the information they require for sustainable construction management. Blatantly seeks to standardise procedures, enable onboarding and training, guarantee compliance, offer reference and troubleshooting assistance, encourage openness and communication, and promote ongoing process and organisational improvement.

During the preparation of this book, AI-assisted technologies were employed to enhance the language, readability, and overall quality of various sections. These technologies included advanced tools for grammar checking, style improvement, and content refinement. The utilization of these tools aimed to ensure clarity, coherence, and a professional standard throughout the manuscript. After the initial improvements made by the AI tools, the author reviewed various sections to ensure that the suggestions were appropriate and aligned with the intended meaning, essence, and context. This review process included verifying the accuracy of the content, maintaining the author's unique voice and style, and making any necessary adjustments to ensure consistency and correctness. In addition, the author edited the AI-enhanced content to refine and personalize it, ensuring that the final version met the highest standards of quality and integrity.

Kampar, Malaysia AbdulLateef Olanrewaju

Acknowledgements

Alhamdullilah. I am very thankful to Rabul l'Allamin (ALLAH (SWT)) for giving me the opportunity and strength to start and complete this project without hindrance. I would particularly like to thank my wife Kafayah and my dear children; Harun, Haneef, and Fatimah, for their relentless patience and understanding during the many long hours that I sat seemingly fascinated in front of my laptop and desktop.

I am deeply grateful to all the respondents who generously provided their time and resources to complete the survey questionnaires that contributed to the data in this book. I also extend my sincere thanks to the anonymous reviewers whose valuable suggestions significantly improved the manuscript. A special acknowledgment goes to Mr. Hasitha Herath of Pivotal International, Dublin, Ireland, for providing some of the essential materials I utilized during the preparation of this book. Within the Faculty of Engineering and Green Technology, I am especially appreciative of the support and encouragement from Associate Professor Zafarullah Nizamani, Ts. Dr. Shalini Sanmargaraja, and Prof. Choon Aun Ng. As a lecturer and mentor, I have been fortunate to engage with many students whose questions and ideas have sparked challenges and reflections in my thinking, greatly influencing my writing. Their curiosity and insights have been invaluable to this work.

Introduction

Objectives of This Book

This book is a collection of research that I conducted, especially in the last 5 years, in the fields of digital construction, quantity surveying, construction management, construction health and safety, and sustainable construction management. Largely, the studies are triggered by the need to develop a master's programme in sustainable construction management, which is an extension of our undergraduate programme, B.Sc. (honours) in construction management at the UTAR. The aim of this book, as will be explained later, is to present sustainable construction management strategies, practices, techniques, methods, and procedures. The contents of this book are determined by the need for sustainable construction management for the construction sector (also referred to as the architecture, engineering construction sector) to contribute to sustainable development goals. Conventional construction management has failed to address the various problems facing the construction industry. Construction projects are completed behind schedules, well over budget, and with high defects. Sustainable construction management is conceived to address these deficiencies and include sustainability dimensions. These problems were translated into objectives, questions, and hypotheses. These are then translated into sustainable construction management objectives. The specific objectives, questions, and hypotheses were further developed into key processes and procedures. Therefore, the ultimate aim of the book is to provide construction companies with an informative operational and procedural understanding of construction functions, an understanding that will enable them to seek problems and develop sustainable construction solutions. The term 'construction company' in this book implies a unit that undertakes construction actions and supports one or two more construction teams. The construction company provides support for their construction teams in various ways, including providing employment, financial security, legal advice, plant and equipment, technical advice, and training. Construction companies are complex and diverse. For example, professional quantity surveying firms, professional engineering firms, professional construction firms, and architectural engineering firms are examples of construction companies. It also includes contracting organisations, subcontracting firms, specialist subcontractors, and trade and work subcontractors. Collectively, construction companies make up the construction sector, or the architectural, engineering, and construction sector, which constitutes about 10% of GDP or GNP in most countries. The book contains practical insights and trends drawing from empirical conclusions that are derived from data collected from experts and practitioners in the construction sector. Furthermore, most of the data is collected from many countries, including Singapore, China, Hong Kong, the UK, India, the USA, Australia, Nigeria, Malaysia, the United Arab Emirates, etc.

General Organisation of This Book

The book contains a total of 14 chapters. Each of the chapters contains a specific aspect of the book developed based on specific objectives, questions, and hypotheses. Each chapter begins with an introduction and ends with a summary. Each has its own unique learning objectives and keywords. The book considered the diversities involved in sustainable construction management. The book's general organisation was created to ensure that most construction companies across all cadres might benefit from the material. Therefore, it is intended to be used by academics, construction managers, clients, project managers, directors, engineers, executives, and other professionals throughout the construction process. Members of the construction team should thoroughly familiarise themselves with the contents of the book. Oversized materials (e.g., maps, figures, tables, and charts) are presented by sectioning the original, beginning at the upper side and continuing from the downside. However, in some cases, the original materials are shrunk to fit in while the eligibility of the content is visible.

Overview of the Chapters

In Chap. 1, a comprehensive overview of the construction sector was provided, setting the stage for a deeper exploration of its evolution and current challenges. The chapter introduced the concepts of sustainable construction and sustainable construction management, highlighting their growing importance in the context of global environmental and social imperatives. A thorough discussion was presented on the multiple justifications for why all stakeholders, from policymakers to industry professionals, should take responsibility for achieving the Sustainable Development Goals (SDGs). In addition, the chapter critically examined the shortcomings of current construction management practices and strategies, identifying key areas where improvement is necessary to meet sustainability criteria. This analysis underscored the urgent need for a paradigm shift in the industry. Finally, the chapter outlined the structure of the book, providing readers with a roadmap of the subsequent chapters and the themes they will cover, ensuring a clear understanding of how the book will address the complexities of sustainable construction management. Chapter 2 explores into the characteristics, size, and overall impact of the construction sector, which, in this book, is considered synonymous with the architecture, engineering, and construction (AEC) sectors. The chapter provides a thorough analysis of the size and characteristics of the construction sector, emphasizing its significant contributions to economic growth and development. These contributions are examined not only in terms of financial impact but also through the sector's influence on employment, infrastructure development, and societal well-being. In addition to this, the chapter offers an in-depth exploration of the duties, rights, and responsibilities of the various players within the sector, shedding light on the complex interplay between different stakeholders, including clients, contractors, architects, engineers, and regulatory bodies. The maturity of clients, particularly in understanding and managing construction processes, is discussed in detail, along with an examination of the unique nature of the construction sector, including its cyclical nature, project-based structure, and the problems it faces. Further, the chapter provides a concise overview of construction contract documents, outlining their importance in defining the scope, responsibilities, and legal obligations of parties involved in a project. Building on the foundation laid in Chap. 1, aspects of sustainable construction are revisited, with additional insights into how sustainability principles are being integrated into the construction sector and the challenges associated with their implementation.

Chapter 3 establishes the foundation for understanding sustainable construction. The chapter begins with a thorough discussion of the principles of sustainability and how they apply specifically to the construction industry. The definition of sustainable construction is explored, along with its key components, which are meticulously unpacked to provide a clear understanding of what constitutes sustainability in this context. A significant portion of the chapter is dedicated to sustainable construction materials and components. Each material is analyzed in terms of its areas of application, with a balanced discussion of its advantages and disadvantages. This analysis helps to illustrate how these materials contribute to the broader goals of sustainability in construction projects. The chapter goes on to describe in detail how construction companies are actively achieving sustainability in their projects. This includes strategies, best practices, and innovative approaches that companies are employing to minimize environmental impact, promote resource efficiency, and ensure long-term economic and social benefits.

To provide a comprehensive perspective, the chapter also offers contextual insights into the historical evolution of sustainability concepts, tracing how the ideas and practices surrounding sustainability have developed over time. For clarity and ease of understanding, sustainability or sustainable development is framed as comprising three fundamental pillars, often referred to as tripods: environmental, economic, and social sustainability. The next three chapters are devoted to an in-depth exploration of each of these pillars, setting the stage for a detailed examination of how they interact and contribute to the overall sustainability of construction projects.

Chapter 4 focuses on the environmental pillar of sustainability within the construction sector. As global awareness of climate change, resource depletion, and environmental degradation continues to grow, the construction industry faces increasing pressure to address the ecological impacts of its practices, strategies, and techniques. This chapter delves into the significant environmental footprint of the construction sector, highlighting the challenges posed by current practices and the urgent need for more sustainable approaches. The chapter begins by detailing the various ways in which the construction sector contributes to environmental issues, such as greenhouse gas emissions, energy consumption, waste generation, and the depletion of natural resources. By examining these impacts, the chapter underscores the critical role that the construction industry must play in mitigating environmental harm and advancing global sustainability efforts. Building on this foundation, the chapter then explores a wide range of methods, approaches, techniques, practices, and strategies that the construction sector can—and must—adopt to make meaningful contributions to the Sustainable Development Goals (SDGs). This includes the adoption of green building standards, the use of renewable and low-impact materials, energy-efficient design, waste reduction strategies, and innovative construction technologies that minimize environmental impact. Throughout the chapter, practical examples and case studies are provided to illustrate how these approaches can be implemented in real-world projects, demonstrating the potential for the construction sector to lead the way in environmental stewardship. The chapter concludes with a call to action, urging construction professionals to embrace these sustainable practices as essential components of their work, thereby ensuring that the industry not only adapts to the demands of a changing world but also actively contributes to the preservation and restoration of our planet.

In Chap. 5, the focus shifts to the economic dimension of sustainable construction practices. This chapter critically examines the alternative methods, techniques, practices, and procedures that can be adopted to deliver construction projects efficiently while maintaining economic viability. The discussion emphasizes that truly sustainable projects must balance environmental and social considerations with economic value, ensuring that these projects remain financially sound without compromising their sustainability goals. The chapter addresses the key economic questions that frequently arise in the context of sustainable construction. For example, what are the costs associated with avoiding construction pollution and waste? Is it economically viable or prudent to implement measures that reduce or eliminate environmental degradation caused by construction activities? These questions are explored in depth, with an emphasis on finding solutions that do not pit sustainability against profitability but rather integrate them in a way that benefits both the environment and the bottom line. By examining various cost-effective strategies, such as resource-efficient design, waste minimization, and the use of sustainable materials, the chapter provides insights into how the construction industry can achieve economic sustainability without sacrificing environmental or social responsibilities. Case studies and examples of best practices are included to illustrate how these approaches have been successfully implemented in real-world projects, offering valuable lessons for practitioners seeking to enhance the economic sustainability of their construction activities.

Ultimately, the chapter advocates for a holistic approach to construction economics, one that recognizes the interconnectedness of environmental, social, and economic factors. By doing so, it demonstrates that sustainable construction is not just about reducing costs or maximizing profits, but about creating long-term value that benefits all stakeholders, including the environment, society, and the economy. Chapter 5 also addresses critical questions such as whether maintenance costs should be a priority in sustainable construction. Detailed answers and considerations are provided, offering insights into the economic implications of maintaining sustainability throughout the life cycle of a construction project. Moving forward, the final pillar of sustainable development—social sustainability—is explored in Chap. 6. This chapter is dedicated to explaining the social aspects of sustainable construction, with discussions primarily based on the findings from my research. Social sustainability in the construction sector involves integrating social factors and assessing the impacts of construction projects and practices on communities, stakeholders, and the workforce. The chapter explores into the social value of construction projects, examining how these projects can contribute positively to society by enhancing quality of life, promoting equity, and supporting community development. It also provides guidance on making informed decisions that prioritize the well-being of the construction workforce, emphasizing the importance of safe working conditions, fair wages, and inclusive practices. Through this chapter, readers will gain a deeper understanding of how social considerations are not only essential for achieving sustainability but also for ensuring that construction projects contribute to broader societal goals. Chapter 7 focuses on the critical concept of net-zero carbon in the built environment. The chapter begins with an in-depth investigation of data related to carbon emissions in the construction sector, highlighting the significant impact that building activities have on global carbon output. As part of this analysis, the chapter examines the growing international commitment to reducing carbon emissions, specifically through initiatives like 'The Buildings Breakthrough.' This global initiative aims to achieve near-zero emission and resilient buildings by 2030, with many countries already signed on to this ambitious goal. The chapter then explores various strategies and measures that can be implemented to achieve near-zero emissions in buildings. These include the adoption of energy-efficient technologies, the use of renewable energy sources, and the implementation of advanced design and construction practices that minimize carbon footprints. The discussion is supplemented with case studies and examples of best practices that demonstrate how these measures are being applied in real-world projects. Additionally, the chapter addresses the financial aspects of sustainable construction, particularly focusing on the up-front capital costs associated with achieving net-zero emissions. It provides insights into how these initial investments can be justified in terms of long-term savings, reduced operational costs, and enhanced building performance, ultimately making a strong case for the economic viability of sustainable construction practices. By the end of this chapter, readers will have a comprehensive understanding of the challenges and opportunities associated with reducing carbon emissions in the built environment, as well as practical guidance on how to implement net-zero strategies in their own projects.

Chapter 8 weaves together the insights and concepts discussed in the previous chapters to establish a comprehensive foundation for sustainable construction management. This chapter theorizes the principles of sustainable construction management, offering a conceptual framework that integrates environmental, economic, and social sustainability into the management of construction projects. It elaborates on how sustainability can be embedded in every phase of a project's life cycle—from planning and design to construction, operation, and eventual decommissioning. The measures proposed are not only explained but also justified, demonstrating their necessity and the benefits they bring to sustainable development goals. Building on this theoretical foundation, Chap. 9 transitions to the action-oriented phase of the book. It presents the results of an international survey that was conducted to identify and analyze the critical success factors (CSFs) for the effective implementation of sustainable construction management. The findings offer practical insights and recommendations that can guide industry professionals in adopting and executing sustainable practices. This chapter serves as a bridge between theory and practice, providing actionable knowledge that can be applied in real-world construction projects to achieve sustainability objectives. Chapter 10 continues the action-oriented focus of the book, shifting from the broader analytical results to a more detailed exploration of specific findings. This chapter is dedicated to presenting the results of the Principal Component Analysis (PCA) conducted on the critical success factors for sustainable construction management. Rather than offering general analyses, this chapter delves deeply into how the PCA identified key components and underlying patterns among the factors that are most crucial for successfully implementing sustainable practices in construction projects. By breaking down these components, the chapter provides a nuanced understanding of the elements that drive sustainability in construction management, offering readers valuable insights that can be directly applied to enhance the effectiveness of sustainable construction initiatives. Chapter 11 delves into the applications, constraints, and benefits of digital technology within the built environment. The chapter emphasizes that the success of sustainable construction initiatives heavily relies on the integration of advanced digital technologies. Tools such as Building Information Modeling (BIM), digital twins, robotics, virtual reality (VR), augmented reality (AR), and artificial intelligence (AI) are highlighted as essential enablers of sustainable practices. The chapter explores how these technologies are being applied to enhance efficiency, reduce waste, and optimize resource use in construction projects. It also examines the constraints that may hinder the widespread adoption of these technologies, including technical challenges, cost considerations, and the need for specialized skills and training. Despite these challenges, the chapter outlines the significant benefits that digital technologies bring to sustainable construction, such as improved project management, enhanced collaboration, and the ability to model and predict environmental impacts accurately. Eventually, the chapter underscores that achieving the goals of sustainable construction is increasingly dependent on the strategic deployment of digital technologies, making their adoption not just advantageous but essential for the future of the industry. Chapter 12 presents a comprehensive overview of global perspectives on sustainable construction. This chapter explores how different regions and countries are approaching sustainability in the built environment, highlighting various strategies, policies, and practices that are shaping the global landscape. It provides insights into how diverse cultural, economic, and regulatory contexts influence sustainable construction efforts worldwide. Additionally, the chapter discusses the essential collaborations required to drive both demand and supply for sustainable construction solutions. It examines how partnerships between governments, industry stakeholders, researchers, and community organizations can enhance the effectiveness of sustainability initiatives. The discussion includes examples of successful collaborations and initiatives that have made significant impacts, illustrating the importance of a coordinated effort in advancing global sustainability goals. By presenting these global perspectives and collaborative strategies, the chapter aims to provide a holistic understanding of how sustainable construction can be promoted and achieved on an international scale. Chapter 13 offers a broad overview of global perspectives on sustainable construction, with a focus on aspects that extend beyond regional building industry impacts. This chapter delves into the complex interplay between economic, social, and environmental challenges that shape sustainable construction practices worldwide. It highlights how these global factors influence the development and implementation of sustainability initiatives.

The chapter emphasizes the importance of considering the broader global context when applying sustainability strategies, acknowledging how global trends, international regulations, and stakeholder dynamics play a crucial role in shaping construction practices. By examining these interactions, the chapter underscores the need for a holistic approach to sustainability that integrates local practices with global standards and insights. Through this comprehensive analysis, Chap. 13 aims to provide a deeper understanding of how global perspectives and cross-cutting issues impact sustainable construction, offering valuable insights for practitioners and policymakers striving to align their efforts with global sustainability goals. Chapter 14 establishes the principles and techniques of sustainable construction management practice and procedure. It provides a detailed exploration of the fundamental practices and procedures necessary for effective sustainable construction management. The chapter addresses the evolving requirements and diverse functions of sustainable construction managers, driven by the expanding and varied needs of the industry. The chapter also highlights the importance of incorporating sustainable construction management firms to meet these demands, considering the wide range of employers and geographic locations involved. This chapter underscores the necessity for adaptable and innovative management approaches that align with both local and global sustainability objectives, ensuring that sustainable practices are effectively integrated across different contexts and projects.

For Whom Is This Book Written

Overall, this book stands as an indispensable resource for students enrolled in a range of professional and degree programs associated with the construction industry. These programs encompass fields such as architecture, building surveying, interior design, civil engineering, construction management, and more. The book offers a thorough and in-depth exploration of the fundamental principles and practices crucial to the construction sector. It covers essential topics including project planning, acquisition, and programming, providing a solid foundation for students engaged in the various stages of construction projects—from conceptualization and design to execution and management. For students, understanding these core areas is not just academic; it is integral to their professional development. The book equips them with the knowledge necessary to navigate the complexities of construction projects and to contribute effectively to their success. It delves into the critical aspects of sustainable construction management, exploring the principles, techniques, and best practices that are vital for ensuring that projects are not only completed successfully but also meet the highest standards of sustainability. By integrating insights into global trends, technological advancements, and the latest industry practices, the book prepares students to address the challenges of the modern construction landscape. It offers practical knowledge that will help them to influence the built environment positively and to excel in their future roles as professionals who drive innovation and excellence in construction. In essence, this book serves as a comprehensive guide that bridges theoretical knowledge with practical application, making it a valuable asset for students aiming to build a successful career in construction. It helps them understand how to effectively manage and execute projects while contributing to broader sustainability goals, ultimately positioning them to make a significant impact in the field.

What This Book Is Designed to Do For You

This book has been developed as a comprehensive reference to the best practices in sustainable construction management. It serves as an inclusive guide that covers processes, procedures, and systems adhering to high standards of both professional and academic competence. The book bridges gaps between various practices, principles, operations, strategies, experiences, and resources, aiming to close the divide between different levels of expertise and access in the industry. The primary objective is to assist construction companies in achieving a harmonious balance by offering guidance to all stakeholders involved in construction projects. For contracting organizations, the book provides valuable insights that can enhance profit margins, boost productivity, and improve reputations. It also addresses the reduction of claims, disputes, fines, and penalties, contributing to more efficient and harmonious project execution. The goal of a sustainable construction manager, as outlined in the book, is to ensure that construction projects are executed in alignment with sustainability objectives and requirements. These objectives encompass a wide range of environmental, social, and economic factors, which are analyzed and integrated into project planning, design, and execution. By doing so, stakeholders can achieve sustainable outcomes that are beneficial for both current and future generations. A sustainable construction manager is tasked with balancing social, economic, and environmental variables throughout the project lifecycle. Their role is crucial in steering projects toward outcomes that are socially responsible, economically viable, and environmentally sound. Essentially, a sustainable construction manager acts as a steward of sustainable development, guiding projects to meet present needs while safeguarding the ability of future generations to meet their own needs. This book, therefore, serves as an essential tool for sustainable construction managers, providing them with the knowledge and strategies required to implement construction projects that adhere to sustainability objectives and criteria. It aims to empower professionals to drive positive change in the construction industry, ensuring that projects are executed in a manner that promotes long-term sustainability and resilience.

Contents

Abbreviations

Several acronyms were used in this book, and it is instructive to identify them to aid in reading the thesis. Following are the key acronyms:

AHP	Analytical Hierarchy Process
AI	Artificial Intelligence
AR	Augmented Reality
BIM	Building Information Modeling
BREEAM	Building Research Establishment Environmental Assessment Method
CBA	Cost-Benefit Analysis
CIOB	Chartered Institute of Building
CO_2	Carbon Dioxide
COBRA	The annual RICS construction, building and real estate research conference
CSFs	Critical Success Factors
CSR	Corporate Social Responsibility
EIA	Environmental Impact Assessment
ESG	Environmental, Social, and Governance
FM	Facilities Management
GBCN	Green Building Council Nigeria
GDP	Gross Domestic Product
GFCF	Gross Fixed Capital Formation
GNI	Gross National Income
GOM	Government of Malaysia
HVAC	Heating, Ventilation, and Air Conditioning
IAQ	Indoor Air Quality
IoT	Internet of Things
ISO	International Organization for Standardization
KBE	Knowledge-Based Economy
KPIs	Key Performance Indicators
LCCA	Life Cycle Cost Analysis
LEED	Leadership in Energy and Environmental Design

PCA	Principal Component Analysis
RICS	Royal Institute of Chartered Surveyors
ROI	Return on Investment
SCM	Sustainable Construction Management
SDGs	Sustainable Development Goals
VR	Virtual Reality

Chapter 1
Nature of the Construction Sector

Abstract The construction sector is a critical pillar of global economies, significantly contributing to GDP and employment while shaping societies through housing, healthcare, education, and infrastructure development. Distinguished by its project-based, temporary nature and intricate stakeholder landscape, the sector encompasses diverse professional competencies and a workforce increasingly including women and educated youth amidst traditional reliance on male, informal, and foreign labour. Market demand is driven by demographic shifts and urbanization, significantly influenced by government policies and cultural contexts. However, the sector faces persistent challenges such as information asymmetry, inadequate working conditions, project delays, and safety vulnerabilities. This chapter argues that enhancing the sector's performance necessitates the widespread adoption of sustainable construction practices. It underscores the crucial need for all stakeholders to understand and fulfill their responsibilities to ensure projects align with sustainability benchmarks and effectively serve evolving societal needs, fostering a more environmentally sound and socially equitable built environment.

Keywords Sustainable performance · Project benchmark

1.1 Chapter Learning Outcomes

The construction market is examined in this chapter. Regardless of the state of development—developed, developing, or undeveloped—the construction industry is significant in the majority of economies. The sector consumed a lot of resources. The sector engages about 10% of the formal sector in most countries and contributes up to 10% of the GDP, or GNI. The sector employs a large labour force from the informal segment. The sector's professionals are diverse, with varying degrees of competencies in the application of technologies. The construction sector comprises a diverse range of professionals with differing levels of competency in the application of technological tools and methods. While the sector is still male-dominated,

A. Olanrewaju, *Sustainable Construction Management: Research and Practice Companion*, Management in the Built Environment,
https://doi.org/10.1007/978-981-96-4910-5_1

the good news is that the number of females is increasing. Many of the workforce members are young and educated. Entry is often not difficult in most countries, except for specialised work. The construction market and its outputs are generally shaped by the culture and system of government of the host countries. The government is the major client in most countries. The clients are generally inexperience, and the information in the sector is asymmetric. In most developed and developing countries, the sector depends on foreign labour, and the site labour force is poorly treated in most countries. After studying this chapter, you will be able to:

- Distinguish the construction sector from other economic sectors.
- Explain the features of the construction sector.
- Explain the competencies of construction parties.
- The rights and responsibilities of the stakeholders in the construction sector
- Understand techniques to design or construct sustainable construction projects

1.2 Introduction

In this chapter, the various aspects of the construction process are examined in the efforts to provide linkage with sustainable construction. Various parties, or stakeholders, are involved in the construction of a project. Construction projects, whether architecture or engineering, follow similar phases or stages, from the simple thought of the clients or sponsors to the final physical object that is ready for use or occupation. Many organisations, including Royal Institute of British Architects, American Society of Civil Engineers, The Chartered Institute of Building, American Institute of Architects (AIA), and Associated General Contractors of America have developed construction processes that are used in many countries. These organisations provide a variety of services, including professional development opportunities, networking events, industry publications, and advocacy initiatives. The demand for construction depends on many factors. For instance, residential buildings are in demand because of an increase in divorce rates, changing lifestyles, children leaving home earlier, a longer life span, and the location of the projects. Roads and railways are needed as easy modes of transportation to connect cities and towns. While most buildings offer an answer to the space requirement, many buildings are built to display royalty, make a statement, and represent power. The design team members must be clear about the value system of the clients or sponsors. When this is not done, the completed projects will be expensive to own or use. The construction process is the collective term for all the activities or actions that are involved in the conceptualization, design, production, and commissioning of a construction project.

1.3 Characteristics of the Construction Sector

Construction means different things to different people. According to ILO [19] construction is defined or explained as:

(i) building, including excavation and the construction, structural alteration, renovation, repair, maintenance (including cleaning and painting) and demolition of all types of buildings or structures;
(ii) civil engineering, including excavation and the construction, structural alteration, repair, maintenance and demolition of, for example, airports, docks, harbours, inland waterways, dams, river and avalanche and sea defence works, roads and highways, railways, bridges, tunnels, viaducts and works related to the provision of services such as communications, drainage, sewerage, water and energy supplies;
(iii) the erection and dismantling of prefabricated buildings and structures, as well as the manufacturing of prefabricated elements on the construction site;

Before we go ahead, it is imperative to posit that, in general, due to the complexity involved in describing and defining construction work, there is a need to agree on whether it is an industry or a sector. These two terms are not the same. An industry, for example, is a collection of related economic activities categorised by the kind of products or services they provide, such as manufacturing, construction, services, mining and quarrying, agriculture, and so on. A sector is a segment of the economy where companies provide the same or a similar good or service. Economies comprised:

- The primary sector includes the extraction and gathering of natural products from the land, such as forestry, mining, and agriculture.
- Processing, manufacturing, and construction are the secondary sectors.
- Financing, entertainment, and retail sales are among the services offered by the tertiary sector.

Whether a nation is developing, developed, or underdeveloped, the construction sector is essential to its progress. This can be attributed to the sector contribution to the GDP, and GNP from fixed capital formation. Construction sector has a significant multiplier effect on other crucial economic sectors in every country. A country cannot see significant progress if its construction sector is inefficient. This is so because the construction sector's primary focus is on developing infrastructure—roads, bridges, trains, buildings, etc.—that raises living standards and promotes economic growth and development. The construction sector contributes between 5 and 10% of GDP in the majority of nations. However, the actual figures would indeed be higher because of poor accounting systems especially in the developing countries. The contribution of the construction activities in the informal segments are exceptionally large globally, though even more so in the developing countries. For instance, in many cases, close to 20% of construction activities occurs in the informal sector. Consequently, in the majority of countries, the outputs of construction may account for more than 20% of GDP or GNP. In the majority of nations, construction output makes up roughly 50% of gross fixed capital creation. While some have demonstrated that the construction sector can be utilised to control the economy, others generally disagree, arguing that the precise amount of its influence is unclear. There are many who contend that the contribution is merely temporary. There are others who contend that the construction sector is drawn to the economy rather than the other way around. Put differently, the construction sector just follows the "path" established by the overall rate of economic expansion; it does not create economic prosperity. The proponents of this conjecture base their argument on the fact that the demand for construction for the "output" of the construction is a derived demand. It is need-based. For instance, if the other sectors do not need a building, would the construction sector produce it?

Whatever stance one takes on the relationship between the construction sector and economic growth, it does not lessen the sector's significance in supplying and maintaining the infrastructure required to promote economic progress. There has been a contention that indigenizing the construction sector is a critical need for a nation to have meaningful and sustained growth since an inefficient construction sector will impede meaningful development in that country. This is an illustration of how infrastructural facilities like buildings, roads, bridges, dams, tunnels, and industries and factories that are outputs of the construction sector are used to stimulate and sustain national economic growth. Whichever country is in a state of development, underdevelopment, or developed, the construction sector is significant. During the COVID-19 pandemic, countries use the construction sector to regulate the economic and boost standard of living. In fact, the construction sector is possibly the only sector that was not completely shut down during the pandemic even as it is difficult not to work 'shoulders to shoulders' on sites. The one major segment included in the quarterly and annual statements of national accounts in most countries is the construction sector. Under the GDP, GNI, and GFCF national accounts, the construction sector is listed twice. GDP, capital formation, and added value are the metrics used to measure the outputs. The outputs of the construction sector are capital or investment goods. These are the goods or outputs that are used to produce other goods and services.

The construction sector is responsible for designing, creating, modifying, repairing, and demolishing engineering projects, buildings, and other structures of a similar nature. Building engineering services and prefabricated component assembly and installation on-site are also included. The features of the construction sector are diverse, and some of them might not be comparable to those of other sector like manufacturing or services. Some of the characteristics are advantages, while others are indeed disadvantages. It is not clear to many whether the construction sector is inclined towards manufacturing or the services sector. To some extent, the construction process is remarkably like the manufacturing process, but at the same time, the success of the construction sector does not depend on specific equipment or a clearly defined process. This feature is like that of the service sector. What is more, the construction sector depends on a large group of individuals and firms to succeed. But like the manufacturing sector, the outputs of construction are often physical products. However, the maintenance aspect of the construction sector is service-based.

Often, physical objects are not delivered. In economic terms, capital is generally classified into two kinds: physical capital and financial capital. However, capital that was produced in the past but is now used in the production of other capital or goods is what is referred to as physical capital. But often, they are simply referred to as capital without the word physical coming before. So, when economists say capital, they mean physical capital. All examples of capital, including machines, buildings, roads, etc., are included in a nation's gross fixed capital formation (GFCF). GFCF is used to stimulate economic activities in any country. The construction sector requires input from some sector, and at the same time, its outputs are inputs to some sectors. For instance, cement, water, oil and gas, timber, etc. are produced from other sectors, but these are used as inputs to construction activities to deliver their outputs. Buildings, roads, and bridges are examples of the outputs of the construction sector that are used as inputs to other sectors of the economy.

Table 1.1 contains some characteristics of the construction sector. In the construction sector, knowing and addressing these characteristics is essential to successful project execution, efficient project management, and ultimately reaching the project objectives and targets. In construction management, clearly defined objectives and measurable targets are essential for guiding project execution and evaluating success. Objectives represent the overarching goals of a project, such as delivering a high-quality building within budget, on time, and in compliance with regulatory and sustainability standards. Targets, on the other hand, are specific, quantifiable milestones that support the achievement of these objectives. For example, reducing construction waste by 20%, achieving a Building Energy Index (BEI) below 100, or completing structural works within 90 days. Together, objectives and targets form the strategic and operational framework for managing construction projects, ensuring that all stakeholders remain aligned in delivering results that meet both client expectations and broader environmental, economic, and social performance indicators.

1.4 Government as the Major Client of the Construction Sector

The government regulates construction activities by deciding when, where, and what to construct, all of which depend on the government's revenues and policies. The construction sector is public-sector dependent. The private sector is profit-driven and will only invest in the sector if the government provides an enabling environment, without which they will not invest. Perhaps a significant difference between the construction sector as a production 'plant' and factory process-based production is that the construction resources, both materials and human resources, are predominantly moved to sites for operation. Although advancements in technology have made it possible to use off-site production on construction, most construction activities and tasks are performed on-site. However, until now, the investment in R&D and

Table 1.1 Summary of the characteristics of the construction sector

Feature	Explanation
Uniqueness	Construction projects are often one-of-a-kind undertakings since they entail the construction of a particular infrastructure or buildings that must satisfy certain needs and goals. The needs and expectations of clients are often not the same even for similar type of construction. Even though projects may share certain characteristics, they are generally distinguished by specific elements including location, design, materials, and client preferences
Complexity	Construction projects can be quite complicated, requiring a variety of interconnected components and specialties. They call for the integration and coordination of several tasks, including project management, engineering, design, engineering, procurement, and scheduling. Several elements, including technical needs, stakeholder participation, regulatory constraints, site circumstances, and the necessity of interdisciplinary collaboration, contribute to the complexity
Long duration	The duration of construction projects can vary, spanning from several months to years, contingent upon the project's complexity and magnitude. The length of time depends on several variables, including the project's scale, construction techniques, site conditions, permissions and approvals, and resource availability. To guarantee the project's successful completion within the expanded period, detailed planning, scheduling, and monitoring are required
Interdisciplinary nature	Numerous engineering specialities are involved in construction projects, including structural, mechanical, electrical, and civil engineering and architecture. To handle particular project components, each profession contributes its knowledge and experience. To accomplish project objectives, various disciplines must work together and communicate well with one another
Resource intensity	Significant amounts of labour, supplies, machinery, and money are needed for construction projects. For a project to proceed, be productive, and be cost-effective, resources must be allocated and managed effectively. In construction projects, availability of resources, procurement, logistics, and resource optimisation are vital factors to consider
Risk and uncertainty	Numerous risks and uncertainties could affect the results of construction projects. Several variables, including unanticipated site circumstances, meteorological conditions, labour disputes, supply chain interruptions, and design modifications, might result in risks. To reduce and handle any hazards, effective risk management and backup plans are crucial
Stakeholder involvement	Several parties are involved in construction projects, each with unique expectations and value systems. The client, investors, design experts, contractors, subcontractors, regulatory bodies, local communities, and end users are a few examples of these stakeholders. For a project to be executed successfully, stakeholder relationships must be managed, concerns must be addressed, and effective communication must be maintained

(continued)

Table 1.1 (continued)

Feature	Explanation
Compliance and regulations	There are numerous rules, guidelines, and standards about construction codes, occupational health, zoning, safety, and environmental effect that construction projects need to respect. To guarantee project legality, safety, and sustainability, compliance with these regulations is crucial
Site-specific considerations	Project sites are intrinsically linked to construction projects. Site-specific elements that affect project planning, design, and construction techniques include topography, soil conditions, accessibility, utilities, environmental restrictions, and local laws
Quality and safety	Safety and quality control must be prioritised in construction projects. For a project to be successful and for the workers' and end users' safety to be guaranteed, construction activities must adhere to strict quality standards as well as safety rules
Capital goods	Construction outputs are capital goods of factors of production because they are required to produce goods and services. For example, buildings are not required for their own sake
Require inputs from other sectors	The construction activities require significant resources from many sectors
Sectoral contribution	Globally, the construction sector contributes about 10% to both GDP and GNI. The sector directly employs up to 10% labour force. The contribution of the informal sector includes, the sector contributes up to a quarter of the GDP and the GNI
Government as the major client	Globally, the government is the major client of the sector with the government's need to provide public services
Resources move to site	The sector has a similar nature to the agriculture sector in this regard. Materials, components and the human capital move to the clients' sites
Paid, before completion	The payment structure is often dictated by the terms of the construction contract and the nature of the project itself. The interim valuation often includes materials and component that are not even delivered to site. Clients are required to pay high mobilization fees. It is customary for clients to fund construction projects before they are finished in order to control problem-risk, guarantee cash flow, and schedule payments according to project advancement
Generate waste	The amount of waste generated by construction activities is huge. Unfortunately, only less than 20% are recycled or reused. Currently, estimates reveal that the construction sector generates 100 trillion tonnes of waste annually
Expensive, large/heavy unit	Construction projects are expensive to own and use. For example, buildings are the most expensive single item on many family balance sheets
Durable products	Construction projects are essentially designed to offer longevity, reliability, and value to clients, providing dependable performance and contributing to sustainability and resource conservation

(continued)

Table 1.1 (continued)

Feature	Explanation
Automation	There is a very low and inadequate integration of technology and machinery to improve and expedite various processes and construction activities
Dominated by large firms	More than 70% of construction projects are executed by large construction companies
Poor investment in R&D	Only less than 5% of construction companies make significant contributions to research and development, and even a few make significant investments in research and development. Only a few construction companies in most countries have established working collaborations with universities and research centres in their countries to support their activities. However, Japan is perhaps outstanding, as large construction companies in Japan often have well-funded research units that spearhead innovation for the companies
Complex price determination mechanism	There are many multiplexities and uncertainties in the pricing techniques of construction projects
Immovable	In general, the construction outputs, including buildings, roads, bridges, dams, etc. are stationary, fixed, or unchangeable cannot and be moved or fixed in place
Efficiency	The construction sector is working within the boundary on the production possibilities curve (PPC) not on the curve. This is a sign of inefficacy and poor productivity
Low-level IT applications	The level of digitalisation in construction is low compared to others sectors, especially in sector manufacturing and services
Poor safety records	The accident rate and casualties are high. More 60% of construction workers have been injured
Labour intensive	The construction sector requires a lot of labour. Therefore, the performance of the construction industry would be impacted by a worker shortage
Service based	There is a deep understanding that, in fact, the success of a construction project depends on the interaction between the parties to the contracts rather than the pieces of equipment, plants, materials, and components. This is regardless of the fact that the final outputs are physical in nature by definition. As a result, many believe that the sector is more similar to service than to the manufacturing sector
Procurement	The dominant procurement strategy in the construction sector is the traditional or conventional procurement strategy. The procurement process involved that activities related to the selection of contractors, contract negotiation, supplier relationship management, and contract administration
Contract	The legally binding agreement for contract is usually between the client or owner and the main contractor that outlines the terms and conditions for the construction project

IT applications has remained small as compared to most of the sectors. Except in a few countries like Japan, where major contractors finance R&D, in most countries, the investment is rarely up to 1% of the total investments in the sector.

Construction work is expensive to construct, own, and use. For those who own a home, this often is the single largest investment they will ever make. Similarly, for those who rent or lease the home, the price they pay periodically for the home is the single payment they make. Most families spend more than 30% of their salary or income on rent to pay mortgages. Yet there is an incidence of foreclosure. The cost of maintenance is also exceedingly high. The number of individuals and businesses that work together to bring one construction into existence is quite large. Compared to most other major sectors, the construction sector is largely labour-intensive in most parts of the world. The construction sector is mostly inefficient. High cases of cost and time overruns have been reported in every economy. The only common difference is the degree of the overruns. Using the production possibilities frontier to explain this, production efficiency is only obtained if a product or service is produced at the lowest possible cost. However, this only occurs at points on the PPF. However, the efficiency of the construction is largely within the boundaries. To simplify, there is a lot of underutilization of the resources. The outputs of the construction sector are uniquely immobile. However, this immobility is generally misinterpreted. For instance, while a building cannot be moved from one place to another, it can move over space. Buildings can be extended, and the height of buildings can be increased. Many individuals and businesses are not only involved in the development process of the buildings; the time needed for the development process is complex and takes a long time. Sometimes, this could take many years, depending on the complexities of the construction.

1.5 Classification of Construction Outputs

The outputs of the construction are quite broad, ranging from buildings to non-buildings. Broadly, the sector can be distinguished into three, including:

1.5.1 Buildings

The construction sector includes the general construction of buildings like residential buildings, office buildings, religious buildings, commercial buildings, social buildings, motels, hostels, storage buildings, farm buildings, hospital buildings, and utility buildings, etc. The building segment is dominated by private clients. The different procurement strategies and funding arrangements also bring both private and public clients together on some of the building construction. In terms of size, the residential building is the largest in most countries. The residential buildings form all types of dwellings, including single-family homes, condominiums, duplexes, apartments, multi-family homes, flats, and townhouses. The decision of which of these to choose from depends on the interests of the available funding arrangements. Those who place higher priority on security and privacy and can afford the operating costs would prefer the single-family home, whereas the condominium arrangement allows for the sharing of operating and maintenance costs. Is it also argued that the resale value of a single-family home is higher as compared to most other types of residences. The owners of the single-family home can also carry out modifications to the buildings without any difficulties, an exercise that might not be easy or even possible with apartments or condominiums. However, the cost of running expensive facilities like swimming pools and playgrounds might be less than condominiums. The level of mechanisation involved with building construction is probably lower in comparison with civil engineering and specialised. In most countries, building construction is handled by indigenous contractors and consultants. Significant parts of modern buildings are based on offsite operations.

1.5.2 Civil Engineering

Civil engineering works are the public infrastructure including roads, streets, bridges, tunnels, railroads, harbours, and other water projects; industrial facilities; irrigation systems; sewage systems; pipelines and electric lines; outdoor sports facilities, etc. Civil engineering is dominated by public clients. In many developing and developed countries, private consortia are involved in the construction of roads, airports, and stadiums under a private–public arrangement in the developing countries. Due to the nature of the civil engineering works, which involve large foreign contractors, although the level of mechanisation is extremely high, civil engineering involves great onsite operation. Civil engineering works take a much longer time to complete as compared to building projects. The costs are usually massive and often involve complex funding arrangements. Consortia involving contractors, investors, banks, and material manufacturers are formed to provide civil engineering work.

1.5.3 Specialised Construction

This mainly involves construction works that are not either building or civil engineering work. The works in this class require specialised skills, responsibilities, plants, or equipment outside the scope of both the building and civil works. This involves an exceptionally large and extensive project. The demolition of existing construction, preparation of mining sites, preparation of mineral properties and sites, rock removal, blasting, drilling, boring, and core sampling for construction, geophysical, geological, building site drainage, drainage of agricultural or forestry land, electrical, plumbing, and other construction installation activities, etc. are some examples of works that fall into this category. One distinctive aspect of specialised construction work may be that it is typically completed on construction sites with very few tiny components assembled there. Although the construction of refineries, sugar plants, iron smelting companies, and similar structures is not civil engineering but rather process engineering, they can also fit into specialised construction activities. The clients involved in the specialised construction are private. The presence of public clients during this phase is extremely limited.

1.6 Players in the Construction Sector

Construction players include clients, architects, civil engineers, structural engineers, services engineers, quantity surveyors, construction managers, facility managers, project managers, main contractors, land surveyors, contractors, subcontractors, suppliers, banks, insurance companies, NGOs, etc. The level of resource [i.e., efforts, skills, and knowledge] correlates with size, type, complexity, client, and project environment. The procurement method has a profound influence on the composition of stakeholders. All players or stakeholders must work together for the construction to be successful. In other words, the members must form a 'team' to ensure that the constructed system meets stakeholder values. Team building is needed to perform the work required by their appointments and contracts. The reasons for working together as a team are to exploit a common goal, target, and commitment to achieve a client value system. In the formation of a team, each of the members must understand their roles and, at the same time, know who to answer and who is answerable to them. These problems must be well defined and explain; otherwise, they could lead to major causes of claims, disputes, and conflict as the construction work advances. The major players are explained next.

1.6.1 Client

Construction is difficult to define or explain, perhaps due to its nature, roles, and experience in the construction business. More than 80% of the construction clients are inexperienced because they have not had formal contact with the construction business. However, in general, a construction client is defined as an entity, individual, group, or organisation that initiates and funds a construction project. They are the entity that engages a construction company to carry out the construction work. The construction works may include new construction, maintenance, renovation, rehabilitation, refurbishment, extension, and conversion of constructed facilities, including buildings (residential or non-residential), civil engineering works, and process engineering work. Clients can also be classified as either vertical construction clients or horizontal construction clients. Primarily, clients procure construction projects to meet their value systems. Construction profession must understand the client needs and expectations. Usually, most clients procure buildings to meet space requirements. While construction clients vary widely, the following are some of the clients found in most countries:

- Private: Homeowners who use contractors for construction or remodelling projects. Most clients in this class work in construction because they require added space to accommodate growing families. Additionally, the clients hired construction businesses to fix structural flaws or to renovate their structures so that they could perform differently.
- Businesses: Organisations in need of construction services to create new office buildings, warehouses, retail spaces, or other commercial facilities, or to renovate, expand, and remodel existing ones. The clients could be partnerships, individuals, groups, etc.
- Developers: Large-scale real estate developers that build hotels, shopping centres, industrial parks, and residential complexes. This is primarily for new construction and renovations. Their internal teams are often in charge of handling repairs and enhancements. Nonetheless, most developers use outsourcing to obtain their maintenance services.
- Government agencies: Local, state, federal, municipal, or council organisations that mandate building projects for the development of infrastructure, including roads, bridges, public housing, hospitals, and schools. These included various construction projects, such as new construction, additions, upkeep, and renovations.
- Non-profit organisations: Nonprofits that depend on building projects to sustain their purposes, such as recreation centres, community centres, schools, hospitals, or other institutions.
- Institutions: Colleges, universities, and schools that build new buildings or extend their campuses are considered institutions. These are often public projects, though they may be privately or publicly held.
- Healthcare organizations: Hospitals, clinics, or lab centres that need to be built to extend their current spaces or construct new infrastructure for the healthcare industry.
- Religious organizations: churches, mosques, temples, and other houses of religion that construct or restore public areas or places of worship.

1.6.2 The Roles of Construction Clients

Because they have a stake in the construction project's success, clients' active participation, decision-making, and communication are essential to meeting their needs and achieving project aims. Clients in the construction sector are crucial to the process. The roles that construction clients must primarily perform during the construction process are as follows:

- Strategic: Projects are owned by the clients; therefore, it is necessary to meet their individual needs and expectations. Therefore, it is the client's responsibility to decide whether a construction project is necessary to fulfil their value systems. The consideration of project sustainability is a part of the value system of contemporary forwarding clients. The project's goals, parameters, and purpose are decided by the clients following their set of values.
- Project Planning: Planning includes deciding the requirements and preferences for a project and how it will be financed or bought. Clients are required to collaborate closely with members of the design team, such as quantity surveyors, architects, engineers, project managers, and construction managers, to create the project plan during the project planning process. The client also makes decisions about the project's budget, schedule, design standards, and functional requirements in addition to honing their needs and wishes.
- Budgeting and Financing: The clients are accountable for arranging the necessary funds to support the projects even for PFI enabled projects. The client should make sure there is enough money accessible for the entire duration of the project. Clients typically get financing from banks, mortgage houses, and other financial organisations, except individual clients, to receive all or part of the funds required to finance the project.
- Decision Making: As a project moves along, its requirements often change. As a result, during the construction process, clients are demanding decisions. A few of the choices could involve examining and accepting shop drawings, specifications, material choices, sustainability standards, design proposals, and construction contracts. To make sure the project is in line with their value system, the clients give the design team and contractors advice and comments. Certain clients need their design professionals' advice.
- Engaging Consultants and Contractors: To help with the project, clients interact and work together with a variety of experts and contractors. This involves appointing the relevant design professionals, such as quantity surveyors, construction managers, architects, engineers, and project managers, to offer the knowledge and aid required for the project to be completed successfully. Modern forward-thinking clients are requesting the design and construction team's sustainable credentials as sustainability awareness grows. This also shows that clients are requesting work with certified sustainable managers. Based on their credentials, record of accomplishment, and experience, the clients assess and choose the most qualified consultants and contractors. Inexperienced clients rely on referrals to choose the contracting firm and design expert.
- Communication and Stakeholder Management: Clients are the principal point of contact for all project participants, regardless of their level of experience and procurement strategy. Design experts, contractors, regulatory bodies, and other stakeholders recommend clients for decision-making when making strategic decisions. The clients make sure that all parties involved in the project are aware of updates, modifications, and choices by effectively easing communication. Clients also manage stakeholder relationships, resolving any difficulties or conflicts that may arise.

- Monitoring Project Progress: Clients check the progress of the building project to ensure that it is according to the predetermined budget, timeline, and aims. Apart from conducting on-site visits to assess the actual state of construction, the client scrutinises the project team's ongoing reports and updates. Project monitoring for clients with little experience is usually handled by a design professional acting on their behalf.
- Approvals and Permits: It is the responsibility of clients to get all required permissions and clearances from government agencies, parastatals, and regulatory bodies. This usually entails checking that the projects adhere to all relevant regulations, including zoning laws, construction codes, environmental standards, and others.
- Acceptance and Handover: Clients are involved in the project's final acceptance and transfer when it reaches that point. They inspect the project, make sure it meets their standards and expectations, and then they may release the last payment to the contractors. Clients also supervise any broken promises and obligations under contracts.

1.7 Classes of Clients

A common classification is based on the contact the client has with the construction market. Based on this yardstick, clients are classified as either experienced or inexperienced.

1.7.1 Experienced and Inexperienced Construction Clients

1.7.1.1 Experienced Construction Clients

Clients are classified as either experienced or inexperienced clients based on their contact with the construction business. In terms of volume of work, most clients are inexperienced. Client built once in their lifetime, and the projects are often residential buildings. Clients with experience in construction are people, companies, and bodies that have worked on construction projects in the past. They have prior experience managing construction projects, interacting with contractors, and going through the construction process. Experienced clientele typically has a certain amount of ability with the industry standards, construction practices, and the intricacies involved in the construction process. Depending on the context, region, and kind of construction projects under consideration, the size of the experienced construction customer sector can change. Individual homeowners, bodies, and companies with experience in construction, such as corporations, government agencies, real estate developers, or construction management firms, may be considered experienced construction clients.

These clients have completed several construction projects over the years. Experienced construction clients often have dedicated units or departments in their organisations with abilities in construction management. The amount of construction activity in each area, the size of the construction industry, and the quantity of ongoing and completed projects are some of the factors that might affect the size of this segment. Experienced clients bring their knowledge, ability, and lessons learned from earlier projects.

1.7.1.2 Inexperienced Construction Clients

Inexperienced construction clients are individuals or organisations that are new to the construction sector or have limited experience with construction projects. Often, they are undertaking their first construction project or have only minimal exposure to construction processes and procedures. These clients lack familiarity with construction terminology, industry practices, and the nuances of managing a construction project. Experienced clients offer fresh insights and perspectives that can highlight the challenges and opportunities faced by newcomers to the construction process. This group of the client include first-time homeowners, small businesses venturing into construction projects, or organisations entering the construction industry for the first time. The size of the inexperienced construction client segment can also vary depending on factors such as location, industry activity, and the specific context being considered. A few examples of the variables that may affect the growth of the inexperienced construction client sector are the degree of new entries in the market, the general demand for construction services, and the regional economy. Not to mention the sheer number of people and organisations that may want to start construction projects but haven't yet done so, the construction industry's development potential must also be considered.

1.7.2 Construction Design Team

The construction business comprises diverse players, including design professionals, construction professionals, clients, and operations teams. This is particularly true for the PPP and PFI initiatives. Regarding their initial interactions with clients and their engagement in the design and construction phases of construction projects, the leading specialists in the field are building engineers, architects, quantity surveyors, and engineers (particularly civil, electrical, and mechanical engineers). A construction design team is a collection of specialists who collaborate and work together to generate the design and plans for a construction project. A design team is sometimes referred to as a design professional or design firm. Those with specific knowledge of various sides of the design and construction process usually make up the team. Numerous specialists are involved in the building process who all have different responsibilities from the moment the clients decide to build, to the time they construct, maintain, renovate, and rehabilitate, to the moment the clients acquire ownership of their created assets. The composition of a construction design team depends on the nature of the project under consideration, the nature of the clients, the location, and prevailing government policies. In general, the key members of construction projects include:

Architects: They are the experts in the creation of architectural concepts and designs. The architects gradually update designs, specifications, and plans during the building phase to satisfy the client and legal requirements. The architects are responsible for general idea and aesthetics of a building or other structure. When designing the architectural blueprints, they consider things like building codes, safety, functionality, and space utilization.

Building engineers: Building engineers make certain that structures can safely support and endure the loads that they will come into contact with during their operation.

Structural Engineers: The design and study of a building's structural system are its main concerns. Structural engineers guarantee that buildings can securely bear the specified weights and withstand a variety of factors, including wind, gravity, and seismic activity. Additionally, they create structural drawings and specifications using pertinent contract documents and architectural drawings.

Civil Engineers: Civil engineers are involved in deciding the strength and strength of the proposed construction, while civil engineers deal with public works such as highways, dams, quays, shipyards and bridges. The civil engineers are concern with project's site-specific requirements, such as grading, drainage, utilities, and other infrastructure requirements. Also, the civil engineer guarantee that the site is proper for building and that the design includes the required site upgrades.

Mechanical Engineers: The building's mechanical systems, including the plumbing, fire safety, and HVAC (heating, ventilation, and air conditioning), are designed by the mechanical engineer. The mechanical engineers ensure these systems meet the project's requirements and follow all applicable rules and standards. Often referred to as "services engineering," mechanical and electrical engineering plays a significant role in contemporary construction, accounting for up to 30% of the entire contract amount for a typical structure.

Quantity surveyors: Cost control, purchasing, and contract disputes in the marketplace and supply chain are all aspects of quantity surveying. On the financial effects of the actions made by other stakeholders and the requirements of the clients, they offer recommendations. As work continues, the quantity surveyors keep an eye on the project and revise the original plans and contractual duties in light of new tasks and modifications.

Electrical Engineers: An electrical engineer's main duty is to design the electrical systems of the building, including the lighting, communication, power distribution, and other electrical installations. The electrical engineers ensure that the electrical design meets project specifications and is both safe and efficient.

Interior Designers: The construction of visually stunning and functional interior spaces is the area of expertise for an interior designer. To design interior spaces, select fixtures, materials, and finishes, and create a cohesive aesthetic that fits the client's preferences and the project's goals, they work closely with the architect and other professionals.

Landscape Architects: A landscape architect's primary responsibility is to create the outdoor environments surrounding a building or other structure. By considering aspects like horticulture, site planning, sustainability, and aesthetics, their integrated design merges with the built environment.

1.7.3 Roles of Design Team

The following are the specific responsibilities of the design professionals in the construction delivery process:

- Examine the client's business case.
- Identify the client's needs.
- Prepare brief
- Establish design criteria.
- Prepare a conceptual design.
- Prepare feasibility and viability studies.
- Conduct value management studies, if needed.
- Conduct a risk assessment throughout the design process,
- Estimate capital costs and construction timetables,
- Prepare the final design specifications and drawings, and arrange for the submission and evaluation of designs.
- Create and implement quality assurance and control standards for the deliverables review process.
- Throughout the design phase, revise construction schedules and capital cost projections.
- Create the final design drawings and requirements in preparation for a tender.
- Acts as the agent for the client.
- Examine contractor submittals for the construction deliverables and suggest accepting them.
- Evaluate modification requests and project expenses for modification orders.
- Visit the location on a regular basis to verify design compliance

1.7.4 Construction Team

The construction team is typically made up of several professionals, including contractors and specialists, who cooperate and work together to guarantee that the project is finished on schedule, within budget, and following the necessary quality standards. The construction team's duties and responsibilities can change based on the project and its organisational structure. To complete the project, the members of the construction team must effectively collaborate, communicate, and coordinate.

Everyone contributes their knowledge and skills. Listed below are some common roles within a construction team:

Construction Manager: From start to finish, the construction manager manages the entire project. All project-related activities, such as scheduling, budgeting, resource allocation, and risk management, must be coordinated and managed by them. For the client, consultants, contractors, specialists, and other stakeholders, use it as the primary point of contact.

Contractors: The contractors oversee converting the specifications and drawings into tangible assets. Following the project plans and specifications, contractors oversee their crews, acquire supplies, and carry out the building activity. They guarantee adherence to occupational health and safety laws, carry out safety inspections, and offer safety training. To guarantee that the job is carried out in compliance with the established standards and regulations, they carry out testing, quality audits, and inspections. They guarantee that corrective measures are implemented and handle any non-conformities. Overseeing daily operations and managing the construction site fall within the purview of the site manager or supervisor. When the work is finished, they ensure that the project's plans, specifications, and schedule are adhered to. The site manager works with subcontractors, oversees safety and quality standards, and handles any issues that arise on the job site.

Architects: Architects handle designing a project and creating detailed construction drawings and requirements. They work closely with the clients to completely understand their goals, translate those demands into a design, and ensure that the finished result satisfies all legal requirements while also being visually pleasing.

Engineers: Engineers with specialised knowledge in their respective domains include structural, civil, mechanical, and electrical engineers. To ensure that the project follows engineering standards and criteria, the engineers perform calculations and analyses, take part in the design creation process, and offer technical guidance during construction.

Subcontractors/specialists: Subcontractors must finish specific building jobs. They are usually mastering in one or more specific crafts, including carpentry, plumbing, electrical, or HVAC installation.

Quantity Surveyor: Quantity surveyors develop cost models during a project to track the project's cost profiles. The quantity surveyors oversee the project budget, check expenditures, and provide financial advice. They also help with procurement and contract management.

1.8 Construction Contract Documents

Construction contract documents are a collection of documents used in the construction sector to specify a project's terms, conditions, and requirements. These documents are meant to reduce the possibility of disagreements or miscommunications by making sure that everyone involved in the construction project is aware of their roles and duties. These documents include:

- Contract Agreement: A legally binding agreement outlining the obligations, terms of payment, and scope of work for both the client and the contractor.
- Drawings: The project design, including floor plans, elevations, sections, details, and requirements, are shown in depth by architectural, structural, mechanical, electrical, and other technical drawings. As a visual aid for contractors and subcontractors, drawings specify the measurements, materials, finishes, and other specifications for the construction.
- Specifications: Specifications are written descriptions of the items, materials, construction techniques, and quality standards needed for the project are found in specifications. To guarantee that the construction work satisfies the required quality and performance standards, specifications supplement designs by offering extra technical information and requirements.
- Schedule of Values: The document lists every task in a construction project and their estimated cost is used to check project progress and payment.
- Change Order Forms: This document is used to record and seek modifications to the first contract.
- Subcontractor Agreements: contracts defining the obligations and payment terms of the contractor and subcontractors for the projects.
- Addenda and Change Orders: this represent any modifications made to the project's budget, schedule, or scope.
- Schedule of Values outlines the amount and timing of payments to the contractor and subcontractors.
- Payment and Performance Bonds: ensure the contractor's performance and payment. With the use of this document, the customer can get paid for services that contractors either did not complete or completed inadequately.
- Insurance Requirements: describes the insurance that the project needs.
- Safety Plan: describes the project's safety standards and protocols.
- Subcontract Agreements: describe the terms of the contractor-subcontractor agreement.

1.8.1 How the Design Team Can Use Drawings to Support Sustainable Construction

- Conceptual drawings: These can be used to discuss and evaluate concepts associated with sustainable construction, such as the planning of green roofs, energy-efficient floor plans, and the orientation of buildings towards the sun.
- Technical drawings: These can be used to define and explain environmentally compliant building supplies, systems, practices, and processes, including air sealing, insulation, and rainwater collection.
- 3D renderings: These can aid in visualising the plan and highlighting its sustainable elements, such as daylighting, shading techniques, and natural ventilation.

- Life-cycle assessments: Throughout their whole lives, drawings can be used to compare and assess how different building systems and materials affect the environment.
- Performance simulations: With the use of drawings, these can be used to predict and model a building's energy and environmental performance, helping to find problem areas and optimise sustainability measures.
- Site analysis: Sketches can be used to evaluate and examine the site, considering elements like wind directions, solar radiation, and the ease of access to natural resources like water and shade.
- Energy-efficient design: Plans and designs for energy-efficient buildings can be created with the help of drawings, which can also be used to specify the locations of windows, insulation, and HVAC systems.
- Material selection: Drawings can be used to plan the choice of materials, considering things like environmental impact, lifecycle cost, and recyclability.
- Water management: Drawings can be used to plan and design water management systems, such as rainwater collecting, greywater recycling, and energy-efficient plumbing fixtures.
- Daylight optimization: By maximising a building's use of natural light, drawings can reduce the need for artificial lighting and energy consumption.
- Indoor air quality: With drawings, the indoor air quality of a building may be planned and developed, including ventilation systems and the use of non-toxic materials.
- Material specification: Engineers and architects have the power to require the use of environmentally friendly and sustainable resources, like recycled and biodegradable materials.
- Energy-efficient systems: Energy-efficient systems for things like lighting, heating, cooling, and ventilation may be specified by architects and engineers to reduce the building's energy usage.
- Indoor air quality: Architects and engineers can specify ventilation rates and non-toxic material usage, among other standards and rules for interior air quality.
- Water management: The water usage of a building can be reduced by specifying water management systems, such as rainwater collecting and greywater recycling, by architects.
- Waste management: By employing waste management strategies like composting and recycling, architects may reduce the impact of their projects on the environment.
- Green roofing: Architects can specify green roofing methods that improve interior air quality, reduce energy use, and support local ecosystems.

1.8.2 Measures that the Design Team Can Use Specifications for to Achieve Sustainable Construction

These specifications can ensure that sustainable design principles are integrated into the building during construction and operation, helping to achieve more sustainable outcomes.

- Material specifications: The design team can specify environmentally compliant and sustainable materials, like those with a tiny carbon footprint and low emissions during construction.
- Energy-efficient systems: The design team has the power to order energy-saving appliances like insulation, lighting, and highly efficient HVAC systems installed.
- Water conservation: The design team can specify low-flow showerheads and toilets, as well as rainwater harvesting systems and other water-saving fixtures and equipment.
- Indoor air quality: Architects can specify the use of elements and practices that promote ideal indoor air quality, such as suitable ventilation, air filtration, and non-toxic coatings.
- Recycling and waste management: Architects might require the implementation of green waste management plans and composting toilets, among other waste management alternatives.
- Certifications and standards: The design team may need the use of certified green building systems and materials, such as those that meet LEED, GBCN, BREEAM, GBI, or green rating tools.

1.8.3 Measures that the Design Team Can Use Contract Agreement to Achieve Sustainable Construction

By the incorporation of these elements in the contract agreement, design teams may certain that sustainable design principles are included in the building project from the concept and that all parties are working together to achieve a sustainable construction.

- Clearly defining sustainability aims and goals: The contract agreement might specify the level of sustainability that is wanted for the building project, including water conservation, energy efficiency, and indoor air quality.
- Materials and systems specifications: A contract agreement might include specifications for sustainable materials and systems, such as water-saving fixtures, energy-efficient HVAC systems, and low-emitting building materials.
- Green certifications: By stipulating in the contract agreement that the building must have green building certifications like Leadership in Energy and Environmental Design, Building Research Establishment Environmental Assessment Method, Green Building Index, China Green Building Evaluation Standard,

Comprehensive Assessment System for Built Environment Efficiency, Sustainable Green Ecosystem Council, Green Mark Certification, Green Building Council of Nigeria or Green Globes, sustainable design and construction practices can be guaranteed.

- Responsibilities: The contract agreement may name certain team members or contractors in charge of implementing sustainable design features, such as renewable energy systems.
- Monitoring and reporting: Regular reporting and monitoring of sustainability performance, including energy use, water use, and indoor air quality, may be required by the contract agreement to help ensure that sustainability targets are being met.
- Project scope: The contract agreement has a description of the project's scope, including its specific sustainability aims and requirements.
- Material selection: The contract agreement can include stipulations about the use of environmentally friendly and sustainable products, such as low-emitting building materials and those with a low carbon footprint.
- Energy-efficient systems: The contract agreement can include provisions about the use of energy-efficient technologies, such as high-efficiency HVAC systems, insulation, and lighting.
- Water conservation: Water-saving fixtures and technology, such as rainwater harvesting systems and low-flow showerheads and toilets, might be included in the contract agreement.
- Indoor air quality: The contract agreement may include provisions about the use of elements and practices that promote ideal indoor air quality, such as suitable ventilation, air filtration, and non-toxic coatings.
- Waste management and recycling: The contract agreement may include details about the use of waste management and recycling equipment, such as composting toilets and green waste management techniques.
- Performance guarantees: Contract agreements may include performance promises like targets for energy use, indoor air quality, and water usage.

1.8.4 Measures that the Design Team Can Use Bill of Quantities to Achieve Sustainable Construction

The design team can ensure that sustainable design concepts are integrated into projects during construction and that the client is aware of the benefits and drawbacks of adopting sustainable construction practices by including the required specifications in the bill of quantities.

- Material specification: The bill of quantities can include a clause requiring the use of and sustainable materials, components, such as low-emitting construction materials and those with a small carbon footprint.

- Energy-efficient systems: The bill of quantities can include specifications for energy-efficient equipment including lighting, insulation, and high-efficiency HVAC systems.
- Water conservation: The BOQ may include specifications for water-saving fixtures and equipment, such as rainwater harvesting systems and low-flow showerheads and toilets.
- Indoor air quality: The BOQ can specify how to use parts and configurations that promote excellent indoor air quality, such as suitable ventilation, air filtration, and non-toxic coatings.
- Recycling and waste management: The BOQ has an overview of how to use recycling and waste management solutions, such as composting toilets and green waste management techniques.
- Quantification of materials and systems: By measuring the quantity of systems and materials needed for the project, the BOQ may give a precise image of the costs associated with sustainable construction.
- Life cycle cost analysis: The bill of materials can be used to perform a life cycle cost analysis, which considers the long-term costs of owning and maintaining the building, such as energy and water usage, maintenance, and replacement prices.

1.8.5 General Conditions in the Contract as a Means to Achieve Sustainable Construction

- Give clear instructions on the goals and requirements of sustainability.
- List the general requirements for sustainable waste management techniques.
- Make recommendations for energy-efficient materials and equipment to use.
- Compel regular reporting and monitoring of developments in sustainability
- Promote collaboration amongst all project participants.
- Encourage material reuse as well as other waste reduction and recycling strategies.
- Identify renewable energy sources and energy-saving equipment to incorporate into an energy-efficient construction strategy.
- Include details about the use of sustainable resources, such as recycled, locally sourced, or certified products.
- Require the use of water-saving measures, like low-flow plumbing and efficient irrigation systems.
- List the non-toxic, low-emitting materials and methods that should be applied to improve the quality of indoor air.
- Make sure the construction site is kept responsibly, with measures taken to prevent soil erosion and preserve fragile ecosystems.

1.9 Summary

The distinctiveness of the construction business is seen in the way every project is distinct from the others in various aspects including timing, choices, needs, expectation, location and use. Projects are transient, and after they are completed, resources are moved to another project somewhere else. The sector is renowned for having a complicated network structure where many project participants have different roles to perform. In addition to supporting other sectors including housing, health care, and education, the construction industry is vital to the economy because it helps carry out government objectives and policies. It is a significant source of employment and job development, especially when it comes to lowering unemployment rates. In addition, the sector has to deal with multiple problems including cost overruns and project implementation delays. The construction sector has a significant role in job development and employment. The sector also has to deal with problems like cost overruns, delays, reworks, inefficiencies, poor communication and collaboration. Contributions from the construction sector, often range from 5 to 10% in many countries. Construction workers make up a sizable portion of work-related mortality and morbidity, making it a high-risk sector in terms of occupational safety and health.

Bibliography

1. Ahn YH, Pearce AR (2007) Green construction: contractor experiences, expectations, and perceptions. J Green Build 2(3):106–122
2. Akintoye A, Skitmore M (1991) Profitability of UK construction contractors. Constr Manag Econ 9(4):311–325
3. Boyd D, Chinyio E (2008) Understanding the construction client. Wiley
4. Brandon PS, Lu SL (eds) (2009) Clients driving innovation. Wiley
5. Charrett D, Potts A (2018) The duty of care in design—can engineers rely on codes of practice? In: The application of contracts in engineering and construction projects. Informa Law from Routledge, pp 101–114
6. Chilombo Chifunda C, Maganaris E, Foster N (2024) Assessing the duty of care architects owe to third parties in the case of failure of professional duty under English Law. Int Acad J Law Soc (IAJLS) 1(3):143–159
7. Clough RH, Sears GA, Sears SK., Segner RO, Rounds JL (2015) Construction contracting: a practical guide to company management. Wiley
8. Cohen AM, Bain JW (1988) Negligence claims in construction litigation. Constr Law 8:3
9. Cooke B (2014) Management of construction projects. Wiley
10. Convery J (1999) Standard form building contracts and duty of care. Mod L Rev 62:766
11. Emmitt S, Ruikar K (2013) Collaborative design management. Routledge
12. Emmitt S (2014) Design management for architects. Wiley
13. Fathalizadeh A, Hosseini MR, Silvius AG, Rahimian A, Martek I, Edwards DJ (2021) Barriers impeding sustainable project management: a social network analysis of the Iranian construction sector. J Clean Prod 318:128405
14. Fewings P, Henjewele C (2019) Construction project management: an integrated approach. Routledge
15. Hays JW (2003) Construction defect claims against design professional and contractors. Constr Law 23:9

16. Haugbølle K, Forman M, Bougrain F (2014) Clients shaping construction innovation. In: Construction innovation, pp 119–133
17. Hughes W, Murdoch JR (2001) Roles in construction projects: analysis and terminology. Construction Industry Publications
18. Holt GD (1997) Classifying construction contractors. Build Res Inf 25(6):374–382
19. International Labour Organisation (ILO) (1937) C062—safety provisions (building) convention, 1937 (No. 62). https://normlex.ilo.org/dyn/normlex/en/f?p=NORMLEXPUB:12100:0::NO::P12100_INSTRUMENT_ID:312207
20. Li A, Zhu Y, Li Y (eds) (2014) Proceedings of the 8th international symposium on heating, ventilation and air conditioning. Springer, Berlin, pp 551–561
21. Ive G, Gruneberg S (2000) The economics of the modern construction sector. Springer
22. Olanrewaju A, Tan SY, Kwan LF (2017) Roles of communication on performance of the construction sector. Procedia Eng 196:763–770
23. Olanrewaju A (2022) An artificial neural network analysis of rework in sustainable buildings. IOP Conf Ser Earth Environ Sci 1101(2):022003
24. Partridge PH, Noletto VA Jr (1988) Construction management: evolving roles and exposure of construction managers and architects/engineers. Am J Trial Advoc 12:55
25. Ruddock L, Lopes J (2006) The construction sector and economic development: the 'Bon curve.' Constr Manag Econ 24(7):717–723
26. Senaratne S, Gunawardane S (2015) Application of team role theory to construction design teams. Archit Eng Des Manage 11(1):1–20
27. Tan Y, Shen L, Yao H (2011) Sustainable construction practice and contractors' competitiveness: a preliminary study. Habitat Int 35(2):225–230
28. Wells J (2007) Informality in the construction sector in developing countries. Constr Manag Econ 25(1):87–93
29. Winch GM (2009) Managing construction projects. Wiley
30. Wong WK, Cheung SO, Yiu TW, Pang HY (2008) A framework for trust in construction contracting. Int J Project Manage 26(8):821–829

Chapter 2
The Nature of Sustainable Construction

Abstract This chapter introduces the concept of sustainable construction and emphasises its critical importance in the construction sector. It provides a foundational understanding of sustainable construction management, outlining the key principles, strategies, and practices that are designed to align construction projects with sustainability targets. The chapter also highlights the need for a dedicated professional discipline in sustainable construction management, noting that it is not yet a core function in the architecture, engineering, and construction sectors. By integrating various techniques and sustainable design elements, the chapter explains how stakeholders, particularly design and construction teams can achieve environmentally responsible, economically viable, and socially beneficial construction outcomes. It draws attention to the challenges faced in implementing sustainable construction practices, such as insufficient stakeholder engagement, poorly executed projects, and the pressures of meeting project timelines and budgets. The chapter also underscores the importance of a flexible, multidisciplinary approach that considers environmental, social, and economic factors at every stage of the construction process. The proposed framework is adaptable to different project contexts, aiming to enhance profitability and productivity while minimizing negative impacts on communities and the environment. Ultimately, this chapter serves as a comprehensive guide for the stakeholders in the construction sector to encourage a systemic approach to sustainable development in construction projects.

Keywords Sustainable development · Sustainability · Construction project

2.1 Learning Outcomes

This chapter serves as an introduction to the book, providing the background to key concepts and terminology referenced throughout. It presents the rationale for why sustainable construction management should be a shared concern across all stakeholders in the built environment. While sustainable construction management is an emerging professional discipline, its effective implementation requires the active

A. Olanrewaju, *Sustainable Construction Management: Research and Practice Companion*, Management in the Built Environment,
https://doi.org/10.1007/978-981-96-4910-5_2

involvement of everyone impacted by construction activities—not just those formally trained in the field. At the time of writing, sustainable construction management has yet to be fully recognized as a distinct function within the architecture, engineering, and construction (AEC) sectors or construction sector. A core aim of this book is to highlight the urgent need for dedicated professionals in sustainable construction management. This call is not intended to replace existing roles but to complement them by addressing responsibilities that lie beyond the traditional boundaries of current AEC disciplines. This chapter introduces the key principles, strategies, techniques, and practices that contribute to achieving sustainability in construction projects. It also explores how these elements are integrated into various stages of design and project execution to support sustainable outcomes. After studying this chapter, you will be able to:

- Justifying the concerns about sustainable development
- Explain construction sustainable development
- Explain sustainable construction management problems
- Explain the focus of each of the chapters.

2.2 Sustainable Development

Sustainability, also known as sustainable development, affects everyone, regardless of whether they reside in industrialised, emerging, or underdeveloped economies, and such make it both a worldwide problems and a global goal. Sustainable development and sustainability are sometimes used synonymously, but while some have defined sustainability as a series of processes and pathways leading to sustainability (e.g., sustainable production, sustainable construction, sustainable materials, etc.), the others have viewed sustainability as a long-term goal (i.e., a more sustainable world). In this text, both are used interchangeably. This is to avoid distraction and confusion, especially as both are intertwin or not same in practice. The concept of sustainable development must be embraced and put into practice if humanity is to survive and thrive [1]. The concept has a similarity with the concept of public goods in economics. Everyone should be concerned about it. Sustainable development has its roots in environmental economics, which was first developed in 1960. Environmental economics is primarily concerned with quantifying the effects (i.e., costs) of economic decisions on the external environment and offering ways to reduce or eliminate these effects, which are typically significant but are frequently unknown at the time when the economic decisions is made. In order to put future development on sustainable paths, sustainable development requires that individuals, families, communities, businesses, nationals, companies, corporations, and international bodies pay close attention to a number of factors, including population growth, the distribution and supply of foods, the survival or extinction of plant and animal species, the supply and consumption of energy, the distribution of income and wealth, industry, and urban settlements. "Sustainable development is the pathway to the future we want for all," says Ban Ki-moon. Sustainable development provides a framework for achieving social justice, environmental stewardship, economic prosperity, and improved governance. In order to achieve sustainable development, it is imperative that social, environmental, and economic aspects be considered while making decisions. For the sake of simplicity, without sacrificing its meaning or generalizability, sustainable development restricts the use of natural resources. Essentially, the public benefits that are obtained should outweigh the costs. Economics, the concept of choice will resonate. Because we all want more that what we can afford, we have to make a choice. In making the choice, we have to considered both the benefits and cost, and the opportunity costs in particular. In a similar consideration, the global consumption is far higher than what the planet could sustainably produce. As a result, the rate of consumption is not sustainable. In a similar vein, far more waste than the planet can handle is produced and released. As a result, some fundamental presumptions must be modified. As of right now, technological advancements are unable to mitigate the effects of climate change. The numerous pledges and allegiances being made during the COP annual summit are not encouraging. The majority of the promises rest on unproven and untested technologies. It is a fact that advances in technology and social structure limit how well we can use the resources that are currently available to us (natural resources). The good news is that social organisations and technologies can work together to create sustainable development. We hope this is an incentive to every individual, company, organization and government.

The concept of sustainable development is built around 3 tripods. To keep the conversation or discourse simple, the three economic, social, and environmental elements of sustainable development are frequently mentioned, despite the fact that there are many indicators. These three are sometimes referred to as the planet, people, and profit. To prevent any one indication from having disproportionate influence over the others, the three should generally be harmonised. In actuality, though, there are situations when the requirements are incompatible with one another. For instance, economic progress necessitates the exploitation of natural resources, which has a detrimental effect on the environment. Additionally, emphasising the social dimension may result in government, corporation, and individual revenue losses that are necessary for economic prosperity. Various people and institutions have defined sustainable development differenttly. It is generally defined as *the development that meets the needs of the present without compromising the ability of future generations to meet their own needs*. This definition was given by the UN World Commission on Environment and Development, also referred to as the Brundtland Commission, in 1987 [1].

The two fundamental ideas behind sustainable development are as follows:

1. The view that "needs," especially the basic needs of the world's poor, should take precedence over all other considerations; and
2. The notion that societal structure and technological advancements have placed restrictions on the environment's capacity to support both the present and future requirements.

The three dimensions of sustainable development are concatenated, not separate. A sustainable future is one in which societal, economic, and environmental factors are balanced for the sake of achieving a higher standard of living. For instance, the provision of food and resources, clean air, and safe drinking water to its populace is contingent upon a thriving ecology. There are multiple arguments for individuals, companies, sectors, industries, corporations, and nationals to be motivated by sustainable developments. The primary motivations are described in Table 2.1.

Table 2.1 Motivation for interest in sustainable development

Motivation	Explanation
Enhanced life quality	Better living circumstances, clean water, healthcare, increased well-being, and more affordable and clean energy are possible outcomes of reaching the SDGs
Financial gains	It explains how implementing sustainable practices can help a person, business, or nation financially. Many sustainable initiatives, such as waste reduction, energy efficiency, and green technologies, can generate income and commercial opportunities
Collaboration	To tackle the problems related to sustainable development, cooperation between individuals, organisations, communities, and governments is obligatory
Legacy	Contributing to the SDGs offers you the opportunity to be a part of a historic movement that will be remembered by future generations
Private commitment	Insist that each person handles their actions. Everyone should know that we are all responsible for the future of our planet and its inhabitants. People should consider the kind of world they wish to inherit for future generations
Inclusivity	Concerns people would ensure that the vulnerable and marginalised people are included in the development process by putting the SDGs into practice
Connectivity	The consideration is that all people have an impact on all others. Not only do governments and organisations need to be concerned about this. Individual actions can all contribute towards the goals, no matter how small. Individuals, corporations, communities, businesses, and sectoral levels can all make contributions
Global agenda	To emphasise that the Sustainable Development Goals (SDGs) are an international agenda that aims to address some of the most pressing problems the world is currently facing, including poverty, hunger, and climate change. Making a beneficial impact on the world for everyone, not just those in one's own country or community, is a requirement of taking part in the SDGs
Innovation and entrepreneurship	Promote the idea that achieving these two elements is necessary to achieve the goals of sustainable development. People should be encouraged to apply their creativity and come up with novel solutions to the problems facing the world today that will last
Environment concern	Present the argument that actions taken to stop climate change, support sustainable construction practices, and practice responsible consumption and note that sustainable production is beneficial for both the global community and nearby ecosystems and wildlife

2.3 Sustainable Construction

The need for sustainable construction has increased since sustainable development was recognised. The biggest danger to the aims of sustainable development is the construction sector. Construction involves greenhouse gases emission. For instance, to meet high demand, a timber processing company can cut down an excessive number of trees. For the company and its workers, cutting a lot of trees makes financial sense since it would lower the cost of producing lumber, which would boost earnings. In addition, customers would benefit since they would be able to buy high-quality timber at a lower price. However, if the tree-cutting process continued without inspection for some time, the company would have to stop operation and put its employees—locals—out on the job market, which would raise the unemployment rate. Not only would this directly affect the company, but other companies that rely on the supplies would have to close their doors to business or face higher price increases. The cost of lumber, for instance, will rise due to increased production requirements and a willingness on the part of the company producing the timber to pay higher salaries, which will result in labour scarcity for other businesses. In the long run, excessive tree-cutting will also cause social discontent and environmental deterioration. Another example would be the growing need for energy, such as electricity and crude oil. While this is necessary to boost the nation's economic prosperity, it will result in significant pollution emissions into the atmosphere, which will negatively affect people's quality of life by deteriorating the air and water. Population growth will be affected, medical treatment costs will rise, and treatment costs for erosion and air and water pollution will rise as well. Another popular construction material is cement, whose usage and manufacture have a big impact on the environment. For example, cement is the main building material used in the construction industry and a necessary input in the manufacture of concrete. However, the emissions of carbon dioxide (CO_2) from cement manufacture would affect human health and the environment.

From the foregoing, it obvious that the construction sector poses a serious challenge to the sustainable development agenda. As an example, multiple studies [3] show that buildings consumed over 40% of global energy, emitted 1/3 of CO_2, used around 25% of harvested wood, released roughly 50% of fluorocarbons, produced 40% of materials for landfills, used 45% of energy for operations, and generated 40% of greenhouse gases. 15% of the world's potentially useful water is consumed by emissions. Five times as many contaminants are found within buildings as outside. The impact is considerably greater when other building-related CO_2 emissions are considered, such as those from the production and transportation of building and demolition materials as Olanrewaju [3] explained. If we consider other energy use related to structures, the amount of energy consumed by buildings increases even more. For instance, the annual energy required to heat and cool a single building is eight to ten times less than the embodied energy in the building's envelope. These figures indicate that a portion of the danger to sustainable growth is posed by the construction sector. Rising sea levels, altered rainfall patterns, erratic weather, melting glaciers, extended wet spells, protracted droughts, floods, landslides, scarcity of water, deforestation, etc. are all effects of climate change that have a big impact on building performance. Even though there are a lot of direct emissions, there are a lot of more indirect emissions. There are three categories for emissions. Scope 1 includes emissions that a person or business produces directly because of operations such as driving a car, burning fossil fuels on-site, and operating boilers. Indirect emissions are covered by scope 2, which includes emissions generated during the generation of heat, steam, or electricity that a business obtains from outside sources. Since they encompass a variety of upstream and downstream activities, scope 3 emissions are the most widespread and difficult to measure, and are produced at many stages of the value chain, including the acquisition and processing of raw materials, transportation, product usage, and product disposal following their useful life. As a result, it is necessary to reconsider how construction are conceived, designed, constructed, operated and approached in practice.

Sustainable construction has been defined as "a construction that meets the needs of the present without compromising the ability of future generations to meet their needs," which is in line with the definition of sustainable development. So, construction that aims to reduce capital expenditure, operating costs, and environmental and community harm while keeping user acceptance is referred to as sustainable construction. It can also be described as a construction project that aims to protect the environment for future usage while also preserving the social, cultural, ecological, spiritual, and economic needs for society's well-being. This concept, in general, makes it possible to measure the components of a sustainable construction. Throughout this book, the phrases "sustainable construction" and "green construction," as well as "energy-efficient construction," are synonymous and occasionally used interchangeably.

The pursuit of sustainable construction has been prompted by worries about resource depletion, pollution, climate change, population growth, and environmental degradation. The book's central claim is that building contributes significantly to climate change. Sustainable construction delivery requirements will result in significant energy optimisation, improved housing durability, waste minimization, improved social impacts, a healthy indoor environment, reduced pollution, consideration of the life cycle and cost of the building project, active user-friendliness, user comfort, and overall satisfaction.

The common features of sustainable construction include:

- optimising energy use;
- minimising CO_2 emission;
- maximising the use of water resources;
- utilising durable construction materials and components;
- using locally sourced construction materials;
- using local labour;
- collecting surface water;
- managing waste effectively;
- minimising all types of pollutions;

- maximising the health and well-being of building occupants;
- fostering and maintaining the local ecology;
- reduce maintenance;
- neighbourhoods;
- site management.

Unlike traditional construction management, sustainable construction management approaches aim to address many environmental, social, and economic problems associated with the conventional construction methods.

2.4 Sustainable Construction Management Problems

Sustainable construction management techniques and practices are purposively designed to address or lessen the many environmental, social, and economic problems that are linked to conventional construction management techniques and practice. The main problems that the sustainable construction management seek to address are contained in Table 2.2. I personally encountered many of the problems on the construction site as a young quantity surveyor in 2000. Some of the are also discussed in the preface of this book. The problems are still present today and are not getting worse. The strategies, practices, techniques and methods of sustainable construction management are developed around these problems.

Sustainable construction management practices aim to create construction infrastructure and structures that are socially and environmentally responsible as well as long-term and economically practical by tackling sustainability problems.

Table 2.2 The problems that sustainable construction management aims to address

Problem	Indication
Resource depletion	Traditional construction methods rely on the exploitation of finite resources, leading to the depletion of such resources. Sustainable practices try to utilise fewer resources by promoting the use of renewable or recycled materials
Energy consumption	Conventional buildings use a lot of energy over their lifetime as well as during construction. A primary aims of sustainable construction are energy-efficient design, construction, and operation to reduce overall energy consumption and promote the use of renewable energy sources
Greenhouse gas emissions	The construction sector is one of the major contributors to greenhouse gas emissions. Sustainable practises aim to minimise carbon footprints through the utilisation of low-emission building processes, sustainable materials, and energy-efficient designs

(continued)

Table 2.2 (continued)

Problem	Indication
Waste generation	Using traditional construction methods usually results in the production of many waste items. Sustainable construction strives to reduce waste through efficient design, construction techniques, and the use of recyclable or recycled resources
Loss of biodiversity and destruction of habitats	Two common effects of construction are habitat destruction and biodiversity loss. Sustainable practices prioritise promoting biodiversity, safeguarding natural habitats, and mitigating adverse effects on ecosystems in and around construction sites
Pollution and water consumption	Water supplies may be negatively affected by construction operations due to pollution and overuse. The primary aims of sustainable construction techniques include water-efficient designs, rainwater collecting, and actions to stop water contamination
Health and well-being	Certain hazardous substances that have an impact on indoor air quality and occupant health may be present in conventional construction materials. Sustainable construction aims to produce healthier interior spaces by using non-toxic, ecologically friendly materials
Inclusivity and social equity	Sustainable construction methods guarantee inclusivity in the planning and execution of construction projects, support local communities, and encourage fair labour practices as a means of addressing social concerns. It promotes universal design and sell create social value for the community where the projects are located
Adaptation to climate change	The principle of sustainable construction recognises the need to adapt to the evolving climate and develop robust structures that can withstand environmental challenges such as extreme weather, rising sea levels, and temperature fluctuations
Life cycle assessments	The conventional construction management put create on the capital cost, thereby neglecting the whole life appraisal. The principle of sustainable construction acknowledges the necessity of adjusting to the changing climate and creating resilient structures that can tolerate environmental stressors including severe weather, increasing sea levels, and temperature variations
Challenges of urbanisation	As transportation congestion, population density, and expanding infrastructure demands are taken care of, sustainable construction practices aim to lessen their adverse effects on the environment

2.5 The Theory of Sustainable Construction Management

To reach sustainability across environmental, social, and economic dimensions, the theory of sustainable construction management presents framework that directs the planning, execution, and administration of construction enterprises. Fundamentally, this theory recognises the built environment's enormous impact on the environment and society at large, and it also recognises that construction projects have long-term effects that go beyond the project's designated area. Essentially, sustainable construction management is about incorporating sustainability objective and requirement into the project's planning and design phases and its construction, operation, and final reuse or decommissioning. Construction Management Enabled Sustainable Development highlights the role of construction management in facilitating sustainable development. Construction management practices enable and support the attainment and achievement of sustainable development goals. In this context, construction management is viewed as a driver of sustainable development, guiding projects towards environmentally responsible, socially equitable, and economically viable outcomes. This all-encompassing strategy makes sure that sustainability factors are integrated into every activity and decision-making process, not just an optional add-on or afterthought. According to this concept, sustainability has several distinct aspects. The main goals of environmental sustainability are to minimise the depletion of resources, reduce waste production and pollution, and mitigate the environmental deterioration caused by construction operations. To reduce the environmental impact of construction projects, this entails implementing techniques like waste reduction, green building practices, energy-efficient design, integration of renewable energy, and sustainable material procurement.

Since social sustainability considers how construction projects affect local communities, labourers, and society, it is equally significant. This covers aspects like encouraging social justice, encouraging community involvement and empowerment, making sure construction workers have safe and healthy working conditions, honouring indigenous rights and cultural heritage, and improving the standard of living for locals. Economic sustainability guarantees that building projects are financially workable and contribute to long-term economic growth, which balances environmental and social concerns. This entails distributing resources as efficiently as possible, reducing lifecycle costs, increasing return on investment, and promoting economic resilience in the face of unpredictability and shifting market conditions. At its core, the philosophy of sustainable construction management acknowledges that to solve urgent global issues like social injustice, resource depletion, urbanisation, and climate change, sustainable practices are not only desirable but also necessary. Through the incorporation of sustainability principles into building methodologies, projects can reduce adverse effects, improve resistance to external disruptions, and boost the welfare of present and future generations.

This text combines two theories together: construction management and sustainable development. Construction Management Enabled Sustainable Development emphasizes the pivotal role of construction management in advancing sustainable development objectives. Within this framework, construction management practices serve as catalysts for transformative change, driving projects towards environmentally responsible, socially equitable, and economically viable outcomes. This is with the primary purpose for achieving sustainable construction target. Through effective project planning, coordination, and execution, construction managers integrate sustainability principles into every stage of the project lifecycle. They prioritize resource efficiency, minimize environmental impacts, promote social inclusivity, and optimize economic benefits. By embracing innovation, collaboration, and stakeholder engagement, construction management becomes a powerful agent of positive change, shaping the built environment in harmony with the principles of sustainable development. The theory of sustainable construction management offers a strong framework for coordinating construction methods with more general sustainability aims, highlighting the interdependence of environmental, social, and economic factors, and directing decision-making procedures towards more accountable and resilient results. Table 2.3 has the key aspect in theorizing sustainable construction management.

In general, resilience-construction, integrating sustainability principles into construction methods, and advancing the welfare of current and future generations are all made possible by the philosophy of sustainable construction management. It

Table 2.3 Key aspects of the theory of sustainable construction management

Key	Explanation
Triple bottom line approach	The triple bottom line approach, which considers social, environmental, and economic concerns when making decisions, is emphasised by sustainable construction management. This strategy aims to reach sustainable results by striking a balance between social justice, environmental conservation, and economic growth
Environmental indicator	Sustainability in construction management places a high priority on resource efficiency and environmental conservation. The primary aim is to minimise the environmental impact of construction activities by the reduction of energy consumption, waste creation, conservation of natural resources, and mitigation of pollution and emissions
Social equity and community engagement	The significance of social equity and community well-being is acknowledged by sustainable building management. It entails interacting with stakeholders, such as workers, marginalised groups, and local communities, to guarantee that building projects enhance society and promote social cohesion, inclusion, and quality of life
Economic indicator	Long-term value creation and economic viability are key components of sustainable construction management. Evaluating the financial viability of sustainable practices and investments in building projects, entails taking lifespan costs, return on investment, and cost–benefit analysis into account
Integrated design and planning	Sustainable construction management encourages an integrated approach to planning and design that incorporates sustainability factors right from the inception of a project. This calls for multidisciplinary cooperation, all-encompassing viewpoints, and the incorporation of sustainable design concepts at every phase of the project's lifecycle
Performance metrics and assessment	To gauge and appraise the sustainability performance of building projects, sustainable construction management uses performance metrics and assessment instruments. This involves tracking developments and finding areas for improvement using metrics including energy efficiency, carbon footprint, indoor environmental quality, and social effect

(continued)

Table 2.3 (continued)

Key	Explanation
Regulatory compliance and certification	Adhering to pertinent rules, guidelines, and certifications about sustainability in the built environment is a crucial part of sustainable construction management. This could entail following environmental laws, building codes, and optional certification schemes
Innovation and continuous development	The field of sustainable construction management is dynamic and ever-evolving, which promotes innovation and constant development. To achieve sustainability goals and solve contemporary issues like resource scarcity, urbanisation, and climate change, it entails embracing innovative technology, materials, and behaviours
Social value	Sustainable construction management to address social inequalities, improve well-being, enhance community resilience, and promote sustainable development. This key involve collaboration between government agencies, businesses, non-profit organizations, and local communities to achieve positive outcomes for society as a whole

signifies a change towards more comprehensive, ethical, and responsible construction methods that put social justice, the environment, and economic growth first.

2.6 Ontology of Sustainable Construction Management

The ontology of sustainable construction management refers to the fundamental nature or essence of sustainable construction management as a field of study or practice. It encompasses the underlying principles, concepts, and relationships that define how sustainable construction is understood, analyzed, and applied in various contexts. The key aspects of the ontology of sustainable construction management are contained in Table 2.4.

To create built environments that are socially just, environmentally conscious, and commercially workable, the ontology of sustainable construction management, taken as a whole, stands for an integrated and holistic viewpoint that aims to advance sustainability practices and principles within the construction sector.

Table 2.4 Key aspect of the ontology of sustainable construction management

Aspect	
Principles of sustainability	The fundamental tenets of ontology are sustainability, encompassing social, environmental, and economic factors. To ensure long-term profitability and resilience in the built environment, sustainable construction management strives to strike a balance between these three pillars
Interdisciplinary nature	A variety of academic fields, including engineering, architecture, surveying, environmental science, economics, sociology, and policy studies, contribute to sustainable construction management. Its ontology reflects the multidisciplinary character of sustainability and the demand for integrated methods to deal with difficult problems
Lifecycle perspective	The ontology acknowledges the significance of considering all stages of a construction project's lifecycle, including planning, designing, constructing and manufacturing, operating, maintaining, and deconstructing. Throughout a construction's and infrastructure's lifetime, it is imperative to minimise environmental effects, maximise social and economic advantages, and optimise resource utilisation, according to the lifecycle perspective
Systems management	Using a system management method, sustainable construction management view the construction projects part of larger social, environmental, and economic systems. The interdependencies and feedback loops between the many elements of the built environment and the larger ecosystem are highlighted by this viewpoint
Ethical and value-based framework	Sustainable construction management's ontology is founded on moral precepts and value-driven judgment. It recognises the underlying moral issues raised by fostering fairness and social justice, respecting ecological boundaries, and striking a balance between the needs of current and future generations
Continuous improvement and innovation	Moral principles and judgments based on values form the foundation of sustainable building management's ontology. It acknowledges the moral questions that are posed when social justice and fairness are promoted, ecological boundaries are respected, and the demands of present and future generations are balanced

2.7 Framework for Sustainable Construction

In construction terminology, a framework supports the structure of a construction projects. It is also defined as a skeleton of a construction. But it carries different meanings in other disciplines. For instance, in computer language, a dictionary defined it as a structure that shows the types of programmes that can be designed as well the interactions among the programmes. To clarify, a framework provides structure and consistency. It allows the process to achieve a particular set of purposes or aims. In general, a framework is an outline, rules, guidelines or skeleton of ideas or concepts formulated, designed or proposed toward achieving a particular set of aims.

A framework can be graphical, conceptual, statistical, and mathematical. A framework is an abstract presentation of real-life object or model. However, some will disagree with these classifications because some believe that model is mathematical expression. In this chapter, the intention is about conceptual framework or guideline. The frameworks developed in this chapter aim to ease designers to incorporate sustainability features into the planning, designing and construction of construction projects. The roles and duties of members of the designed professional are contextualised into the various stages of the construction plan of work and the three main components of sustainability. These were developed based on the Institution of Civil Engineers (ICE), the Construction Industry Council (CIC) Scope of Services, the Construction Industry Research and Information Association (CIRIA) Guidelines, the American Society of Civil Engineers (ASCE) Guidelines, the Project Management Institute (PMI) Project Management Body of Knowledge (PMBOK), the Royal Institute of British Architects (RIBA) plan of work 2020 and the Chartered Institute of Building (CIOB) guidance notes. These plans of work and guidelines, though not specific to sustainable construction, provide valuable frameworks and best practices for managing civil engineering projects, ensuring that they are delivered efficiently, safely, and in accordance with industry standards and regulatory requirements. This is contained in Table 2.5. The three main components of sustainability are embedded in the phases of the construction process.

2.8 Epistemology of Sustainable Construction Management

The sustainable construction management epistemology refers to how knowledge about sustainable construction is acquired, confirmed, and applied within the built environment or the surveying, architecture, construction and engineering (SACE) sector. It involves understanding the sources, methods, and frameworks through which knowledge is generated, organized, and given, and the underlying assumptions and paradigms that shape our understanding of sustainability in construction management. Table 2.6 contains the key aspects of the epistemology of sustainable construction management:

The epistemology of sustainable construction management is characterized by a dynamic and interdisciplinary approach to knowledge generation, validation, and application, guided by empirical research, case studies, standards, ability, and ethical considerations. It reflects a commitment to advancing understanding and practice in support of sustainable development goals within the built environment.

Table 2.5 Sustainable construction management plan of work/matrix

Stage	Environmental	Economics	Social	Comments
Strategic appraisal	In the strategic definition phase, we concentrate on the smooth integration of environmental requirements into critical areas such as briefing and construction business cases. We carefully choose and consider sustainability metrics when building business cases, considering both short- and long-term environmental effects. To shape the project through stakeholders' collective consciousness, we simultaneously involve stakeholders in the construction briefing to understand their values and environmental goals. By taking this proactive approach, we can make sure that environmental responsibility is integrated into the construction plan from the opening, highlighting our dedication to sustainability as not just a part but an essential driver of it	During the strategic definition stage, align cost optimisation criteria with the value systems of the client or users by seamlessly incorporating them into the creation of business cases, briefs, and other project criteria. This entails a thorough assessment of cost-related variables, guaranteeing that economic sustainability is a primary factor in decision-making processes from the start	We demonstrate our commitment to social sustainability at the strategic definition process by incorporating community benefits into crucial documents like business cases, briefings, and other project requirements. This integration represents a thorough endeavour to match the project's objectives with the client's and end users' values, not just a procedural process This entails considering how the project will affect the larger community, promoting favourable social outcomes, and enhancing the general well-being of individuals who will be affected both directly and indirectly Engaging with stakeholders during the construction briefing is an opportunity to gain insight into their aims and societal values. It is emphasised by this proactive strategy that social sustainability is an essential part of the project's DNA rather than an add. It reflects our dedication to building resilient and inclusive communities by designing areas and structures that serve practical purposes while also strengthening the bonds that bind society together. Prioritising community benefits in the strategic definition stage paves the way for a project that crosses conventional limits and appears as a model for socially conscious and sustainable development	While this stage is critical, sustainability is not well addressed in the previous texts. The traditional procurement is also cited as a problem
	In the strategic appraisal stage of construction development, sustainability is a top priority. This requires a comprehensive strategy that aims to maximise economic efficiency, improve social well-being, and minimise environmental impact Reduced pollution, waste reduction, efficient cost management, and decreased carbon emissions are all part of the key sustainability requirements. A sense of community and inclusivity are fostered by the construction's simultaneous efforts to support neighbourhood and cultural integration Achieving these sustainability goals requires making critical strategic decisions about finance means, design and construction team makeup, orientation, and site selection. While cautious team selection guarantees knowledge in line with sustainable standards, thoughtful site selection and orientation can help preserve the environment In addition, the utilisation of environmentally sustainable building techniques and economical finance models can efficiently lower pollution, waste production, crowded living conditions, and upkeep expenses. This complete strategy has a favourable impact on social dynamics by going beyond the physical features of construction. It lowers the rate of crime, encourages intercultural harmony, builds friendships, improves household well-being, and eventually builds a community that is peaceful and happy Integrating these sustainability indicators into the strategic definition phase, in essence, signifies a commitment to creating spaces that resonate with long-term well-being and positive community impact, while also supporting economic viability and harmonising with nature. This goes beyond simply meeting construction goals			

(continued)

Table 2.5 (continued)

Stage	Environmental	Economics	Social	Comments
Briefing	*Site selection for waste reduction and environmental preservation* Examine how careful site selection can drastically reduce pollution, waste production, and environmental deterioration. Analysing the effects of deliberate site selections on resource conservation and natural ecosystems is part of this *Construction orientation for environmental mitigation* Examine how construction orientation affects environmental effect mitigation. Examine how waste production, pollution, and environmental damage can be decreased by strategically aligning construction with natural factors *Team selection for sustainable design and construction* Analyse how waste reduction, CO_2 emissions, and pollution are affected by the choice of design and construction teams. Examine the advantages of working with professionals to achieve environmentally sensitive design and construction processes, such as LEED specialists *Financial methods (e.g., BREEM) for pollution reduction* Examine how construction financing, by utilising the knowledge of professionals, to reduce waste, mitigate CO_2 emissions, and regulate pollution. Examine how sustainable financing influences ecologically friendly building techniques This thorough investigation conducted during the 'Preparation and Brief' phase offers insights into how strategic choices regarding team composition, construction orientation, site selection, and financing options all work together to advance environmental sustainability goals in construction projects	*Site selection for enhanced construction economy* Examine how the construction industry is affected by wise site choices. Examine how careful site choice can reduce expenses, simplify coordination, and improve project efficiency *Construction orientation and economic advantages* Examine the effects that construction orientation has on the construction industry. Analyse the financial benefits of cutting expenses and increasing resource efficiency while building in harmony with the environment *Design and construction teams' role in construction economics* Examine how construction economics is directly impacted by the makeup of design and construction teams, including quantity surveyors and construction managers. Examine their contributions to overall economic sustainability, effective project management, and cost optimization *Financial methods for cost reduction* Analyse how the construction finance approach lowers costs. Examine how sustainable financing options can reduce project costs overall and guarantee long-term financial viability during the building phase This in-depth analysis conducted during the 'Preparation and Brief' phase offers insightful information about how strategic choices about building orientation, site selection, and team composition, and financing options all work together to improve the financial viability of construction projects	*Site Selection for Community and Household Well-being* Examine how thoughtful site choice enhances the quality of life for households and the community. Examine the benefits for facilities, accessibility, and overall quality of life *Construction orientation and its influence on community and house well-being* Analyse how construction orientation contributes to the improvement of household and community well-being. Examine the favourable effects of intelligent orientation on elements such as sunlight, greenery, and general livability *Design and Construction Teams' Contribution to Stakeholder Well-being* Examine how the comfort, happiness, and general well-being of the community are directly impacted by the makeup of design and construction teams, which includes architects and project managers. Think about how they can help create environments that encourage inclusivity and social harmony *Financial methods for enhanced integration* Examine how the construction finance model promotes community integration. Examine how sustainable funding strategies can improve community activities, build social ties, and advance general well-being This thorough investigation during the 'Preparation and Brief' phase offers insightful information about how strategic choices about team composition, construction orientation, site selection, and financing options all work together to promote social sustainability and improve the quality of life for communities and households engaged in construction projects	Like the strategic definition phase, sustainability is not often addressed. The lack of collaboration between the design team is cited as a problem

(continued)

Table 2.5 (continued)

Stage	Environmental	Economics	Social	Comments
Concept design	*Reducing embodied energy* Investigate methods for reducing the embodied energy in building supplies and parts. Think about using recycled or locally obtained materials, low-impact manufacturing techniques, and sustainable sourcing *Optimizing operational energy efficiency* Examine strategies for improving the energy efficiency of operations. To reduce the building's long-term energy consumption, this entails incorporating passive design ideas, energy-efficient technologies, and renewable energy sources *Mitigating waste generation* Evaluate ways to cut waste at every stage of the building and operation process. Consider design decisions that emphasise waste-free materials, investigate recycling possibilities, and use building techniques that reduce waste production overall *Controlling pollution impact* Identify ways to reduce pollution when a building is being built and when it is being used. This includes making environmentally friendly material choices, putting pollution avoidance measures into action when building, and adding systems that lessen pollutants produced during operation *Minimizing energy and water consumption* Investigate innovative methods for reducing water and energy usage. To do this, energy-efficient appliances, water-saving fixtures, and intelligent systems that optimise resource utilisation across the building's lifecycle must be integrated To address environmental sustainability comprehensively	*Economizing construction costs* Identity for ways to save building expenses without sacrificing quality. Value engineering, economical material selections, and construction techniques that improve cost-effectiveness without sacrificing structural integrity should be considered *Minimizing maintenance costs* Identify design options that reduce long-term maintenance costs. This entails making use of long-lasting materials, and implementing low-maintenance construction technologies, and guaranteeing accessibility for regular maintenance and repairs *Facilitating future modifications with cost efficiency* Make future adjustments, such as renovations or alterations, to maximise cost-effectiveness. Think about adaptable systems, modular building methods, and flexible design components that enable simple alterations without having a significant budgetary impact By carefully preparing for future alterations, minimising maintenance costs, and managing construction costs effectively	Enhancing Household and Community Wellbeing Examine design approaches that enhance the welfare of families and the larger community. Consider components that improve accessibility, cultivate a feeling of community, and put inhabitants' comfort and health first *Promoting happiness through design choices* Examine how design decisions can enhance occupant satisfaction. This entails designing areas to take full advantage of natural light, encourage indoor-outdoor flow, and include components that enhance the quality of life *Cultivating a positive community image* Determine how the design idea enhances the good reputation of the neighbourhood. Give careful thought to community areas, landscaping, and architecture features that improve the neighbourhood's overall character and image By placing a high priority on household well-being, encouraging happiness through deliberate design decisions, and fostering a positive community image	

(continued)

Table 2.5 (continued)

Stage	Environmental	Economics	Social	Comments
Detailed design	*Reducing embodied energy* Investigate innovative ways to reduce embodied energy in building supplies and parts. Reevaluating material selections, implementing sustainable sourcing methods, and giving low-impact manufacturing processes top priority are all necessary to achieve this *Optimizing operational energy efficiency* Examine innovative methods to improve operational energy efficiency. This entails integrating ultramodern technology, passive design ideas, and renewable energy sources to drastically lower the building's overall energy usage *Mitigating waste generation* Use improved techniques to cut waste even more during the building and operation phases. This could entail innovative recycling techniques, the ideas of the circular economy, and building strategies intended to reduce waste production overall *Controlling pollution impact* Create plans to actively reduce pollution while buildings are being built and operated. Consider innovative materials with less of an influence on the environment, strict procedures for preventing pollution, and innovative technologies that minimise operational pollutants *Minimizing energy and water consumption* Investigate innovative methods for reducing water and energy usage. This means using the newest water- and energy-saving fixtures, appliances, and intelligent systems that maximise resource utilisation throughout the building's lifetime Intending to minimise waste and pollution, optimise energy and water consumption in building projects, reduce embodied energy, and improve operational efficiency	*Economizing construction costs* Investigate creative ways to reduce building expenses without sacrificing quality. This entails going back to value engineering, making effective material selections, and using construction techniques that increase the economy without sacrificing structural integrity *Minimizing maintenance costs* Examine design options that reduce long-term maintenance costs in a planned manner. This entails making use of long-lasting materials, implementing low-maintenance construction technologies, and guaranteeing accessibility for regular maintenance and repairs *Facilitating future modifications with cost efficiency* Make sure to optimise cost efficiency while planning for future adjustments, such as refurbishments or renovations. Think of easily adjustable systems without having a significant negative influence on the economy, as well as flexible design features and modular building processes To ensure a well-balanced and financially workable construction project	*Image enhancement through thoughtful design* Examine architectural features that enhance the building's favourable perception. Consider landscaping, public areas, and architectural aspects that improve the building's overall recognition and visual appeal *Recognition and community identity* Examine how the design might enhance identification and recognition within the community. This entails designing areas that are consistent with the neighbourhood's ideals, encouraging a sense of community, and adding features that capture the neighbourhood's character *Comfort-centric design choices* Consider the positive effects that design choices can have on community comfort and resident well-being. Think of features that improve the quality of the air inside the building, make the most of natural light, and encourage general well-being both inside and outside the structure *Neighbourhood cohesion and integration* Examine design approaches that foster community cohesion. To build a stronger feeling of community, this entails creating public areas, taking accessibility into account, and encouraging social contact possibilities By improving the building's image, encouraging community recognition, putting comfort first, and creating neighbourhood togetherness	Energy efficiency components are considered. While sometimes it affects economics, social aspects are not considered

(continued)

Table 2.5 (continued)

Stage	Environmental	Economics	Social	Comments
Technical design	*Minimizing embodied energy in detail design* Investigate minute details to reduce the embodied energy in building supplies and parts. To lessen the overall environmental impact, consider creative design solutions, efficient manufacturing techniques, and sustainable sourcing *Enhancing operational energy efficiency* Identify advanced technological ideas to improve energy efficiency in operations. This entails adjusting building systems, incorporating ultramodern technology, and maximising the structure's total energy performance *Strategies to reduce waste and pollution* During the construction phase, implement exact technological steps to minimise pollution and waste generation. This could involve the use of cutting-edge building techniques, recycling programmes, and pollution control measures that are incorporated into the technical requirements *Optimizing energy and water consumption* Employ precise technological measures to reduce waste production and pollution throughout the construction period. This could entail using state-of-the-art construction methods, recycling initiatives, and pollution control strategies integrated into the technical specifications By optimising energy and water consumption in construction projects, minimising embodied energy, improving operational efficiency, lowering waste and pollution, and reducing waste	*Value engineering in detail design* Investigate value engineering options to save expenses without sacrificing quality. This entails a thorough examination of the design components and requirements to pinpoint affordable substitutes and effective fixes *Efficient material specifications* Examine ways to save costs on materials without sacrificing durability or function. Choose materials that strike a compromise between cost and quality to make sure you make a wise choice that supports the project's objectives for economic sustainability *Streamlined construction processes* Use construction methods that increase productivity without sacrificing quality or safety. Provide technical information that facilitates efficient workflows, cutting down on construction time and, hence, overall expenses *Lifecycle cost considerations* While defining the specifications, consider the lifecycle costs of the systems and materials. To ensure a thorough economic analysis, consider not only the initial costs but also ongoing maintenance and operating expenses Using value engineering, the choice of affordable materials, the optimisation of construction procedures, and the careful consideration of lifecycle costs in the design and specifications	*Architectural Aesthetics for Positive Image* Include design components that help create a favourable impression of the building. landscaping, external elements, and detailed architectural aesthetics that improve the building's overall recognition and visual appeal *Recognition through innovative design features* Include innovative design elements that help the building become known in the neighbourhood. Think about distinguishing features that make a statement and serve as markers that help people develop a sense of who they are *Comfort-centric design choices* Put comfort first when creating the requirements and convoluted design. Make necessary adjustments to lighting, HVAC systems, and interior design aspects to create areas that put inhabitants' comfort and well-being first *Neighborhood integration in technical specifications* Include technical requirements that ease the integration of neighbourhoods. Describe the public areas, features for people with disabilities, and architectural components that support a friendly and connected neighbourhood environment By improving the building's image, encouraging recognition, placing a premium on comfort, and fostering neighbourhood togetherness through detailed design and specifications	Only technology-based solutions are considered

(continued)

Table 2.5 (continued)

Stage	Environmental	Economics	Social	Phase	Comments
Assembling	*Strategic site management for efficiency* Put into effect strategic site management techniques that maximise energy efficiency, waste minimization, and environmental impact overall. Reducing the number of resources used on-site, involves effective scheduling, waste management strategies, and sustainable construction techniques *Sustainable material procurement strategies* Purchasing materials should be done sustainably by choosing materials with less impact on the environment, and recyclability and ethical sourcing. Reduce the embodied energy and pollutants associated with materials by working with suppliers who follow environmentally responsible practices *Competent subcontractors and specialists engagement* Make certain that qualified subcontractors and experts with knowledge of sustainable building techniques are engaged. Work together with experts who place a high priority on waste minimization, energy-efficient solutions, and ecologically friendly construction practices *Innovative construction technologies* Investigate and use innovative construction solutions that diminish waste, maximise environmental effects, and improve operational energy efficiency. This could use modular building, prefabrication, and sophisticated on-site production procedures *Water and energy-efficient on-site practices* Put into action on-site procedures that prioritise energy and water conservation. To reduce operational energy consumption during construction, use water-saving techniques, and energy-efficient machinery, and give priority to renewable energy sources Broadly, this stage addresses environmental sustainability holistically through site management, sustainable material procurement, hiring qualified subcontractors, utilising innovative technologies, and on-site implementation of water and energy-efficient practices	*Strategic material selection for cost efficiency* Maximise cost effectiveness by carefully choosing your materials. Consider a material's longevity, long-term worth, and economic impact while evaluating it. Make economical options your priority while supporting quality in both on- and off-site production *Claim avoidance through robust contracts* Reduce the likelihood of claims and disagreements by creating strong contracts that include obligations, deadlines, and channels for resolving conflicts. Throughout the construction process, a well-structured contract can help ensure financial sustainability and prevent needless costs *Proactive risk avoidance measures* Take preventative action to minimise hazards that can result in unexpected expenses. This includes creating risk mitigation plans, conducting in-depth risk assessments, and putting backup plans in place to guarantee financial stability throughout the production and construction stages *Reduction of variations through detailed planning* Reduce variances and alterations throughout construction by using thorough planning and communication. Create a thorough strategy from the commencement that outlines the project's needs and specifications in detail to lessen the possibility of expensive adjustments and modifications During the 'Manufacturing and Construction' phase, this improved method seeks to attain economic sustainability through deliberate material selection, proactive risk management, avoidance of claims, and variance minimization-all of which support financially feasible and cost-effective construction processes	*Effective conflict resolution strategies* Use efficient conflict resolution techniques to resolve problems quickly and stop arguments from getting worse. Encourage team members to share their concerns and work together to create a culture of open communication. Make sure that disputes are settled fairly and quickly *Enhanced site communications* Make open lines of communication a top priority while on the job. Provide a venue for regular communication, use digital technologies to provide updates in real-time, and make sure everyone in the team is informed. This increases overall site communication, fosters a collaborative atmosphere, and reduces miscommunication *Productivity-driven site culture* Establish a site culture that is focused on productivity by establishing clear standards, offering sufficient training, and praising and rewarding exceptional work. Promote cooperation among employees, give them more authority, and put effective processes in place to increase output overall and support social sustainability *Training for conflict prevention* To avoid confrontations and arguments, train staff members who will be working on-site. Give the group tools for resolving disputes, encourage inclusivity and diversity, and establish a welcoming workplace that prioritises the health of all those engaged in the building process By promoting an inclusive and productivity-driven site culture, improving site communications, and implementing effective dispute resolution, this enhanced approach within the 'Manufacturing and Construction' phase seeks to address social sustainability holistically		Currently, most attention is paid to this stage

(continued)

Table 2.5 (continued)

Stage	Environmental	Economics	Social	Phase	Comments
Handover and operation	*Comprehensive commissioning plan* Create an extensive commissioning plan that explains each household's obligations in detail. This plan should include testing, safety, security, emergency readiness, and operational rules and outlining the procedures for a smooth transfer *Critical equipment list integration* To prioritise and maximise the operation of crucial systems, incorporate an essential equipment list into the commissioning strategy. This focused strategy reduces energy use and possible waste by ensuring that vital equipment runs as efficiently as possible *Holistic testing approach* Adopt a comprehensive testing strategy that includes every facet of the building's operation in testing. This covers waste management, energy systems, and pollution prevention techniques. Extensive testing guarantees that every part function sustainably for the environment *Comprehensive operation manuals* Create comprehensive operation guides that instruct users on energy-saving procedures and sustainable behaviours. To encourage ecologically responsible building use, including instructions on waste management, energy-saving techniques, and pollution control measures *Training for sustainability* During the handover period, conduct training workshops for homes on sustainable behaviours. This contributes to a more environmentally friendly building operation by teaching tenants about waste reduction, pollution prevention, and energy-efficient usage A commissioning plan, crucial equipment integration, thorough operation manuals, testing, and training for sustainable practices are all part of this enhanced strategy, which is part of the 'Handover and Close Out' phase and attempts to address environmental sustainability holistically	*Energy-efficient commissioning plan* Create a commissioning strategy that gives energy efficiency a top priority for the duration of the building's operation. This entails putting smart technologies into practice, testing and optimising systems to reduce energy waste, and making sure that every part performs at its best *Waste reduction strategies* Include waste management techniques in the commissioning plan. This entails choosing and put into action strategies to reduce operational waste, encourage recycling programmes, and embrace behaviours that support the operation of buildings in a way that is more economically sustainable *Pollution control measures* Put pollution control measures into action following the commissioning plan. This entails following environmental standards, testing and improving systems to reduce pollutant emissions, and making sure the building's operational phase is in line with aims for economic sustainability *Optimizing energy consumption* Provide tactics in the commissioning plan for maximising energy usage. To reduce operating costs, this entails teaching stakeholders about energy-efficient behaviours, implementing energy-saving technologies, and offering recommendations for the operation of sustainable buildings *Water conservation initiatives* Incorporate water-saving measures into the commissioning plan. During the building's operational phase, outline steps to minimise water consumption, put effective water management systems in place, and encourage sustainable water usage practices Comprehensively addressing economic sustainability is the goal of this refined approach during the 'Handover and Close Out' phase. It includes an energy-efficient commissioning plan, waste reduction strategies, pollution control measures, energy consumption optimisation, and water conservation initiatives	*Personalized comfort settings* Include elements in the commissioning plan that let homeowners customise their living environments. This could include lighting preferences, HVAC system adjustments, and other environmental settings that enhance personal comfort *User-friendly operation manuals* As part of the commissioning plan, provide easily navigable operation manuals. Clearly stated guidelines on how families can maximise comfort feature, making sure they are knowledgeable and able to quickly operate the building's systems to improve their level of happiness overall *Efficient communication channels* Provide effective routes of communication in the commissioning plan to handle household input and concerns. Encourage a prompt response to questions and the development of a caring atmosphere that puts residents' happiness and well-being first *Community engagement initiatives* Incorporate community participation programmes into the plan for commissioning. To promote a feeling of community and contribute to social sustainability by improving the general satisfaction and well-being of inhabitants, organise forums or activities that bring homes together *Education on sustainable practices* Incorporate instructional elements that educate households about sustainable habits within the commissioning plan. Encourage the adoption of behaviours and practices that support social sustainability, such as eco-friendly living, community service, and a shared dedication to a cosy and fulfilling home Through the prioritisation of individualised comfort, user-friendly communication, community participation, and education on sustainable practices within the commissioning plan, this improved approach within the 'Handover and Close Out' phase seeks to address social sustainability holistically		This is seldom identified as an opportunity to reaffirm sustainability in construction delivery

(continued)

Table 2.5 (continued)

Stage	Environmental	Economics	Social	Phase	Comments
Use and operation	*Efficient usage practices* Encourage tenants to use resources efficiently. Encourage eco-friendly behaviour by using user-friendly guidance and educational activities to promote responsible use of energy and water, waste reduction, and pollution prevention *Strategic retrofitting measures* Reduce operational energy consumption and improve energy efficiency by implementing strategic retrofitting initiatives. This could entail installing energy-saving equipment, improving insulation, and using intelligent technologies that maximise resource utilisation *Sustainable extension planning* Give sustainable design concepts top priority while arranging additions. Reduce CO_2 emissions and energy consumption by using energy-efficient materials, renewable energy sources, and technology that supports environmental sustainability goals *Refurbishment for Sustainability* Refurbishment projects can be made more sustainable by using eco-friendly materials, increasing energy efficiency, and making waste reduction techniques. Try to prolong the life of construction materials to reduce the impact replacements have on the environment *Regular maintenance for efficiency* Stress routine maintenance procedures that improve operational effectiveness. In line with environmental sustainability goals, proactive maintenance can minimise energy waste, lessen the need for major repairs, and extend the life of building components *Water conservation initiatives* When the system is being used and run, conservation measures are in place. To reduce total water consumption, promote water-saving behaviours, install water-efficient fixtures, and think about sustainable landscaping choices Through the promotion of efficient consumption, strategic retrofitting, sustainable extension planning, refurbishment for sustainability, routine maintenance, and water conservation efforts, this refined approach under the 'Use and Operation' phase seeks to completely address environmental sustainability	*Proactive maintenance practices* To save money on repairs and extend the life of building components, adopt proactive maintenance techniques. Saving money on operations can be achieved through planned maintenance schedules, prompt repairs, and routine inspections *Cost-efficient retrofitting measures* Use affordable retrofitting techniques to improve operational performance and energy efficiency. Make enhancements that have a good return on investment your priority. Examples include smart technologies that maximise resource consumption, better insulation, and energy-efficient lighting *Economical extension planning* Prioritise inexpensive design concepts while organising additions. To save operating costs, by examine inexpensive materials, investigate low-cost energy-efficient options, and make sure the extension complies with the objectives of economic sustainability *Budget-conscious refurbishment* Incorporate cost-effective renovation techniques that support long-term economic viability. Choose affordable materials, give maintenance a higher priority than major replacements, and implement refurbishing techniques that result in long-term operational savings *Energy consumption optimization* Maximise energy use by taking economical actions. To cut operating expenses, install energy-efficient equipment, carry out energy audits to find areas for improvement, and train tenants on energy-saving techniques *Water consumption efficiency* During the use and operation phase, place special emphasis on water consumption efficiency. Implement economical strategies that reduce water usage and lower total operating costs, such as water-saving fixtures, effective irrigation systems, and landscaping techniques Through the promotion of proactive maintenance, economical retrofitting, cost-conscious extension planning, economical refurbishment, energy consumption optimisation, and water consumption efficiency, this refined approach within the 'Use and Operation' phase seeks to comprehensively address economic sustainability	*Community engagement initiatives* Urge customers and building occupants to participate in the community. Encourage a feeling of community by organising activities, sharing areas, and working together on projects that enhance the general integration and the well-being of the community *Comfort-centric design considerations* Encourage the implementation of user-friendly design principles that put comfort first. This could entail designing warm and welcoming public areas, offering functional and beautiful spaces, and adding features that improve residents' general comfort *Promotion of psychological welfare* Encourage psychological well-being by establishing surroundings that place a high value on mental health. Promote stress-relieving activities, providing users with access to green spaces, and incorporate natural components into the building to enhance their psychological well-being *Satisfaction-oriented amenities* Offer facilities and services that prioritise customer happiness to improve building users' overall experience. Think about adding leisure areas, wellness initiatives, and assistance services that enhance residents' happiness and fulfilment *Crime prevention measures* Work together with the local police to put crime prevention measures in place both inside and outside the facility. This could involve installing security measures, community policing programmes, and well-lit walkways to make the area safer and lower the chance of crime *Disease prevention strategies* Advocate for behaviours that help prevent illness. Establish hygienic practices, offer enough ventilation, and encourage behaviours that are health-conscious to cultivate a population that is resilient to illness within the building Through the promotion of community participation, comfort-centric design, psychological well-being, satisfaction-oriented facilities, crime prevention, and disease prevention techniques, this enhanced approach under the 'Use and Operation' phase seeks to holistically address social sustainability		Few households are doing this with the Do It Yourself (DIYS) approach. Yet it is only about energy efficiency

Table 2.6 Aspect of epistemology in sustainable construction

Aspect	Explanation
Interdisciplinary knowledge	Sustainable construction management draws on knowledge from diverse disciplines, including engineering, architecture, environmental science, economics, sociology, and policy studies. Its epistemology acknowledges the importance of integrating insights from multiple disciplines to address complex sustainability challenges in the built environment
Empirical research	Empirical research plays a significant role in advancing knowledge about sustainable construction management. This includes studies investigating the environmental, social, and economic impacts of construction practices and evaluations of the effectiveness of sustainability interventions and policies
Case studies and best practices	Case studies and best practices provide valuable insights into real-world examples of sustainable construction projects and initiatives. They serve as crucial sources of knowledge about successful strategies, innovative approaches, and lessons learned in sustainable construction management
Standards and certification	Standards and certification systems contribute to the epistemology of sustainable construction management by setting up criteria and benchmarks for sustainable building practices. These frameworks help to codify knowledge and provide guidance for practitioners
Ability and professional experience	Ability and professional experience play a significant role in shaping the epistemology of sustainable construction management. Practitioners, researchers, and industry professionals contribute valuable insights and perspectives based on their hands-on experience and practical knowledge of sustainability principles and practices
Ethical considerations	The epistemology of sustainable construction management acknowledges the ethical dimensions of knowledge production and application. It recognizes the importance of ethical conduct, transparency, and accountability in research, decision-making, and practice, particularly about sustainability goals and values
Continual learning and reflection	Sustainable construction management embraces a culture of continual learning and reflection. Its epistemology emphasizes the importance of critical thinking, reflexivity, and open dialogue in advancing knowledge and understanding of sustainability issues within the construction industry

2.9 The Duty of Care of Sustainable Construction Managers

Tort laws are a branch of civil law that deal with situations where one person's actions cause harm or loss to another person, either intentionally or through negligence. In simpler terms, tort law allows individuals to seek compensation when they have been wronged or injured—physically, emotionally, or financially by someone else's wrongful act (a "tort"). The main goal is to restore the injured party to the position they were in before the harm occurred. Common types of torts include; a) negligence—e.g., a contractor failing to follow safety protocols and causing an accident, b) Intentional torts—e.g., assault, trespass, or defamation, and c) Strict liability—e.g., harm caused by using hazardous materials, regardless of intent or precautions. In construction: Tort law could apply when: a) A building defect causes injury to a third party, b) Unsafe site conditions harm passers-by and c) Poor workmanship results in property damage. Duty of care is the first element in a tort law. The others are breach, causation and damages. To establish a wrong doing, wrongful act or omission that support compensation, the plaintiff must proof with material evidence that defendant own them a duty of care. The other components that need to be proven will depend on the kind of tort. For instance, strict liability torts do not require evidence of fault or carelessness, but intentional torts do require proof of intent to cause harm. It is crucial to remember that the plaintiff has the burden of proof to prove each component of the tort. Duty of care exist for the professionals in a construction contract. To establish a breach of duty, it must be demonstrated that the defendant's actions fell below the level of care that a reasonable person would have used in identical circumstances. The duty of care is a legal requirement imposed on persons or organisations that requires them to act with the care and caution that a reasonable person would use in similar circumstances. A duty of care might arise in a variety of scenarios, such as sustainable construction managers providing competent care to their clients. The level of care required in a given situation will be determined by the circumstances of the case. For example, a sustainable construction manager's duty of care would be evaluated in comparison to the level of care provided by other sustainable construction managers in their field. If a sustainable construction manager fails to fulfil their duty of care and causes harm or loss to construction stakeholders or their property, they may be held legally accountable for their carelessness. The injured party may be entitled to compensation for their losses, including medical expenditures, lost wages, and pain and suffering.

The components of the duty of care in tort law are:

- The first element is the defendant's duty of care towards the plaintiff. A duty of care is a legal requirement to use reasonable care to prevent injury to others.
- Foreseeability: The plaintiff's harm must have been foreseeable by the defendant during the claimed breach of duty. The test for foreseeability is whether a reasonable person in the defendant's situation would have predicted the injury as a probable result of their actions or omissions.
- The defendant broke their duty of care by failing to fulfil the standard of care expected of a reasonable person in their position under the given circumstances.
- The plaintiff must prove that the defendant's breach of duty caused the loss they suffered. There must be a causal link between the defendant's actions and the plaintiff's injuries.
- The plaintiff must have experienced damage, such as physical injury, emotional distress, or financial loss, due to the defendant's breach of duty.

Individuals must take reasonable care to prevent inflicting injury to others, and if they fail to do so, they may be held accountable for any resulting harm. To establish a claim for breach of duty of care, the plaintiff must prove that the duty existed, that the defendant breached their duty by failing to satisfy the standard of care, that the injury suffered was foreseeable and caused by the breach of duty, and that the plaintiff incurred some type of loss.

2.10 Duty of Care of Professional Sustainable Construction Managers

Professional sustainable construction managers supervise the design, construction, and operation of buildings and infrastructure projects with a focus on sustainability requirements and objective. The duty of care of sustainable construction management includes a variety of tasks aimed at fostering environmentally responsible and socially conscious construction processes while protecting the health, safety, and well-being of all stakeholders.

The duty of care sustainable construction managers compels them to operate with the competence, knowledge, and care that a reasonably competent and experienced sustainable construction manager would exercise in similar circumstances. Professional sustainable construction managers have a duty of care to operate in the best interests of their clients and guarantee that their work meets the highest standards of professionalism and competence, with a primary focus on sustainable construction management practices and principles.

The duties of care of professional sustainable construction managers will include:

- Ensure that design and construction methods meet or exceed appropriate environmental standards and laws, such as LEED, GBI, Green Star, BREEAM.
- Incorporating sustainable materials and technologies in the design and building process, such as renewable energy sources, green roofs, or water-saving systems.
- Overseeing the maintenance and operation of the building or infrastructure project to ensure continuous sustainability.
- Advising their client on best practices for sustainable construction and operation, as well as any difficulties that may occur during the design or construction phases.
- Sustainable construction managers are accountable for reducing the environmental impact of construction projects. This includes decreasing resource consumption, effectively managing waste, and using eco-friendly building materials and technology to reduce pollution and protect natural resources.
- They guarantee that construction activities follow applicable environmental regulations, building codes, and sustainability guidelines. This includes remaining current with local, national, and international environmental legislation and ensuring that construction techniques meet these standards.
- They create risk management methods to prevent environmental dangers such as soil contamination, air and water pollution, and habitat destruction, and they take proactive steps to mitigate any consequences.

- They promote social responsibility within construction projects, considering the well-being of communities, workers, and future generations. This includes interacting with local stakeholders, addressing social issues like equal employment opportunities and community development, and promoting inclusive and ethical practices throughout the project's lifecycle.
- Sustainable construction managers prioritise worker and occupant health and safety throughout the construction process. To avoid accidents, injuries, and occupational health concerns, they put in place safety protocols, offer proper training and protective equipment, and keep the workplace safe.
- Sustainable construction managers also owe a duty of care to third parties who may be impacted by their activity, such as neighbouring property owners or the general public. In some situations, sustainable construction managers may be held accountable for these parties' damages if they fail to fulfil their duty of care.

2.11 Summary

In this book, we have laid the groundwork for a comprehensive approach to sustainable construction, aimed at guiding stakeholders—especially the design and construction teams—in aligning their practices with sustainability criteria. This preliminary framework or typology is inspired by the Institution of Civil Engineers (ICE) Conditions of Contract, The Construction Industry Council (CIC), the Construction Industry Research and Information Association (CIRIA) Guidelines: The American Society of Civil Engineers (ASCE) Guidelines, and CIOB Planning Protocol 2021 (CIOB PP21), guide to managing Safety–Critical Elements (SCEs) in building construction and RIBA Plan of Work 2020 but extends beyond them by incorporating a wealth of theories and empirical evidence. The underlying premise is that achieving sustainability in construction delivery is a multidisciplinary endeavour. It involves a complex interplay of processes, procedures, and stakeholders, each playing a critical role at various stages of construction development—from the initial concept through design, construction, operations, and ultimately disposal. This chapter dig into the reasoning behind this book, explaining the necessity of introducing a sustainable construction framework. The primary goal is to enhance the profitability and productivity of contractors while minimizing the negative impacts of construction activities on communities, residents, users, and the environment. Sustainable building practices are intended to offer long-term benefits, not only in terms of environmental stewardship but also by contributing to the economic viability of the construction industry. However, despite the clear benefits, there have been numerous instances where sustainable construction projects were poorly planned or hastily executed, leading to suboptimal outcomes. These failures often arise due to a lack of comprehensive understanding of sustainability principles, insufficient stakeholder engagement, or the pressures of meeting deadlines and budgets. The framework presented in this book seeks to address these challenges by providing a structured approach that emphasizes careful planning, stakeholder collaboration, and adherence to sustainability criteria throughout the entire construction lifecycle. Moreover, this book recognizes that sustainable construction is not a one-size-fits-all solution. Different projects may require tailored approaches depending on the specific environmental, social, and economic contexts in which they are situated. Therefore, the matrix or typology we propose is flexible and adaptable, allowing for customization to meet the unique demands of each project. In summary, the book serves as a guide for stakeholders in the construction industry, offering a pathway to achieving sustainability in construction delivery. It stresses the importance of a holistic approach, where sustainability is integrated into every stage of the construction process, from planning and design to execution and beyond. By following this framework, the construction industry can contribute to the broader goals of sustainable development, ensuring that buildings are not only functional and cost-effective but also environmentally and socially responsible.

Bibliography

1. Brundtland Commission (1987) World commission on environment and development, our common future. Oxford University Press, Oxford and New York
2. Ibiyeye AI, Shamang KJ, Malachy G (2024) Awareness of building sustainability: a case of architectural students of Kaduna state university. Soc Sci Hum Open 10:100927
3. Olanrewaju AL (2022) An artificial neural network analysis of rework in sustainable buildings. IOP Conf Ser Earth Environ Sci 1101(2):022003
4. Olanrewaju A, Tan SY (2018) An exploration into design criteria for affordable housing in Malaysia. J Eng Des Technol 16(3):360–384
5. Ndumbe SI, Velikov P (2024) Government strategies on cybersecurity and how artificial intelligence can impact cybersecurity in healthcare with special reference to the UK. In: Cybersecurity and artificial intelligence: transformational strategies and disruptive innovation. Springer Nature Switzerland, Cham, pp 217–236
6. MacRoberts (Firm) (2015). MacRoberts on Scottish construction contracts. Wiley
7. Wang J, Ding Z, Zou L, Zuo J (eds) (2013) Proceedings of the 17th international symposium on advancement of construction management and real estate. Springer Science & Business Media

Chapter 3
Foundations of Sustainable Construction

Abstract This chapter provides a foundational understanding of sustainable construction, addressing the need for a shift away from conventional building practices that are resource-intensive and environmentally damaging. The construction sector is under increasing pressure to mitigate its impact on the environment while ensuring social and economic sustainability. Sustainable construction is positioned as a solution to these challenges, emphasising techniques, principles, and strategies that promote long-term ecological balance and resilience in the built environment. The chapter explores key sustainable practices, including the use of green materials, energy-efficient design, waste reduction, and water conservation, highlighting their role in enhancing the sustainability of construction projects. It also examines the importance of certification systems that validate the sustainability of materials and practices. By integrating these practices throughout the construction process from design to operation, stakeholders can achieve improved environmental performance, reduced carbon footprints, and increased project longevity. Additionally, the chapter discusses the benefits, prospects, and challenges associated with sustainable construction, offering a roadmap for overcoming obstacles and fostering a greener, more resilient construction industry. Ultimately, this chapter equips readers with the knowledge and tools necessary to contribute to the transformation of construction practices for a sustainable future.

Keywords Sustainable performance · Project benchmark · Sustainable · Development · Resource management

3.1 Learning Outcomes

This chapter lays the foundation for sustainable construction. Conventional construction has a host of problems, that require paradigm shift if the construction sector is to provide value added service; to be an asset rather than a problem to economic growth and development. The conventional construction are not only expensive to own and operate, but they also pose threat to the environment. The current resilience

A. Olanrewaju, *Sustainable Construction Management: Research and Practice Companion*, Management in the Built Environment,
https://doi.org/10.1007/978-981-96-4910-5_3

of construction projects against the growing impacts of climate change is poor. Sustainable construction is conceived to address the shortcomings that are associated with conventional construction. This chapter addresses the multiple techniques, principles, strategies, and practices that incorporate various construction projects to achieve sustainable construction projects. Aspects that apply to sustainable design are explained. After studying this chapter, you will be able to:

- Demand for sustainable construction
- Explain the basic principles of sustainable construction.
- Explain green or sustainable certifications.

3.2 Introduction

The global construction sector stands at the crossroads of unprecedented challenges and transformative opportunities, calling for a paradigm shift towards sustainability. As the demands for infrastructure grow, so does the imperative to construct projects that minimise environmental impact, enhance social well-being, and ensure economic viability. In this context, the foundation of sustainable construction emerges as a critical cornerstone for reshaping the future of the built environment. The urgency to adopt sustainable practices in construction stems from the recognition that the built environment significantly influences the world's ecological balance and societal resilience. Traditional construction methods, often characterised by resource-intensive processes and high carbon emissions, have led to environmental degradation, and contributed to climate change. The pressing need to address these challenges has given rise to a comprehensive approach that considers not only the product but the entire life cycle of a structure. The foundation of sustainable construction represents a commitment to a future where the built environment coexists harmoniously with nature and society. This chapter will delve deeper into the key principles, innovative practices, and procedures10 that underscore the foundations of sustainable construction. Through an exploration of the multifaceted aspects of sustainable construction and the principles of sustainable construction, the most pressing challenges constraining the achievement of sustainable construction goals were discussed. The chapter also outlines ways for the public to contribute actively to the transformation of construction practices for a greener and more sustainable tomorrow.

3.3 Principles of Sustainable Construction

The principles of sustainable construction revolve around creating built environments that follow environmental and social ethics. The principle of sustainable construction involves the basic or fundamental requirements and core concepts that are required to achieve the sustainability goals of construction projects. It is essential to provide a basis for value-added decision-making and guiding actions. Sustainable construction principles aim to minimise the negative impacts of construction and promote long-term resilience. Sustainable construction energy efficiency, application of sustainable materials, water conservation, waste management, site management, protection of biodiversity, adaptive design, healthy indoor environments, social equity and inclusivity, whole-life appraisal, community engagement, environment certification and compliance, innovation and technology, transportation, and mobility.

3.3.1 Energy Efficiency

The amount of energy that construction projects consume during construction and operation is huge. Globally, construction projects consume about 40% of the total energy consumed. About 70% of this is attributable to buildings. Energy efficiency in construction projects refers to the implementation of design, materials, and construction practices that aim to minimise energy consumption and optimise the performance of buildings and infrastructure. Therefore, construction projects must be designed and constructed to maximise energy efficiency, reduce energy consumption, and incorporate renewable energy sources to decrease reliance on non-renewable energy. The goal of energy efficiency is to reduce the amount of energy needed for various aspects of a construction project, from the construction phase to the time the construction is completed and out of operation. This is a strategic approach to creating sustainable built environments. By prioritising energy efficiency in construction projects, stakeholders can reduce the environmental impact of buildings, lower operational costs, and contribute to a more sustainable and resilient built environment. The primary aspects of energy efficiency in construction projects include:

I. Design Considerations

Passive Design: Buildings can minimise their demand for artificial lighting, heating, and cooling by implementing passive design principles, which enhance natural light, ventilation, and insulation. Achieving sustainable structures requires careful consideration of how buildings are positioned to maximise natural sunlight. Building orientation can maximise solar exposure, lowering the need for artificial lighting and heating.

II. Building Envelope

Projecting and shading the activity within the buildings is one of the building envelope's main purposes. Therefore, minimising heat transfer and keeping a suitable interior temperature can be achieved by installing high-quality insulation on walls, roofs, and floors to reduce energy usage. In the end, this would contribute to the decreased requirement for air conditioning or heating systems. To improve thermal performance, energy-efficient windows and doors with good sealing and insulating qualities should be chosen.

III. Effective Heating

It is advised that HVAC systems be fitted with smart controls for best performance and that they be suitably scaled for the building.

IV. Energy-Efficient Lighting

Whenever feasible, the design team should avoid using artificial lighting. This can be achieved by implementing daylighting techniques and adopting energy-efficient lighting technology, such as LED lamps, compact fluorescent lamps (CFLs), Halogen Incandescent Lighting.

V. Renewable Energy Integration

Construction companies should use on-site renewable energy sources, like solar panels, especially during construction to provide clean energy and lessen reliance on grid-supplied electricity. This will have a significant positive impact on site profitability and productivity, not to mention helping to maintain a clean atmosphere.

VI. Smart Building Technologies

Though this might not apply to all building types, there will be a lot of possibility to enhance energy efficiency using sensors and controls to monitor and optimise energy usage in real time.

VII. Energy-Efficient Appliances and Equipment

Energy-efficient appliances and equipment can help reduce energy consumption throughout the building's operation and use, as the building's appliances and equipment consume a significant amount of energy.

VIII. Water Heating Efficiency

The cost of heating water is huge in many nations. Consequently, employing cost-effective and environmentally compliant water heating systems-such as energy-efficient water heaters and appropriately insulated hot water pipes will help maintain a clean atmosphere.

IX. Life Cycle Analysis

High carbon emissions are stored in construction materials. Thus, to assess the environmental impact of various building materials and techniques, construction companies and their clients need to work together to minimise the carbon emission in construction projects.

X. Commissioning and Monitoring

Over 90% of the life of a construction project requires maintenance. As a result, the construction parties should put commissioning procedures into place to guarantee that systems function as intended, and routinely check and repair equipment to quickly detect and resolve issues related to energy efficiency.

XI. Occupant Empowerment

Building occupant behaviour is crucial for energy and maintenance management. Consequently, encouraging energy-saving practices and teaching tenants about energy-efficient practices would support sustainable construction management.

XII. Local Climate Considerations

Sustainable construction management practices are being undermined by imported design, materials, practices, and components that are not considered locally. To improve energy efficiency, the design team should take the local environment into account while designing buildings. This includes taking prevailing winds, humidity levels, and temperature fluctuations into account.

3.3.2 Materials Selection

Reliable sourcing, production, use, and recycling of materials minimises their environmental impact throughout their life cycle, making them sustainable. The entire sustainability of the construction and manufacturing processes is enhanced by these materials. When choosing sustainable materials, the design team should consider features like labour exploitation, resource efficiency, recyclability, and minimal environmental effects during extraction, production, transportation, and use. Resilient and ecologically conscious constructions require the use of sustainable materials. The choice of these materials encourages a more environmentally conscious and circular approach to the built environment and adds to the overall sustainability of construction projects. To support the goals of sustainable construction, specifiers and contractors should be aware of the traits that sustainable materials exhibit:

I. Renewable Resources

To minimise the depletion of natural ecosystems, sustainable materials must be sourced from renewable resources that can be regenerated quickly.

II. Recyclability

To reduce the quantity of construction wastes are dumped in landfills, the materials should be readily recyclable or reused at the end of their life cycle. This need is frequently linked to the waste management cycle.

III. Low Embodied Energy

When the materials are extracted, processed, manufactured, transported, and installed in building projects, they should have minimal negative effects on the environment.

IV. Non-Toxic

A lot of construction supplies and parts are dangerous for users and tenants. Toxic construction materials are sources of many sick building syndromes. Sustainable materials, however, ought to be devoid of or contain little dangerous chemicals. Construction materials lessen environmental pollutants and encourage healthier indoor settings.

V. Durability

Durability is a need for buildings to minimise materials and disturbances. The choice of materials can help achieve this. Because sustainable materials are so robust, there is less need for invasive repairs, and the environmental impact is reduced overall.

VI. Local Sourcing

Sustainable materials are procured locally whenever possible to lower transportation-related energy consumption and boost local economies, all while reducing carbon emissions and creating jobs for the community. Sustainable materials guarantee the preservation and conservation of natural resources. Sustainable development can be promoted while lowering the extraction and use of scarce resources using sustainable materials. Table 3.1 lists a few examples of sustainable materials.

3.3.3 Water Conservation

Most countries have expensive and limited freshwater supplies, mostly because of inadequate harvesting practices. To minimise water use in buildings, sustainable construction places a strong emphasis on the application of water-efficient design and technologies, considering both indoor and outdoor water usage. Using techniques, materials, systems, and technologies to minimise water use, lessen wastewater production, and increase overall water efficiency during the building process and project lifetime is known as water conservation in sustainable construction. Stakeholders can improve water efficiency, lessen their influence on the environment, and encourage the wise use of water resources in the built environment by including water conservation methods. The following are the main elements of water conservation in sustainable construction:

I. Site Planning and Design

Rainwater Harvesting: This incorporates rainwater harvesting systems to capture and store rainwater for use in construction activities or for landscaping purposes.

Penetrable Surfaces: The site should be planned and designed to use permeable surfaces, such as permeable pavements. Such a design would enable rainwater to infiltrate the ground, thereby reducing runoff and promoting groundwater recharge.

Landscaping Design: Plan landscaping to include native plants, which are resistant to drought and require less water for their survival.

Table 3.1 Examples of sustainable materials

Material	Nature
Bamboo	A resource that renews quickly and can be utilised to make flooring, furniture, and structural components, among other features in the building
Recycled steel	Steel made from recycled scrap metal minimises energy use and lowers the need for new raw materials
Reclaimed wood	Salvaged wood from abandoned structures, shipping pallets, and other sources gave materials that would otherwise be thrown away a second chance at life
Recycled concrete	Crushed concrete from destroyed buildings is used as aggregates in freshly mixed concrete, minimising waste, and the requirement for virgin materials
Cork	Cork is a renewable and sustainable substance that is harvested from the bark of cork oak trees and utilised in flooring, insulation, and other applications without causing harm to the tree
Recycled glass	The glass that has been crushed and processed from industrial or post-consumer sources to make concrete, tiles, and countertops made of glass
Hempcrete	An environmentally friendly building material made of hemp fibres, lime, and water that has good insulating qualities
Soy-based insulation	Soybean oil insulation is a renewable resource that provides an eco-friendly substitute for conventional insulating materials
Recycled plastic	Plastics made from recycled materials—such as bottles or packaging—that are used to make a variety of products, like outdoor furniture and building materials
Wool	Natural wood fibres are utilised instead of synthetic materials for insulation, offering fire resistance and temperature control
Straw bales	A renewable and naturally occurring building material made from agricultural waste that is used to insulate walls
Natural stone	Stone, like granite or limestone, that is responsibly and locally obtained and mined is utilised in building because of its strength and visual attractiveness

II. Efficient Water Use During Construction

Effective Irrigation Techniques: To reduce water consumption during landscaping and site development, parties should employ effective irrigation techniques like drip irrigation or soil moisture sensors.

Water-Efficient Equipment: To minimise water waste throughout various building processes, parties should use tools and equipment that are water efficiency.

III. Water-Efficient Building Systems

Low-Flow Fixtures: Parties should install low-flow taps and faucets, showerheads, and toilets for the construction site activities as well as for the completed building to reduce water consumption.

Graywater Systems: The design team should ensure that the design incorporates graywater systems to reuse lightly used water for construction site activities as well as for completed buildings for non-potable purposes such as irrigation or flushing toilets.

Water-Efficient Appliances: Water-efficient equipment, like washing machines and dishwashers, should be selected by the design team for locations where the building's operation requires the usage of water.

IV. Water Recycling and Treatment

Water Treatment Systems On-Site: To recycle and reuse water for non-potable uses or construction, the contractors need to install water treatment systems on-site.

Contractors should treat wastewater from construction activities using innovative water reclamation technologies so that it may be safely disposed of or reused.

V. Water Monitoring and Management

Smart Metering: When feasible, the design team should include smart metering systems to track water usage, identify leaks, and optimise water use instantly.

VI. Material and Components Selection

Water-Efficient Materials: To improve the projects' overall water efficiency, the design team should specify building materials and components that demand less water during manufacture or manufacturing.

VII. Compliance with standards and codes

Compliance with Water requirements: To show a dedication to water conservation, the design should meet or beyond local water efficiency laws and requirements.

VIII. Life Cycle Assessment

Water Impact Consideration: To assess the overall water impact of construction materials and processes, incorporate water considerations into life cycle evaluations.

3.3.4 Waste Reduction and Recycling

The problems of waste generation in the global construction industry is both significant and escalating. According to the World Bank, construction and demolition (C&D) activities contribute approximately 2.2 billion tonnes of solid waste annually, with this figure projected to rise to 2.59 billion tonnes by 2030. If the current practices persist. This figure is projected to triple by 2030 if current practices persist. In certain countries, construction waste accounts for up to 40% of total solid waste, positioning it as a key area for intervention. Beyond its environmental impact, unmanaged waste drives up project costs, causes delays, and exposes stakeholders to regulatory penalties. However, the goal should not only be to manage waste post-production but to prevent its generation from the outset through smart design, accurate procurement, prefabrication, and effective labour management. The construction sector is both a catalyst for development and a major contributor to environmental challenges. In addition to solid waste, construction sites are sources of air pollution, water contamination, and noise pollution. This is not just a waste problem; it is a systemic problem. The traditional waste management approaches of dispose, dump, and forget are no longer feasible. Landfills are reaching capacity, regulatory pressure is increasing, and waste represents not only an environmental burden but also a significant economic loss. Every tonne of material wasted signifies money lost, resources depleted, and additional carbon released into the atmosphere. Construction stakeholders must rethink their approach to waste, shifting from a focus on prevention and reduction to creating circular, market-driven ecosystems in which materials are repurposed, reused, and revalued. The construction sector must evolve from the conventional linear "take-make-dispose" model and the traditional waste hierarchy toward a more circular, market-oriented ecosystem that prioritizes material recirculation and value retention. Envision a future where surplus steel is exchanged via digital platforms, reclaimed timber is recertified for new projects, and "waste" is transformed into a valuable resource for others. Achieving this requires more than just innovation; it requires a cultural shift. We need clients to trust reused materials, governments to reform accreditation systems, and companies to adopt digital tools for material tracking and transparency. Transparent information must be shared. Stakeholders must ask not only, "What does it cost?" but also, "What does it waste?" If a project generates 100 tonnes of waste, and the average material cost is $300/tonne, the potential financial loss (excluding disposal costs) is $30,000 in material alone. Quantity surveyors must sharpen their cost management skills. Construction waste management must evolve into construction resource management. This is not just to reduce harm, but to design value environmentally, economically, and socially into the projects. The constriction parties should adopt practises that promote recycling, reusing, and minimising the number of materials sent to landfills will help minimise waste from construction and operations. To reduce the negative effects of construction operations on the environment and encourage resource efficiency, systemic waste management techniques are essential parts of sustainable construction practices. Stakeholders may limit the environmental impact of building projects, encourage responsible resource management, and support a circular economy by incorporating waste reduction and recycling initiatives into sustainable construction practises. In sustainable construction, some crucial tactics for waste reduction and recycling include:

I. Waste Management Plan

At the start of the construction project, the construction should create a comprehensive waste management plan. Strategies for waste minimization, recycling, and appropriate disposal during the building process.

II. Source Reduction

Optimise material consumption to reduce waste at its source. Placing precise orders for components and materials. Preventing the over-specification of materials. Source reduction would aid in preventing the creation of needless waste.

III. Recyclable and Reusable Materials

Reusable or recyclable materials should be given priority by the design team. To lessen the need for new raw materials, the design team should choose building materials with a larger percentage of recycled content.

IV. On-Site Sorting and Collection

Construction waste should be sorted and collected on-site by the contractors. To aid in effective recycling, constructors provide distinct bins or skips for various commodities, including concrete, glass, metal, wood, and plastics.

V. Construction Waste Recycling Facilities

It is recommended that contractors collaborate with reputable local construction waste firms to guarantee that the items they gather are appropriately handled and kept out of landfills.

VI. Deconstruction Instead of Demolition

When demolition existing structures, consider deconstruction as an alternative to standard destruction. Disassembling a structure piece by piece to save resources for future use minimises waste. This process is known as deconstruction.

VII. Material Recovery and Salvage

When it comes to remodelling or demolition projects, contractors should gather and rescue items that can be applied to new or ongoing projects. It could be necessary to get the client's and design team's agreement before using salvage materials in projects. The government must establish rules and certifications on this, and the building industry must inform its clients of this. Most of the new material and component approvals in most nations.

VIII. Concrete Recycling

The need for virgin aggregate can be decreased by using recycled concrete as aggregate in fresh concrete.

IX. Wood Recycling

Pallets, crates, and leftover building supplies are examples of salvaged and recycled wood that can be utilised for a variety of projects, such as mulch or biomass energy production.

X. Metal Recycling

Recycling metal helps preserve natural resources by lowering the requirement for energy-intensive raw material extraction. It is recommended that contractors segregate and recycle metal waste, including aluminium and steel.

XI. Plastic Recycling

Programmes for recycling plastic building materials, such as insulation, pipes, and packaging, should be put in place by contractors. When feasible, use recycled plastic materials.

XII. Documentation and Reporting

Throughout the whole construction project, contractors should identify, review, record, and monitor their efforts to reduce waste and recycle materials. The success of wastes diversion and recycling programmes can be reported on to guide future initiatives.

XIII. Government Regulations and Certification

The design team and contractors must abide by local recycling and waste management laws. Aim for certifications that highlight sustainable construction methods, such as recycling and waste minimization, such as LEED, Nigeria Green Building Council, Green Mark Certification, DBJ Green Building certification, Three-Star System, The Green Building Evaluation Label (GBEL) BREEAM, and GBI.

XIV. Life Cycle Assessment

To examine the environmental impact of construction materials and processes, including waste creation, the design team should do life cycle evaluations. Making educated decisions on the choice of building materials and techniques is possible with the help of this knowledge.

3.3.5 Site Selection

One of the clients' roles in a construction contract is to be the owner of the land on which the project will be constructed. But whether the clients already own the land, or the decision is based on the design team's proposal, parties should choose construction sites that have the least negative environmental impact possible, considering features like ecosystems, biodiversity, and closeness to already-existing infrastructure. When choosing a site for sustainable development, environmental, social, and economic considerations must be carefully considered to reduce the negative effects of construction on the neighbourhood and to support long-term sustainability. The primary components of managing and choosing sites for sustainable projects differed from that of the conventional projects. The primary considerations when choosing a site for sustainable construction:

I. Ecological Impact Assessment

A comprehensive ecological impact assessment should be carried out by the clients and the design team to comprehend the site's natural features, biodiversity, and ecosystem. The assessment's primary goal was to identify areas that needed to be mitigated or conserved while development was underway.

II. Brownfield Redevelopment

The groups must think about repurposing previously built land to boost metropolitan areas, lessen the strain on greenfield sites, and lessen the environmental damage brought on by new construction.

III. Proximity to Infrastructure

The locations that are closest to the current infrastructure-such as utilities, services, and public transportation-should be preferred by the teams. The main benefit of this is that it will lessen the requirement for new infrastructure construction and the environmental effects of providing transport infrastructure.

IV. Flood Risk and Climate Resilience

When choosing a location, the teams had to evaluate the risk of flooding and take the possible effects of climate change into account. The teams should also consider the local environment while designing and planning for climate-resilient building to reduce the hazards associated with extreme weather occurrences.

V. Community and Stakeholder Engagement

To learn about the wants and worries of stakeholders and the local community, the teams should interact with them. The local community's survival, as well as that of the local flora and fauna, should be threatened by the projects. The social and cultural features of the location should be considered, and community input should be included in the decision-making process.

VI. Access to Green Spaces

The sites that facilitate access to green spaces and foster a sense of connection with nature should be highly prioritised by the teams. To improve the general well-being of residents, landscaping and green infrastructure should be included in the design.

VII. Historical Preservation

Sites of historical or cultural significance should be respected and preserved. When it's feasible, incorporate old buildings into new projects or make sure that these assets are respected and protected during construction.

3.3.6 Site Management

In sustainable construction, the objective is to reduce the negative environmental effects of construction, enhance community well-being, and support sustainability in general. These are the main elements of site management in environmentally compliant buildings:

I. Erosion and Sediment Control

To stop soil erosion and water flow during construction, the teams should effectively apply erosion and sediment control techniques. This covers the application of sediment basins, erosion control blankets, and silt fences.

II. Waste Management

Teams should create a thorough waste management strategy that focuses on recycling, reusing, and reducing construction waste. To keep materials out of landfills, the teams should put in place processes for on-site sorting and collection.

III. Energy Efficiency during Construction

To reduce energy usage, the teams should optimise the construction operations. To optimise energy efficiency during construction, the site should employ energy-efficient tools and technology.

IV. Water Conservation

Throughout construction, the teams should follow water-efficient procedures, such as controlling dust and using less water, and they should think about using recycled water for building.

V. Noise and Air Quality Control

During construction, the teams should employ equipment with noise-reducing characteristics, regulate noise and air pollution, and execute dust control procedures to maintain air quality.

VI. Wildlife Protection

During construction, the crews should take precautions to conserve the local animals, particularly in ecologically sensitive places. To reduce disturbance, this may specifically entail establishing buffer zones, building animal crossings, or sticking to strict construction schedules.

VII. Health and Safety

Using personal protection equipment, training initiatives, and sustainable safety practices, the teams should place a high priority on the health and safety of construction workers.

VIII. Site Layout

To minimise disturbance to the natural terrain and existing ecosystems, the teams should optimise the layout of construction sites. This could entail considering how to preserve natural features and make effective use of available space.

IX. Green Certification

To manage the site sustainably, the teams should work towards green building certifications like LEED, Nigeria Green Building Council, Green Mark Certification, DBJ Green Building certification, Three-Star System, The Green Building Evaluation Label (GBEL) BREEAM, and GBI. These certifications include standards for sustainable site selection and construction management techniques.

X. Post-Construction Site Restoration

The teams should devise a plan for restoring the property after construction, which should include landscaping, restoring the habitat, and demolishing any temporary buildings. The primary goal of this is to close the site in a way that supports the health of the environment and the community.

3.3.7 Biodiversity and Ecosystem Protection

Through the preservation of natural habitats, the incorporation of sustainable areas, and the application of landscaping techniques that support native ecosystems, biodiversity can be protected and enhanced. The preservation of biodiversity and the general health of natural habitats are the goals of biodiversity and ecosystem protection in sustainable building, which include methods and techniques designed to reduce the adverse effects of construction operations on nearby ecosystems. When constructing sustainably, some key factors and methods for protecting ecosystems and biodiversity could be:

I. Site Assessment and Planning

To locate and map current ecosystems, hotspots for biodiversity, and sensitive habitats, the teams should conduct in-depth site surveys. The assessment's findings can help in planning construction sites and preventing the destruction or disturbance of highly valuable ecological areas.

II. Green Infrastructure and Landscape Design

It is recommended that the teams incorporate components of green infrastructure into the construction project. Permeable surfaces, rain gardens, and green roofs are a few examples of components that are excluded. These characteristics would promote biodiversity, aid in the formation of habitats, and increase the built environment's overall resilience.

III. Protected Areas and Buffer Zones

The teams ought to create buffer zones and protected areas around wildlife habitats, waterways, and delicate ecosystems. These areas would offer safe havens for the local species and aid in protecting natural environments from problems brought on by construction.

IV. Wildlife Corridors and Crossings

To connect fragmented ecosystems, the teams should design animal crossings and corridors. This promotes species mobility and preserves genetic variety within populations, lessening the isolation of wildlife communities.

V. Habitat Restoration and Enhancement

As part of the building project, the teams should take up habitat enhancement and restoration projects. To make up for any habitat disturbance brought on by building, this could entail re-establishing wetlands, growing natural flora, or developing wildlife-friendly features.

VI. Sustainable Construction Practices

The teams ought to use building techniques that cause the least amount of disruption to ecosystems. This covers the use of temporary construction mats to protect soil and plant, as well as low-impact construction methods including reduced site grading.

VII. Non-Toxic Construction Materials

It is recommended that the construction teams choose non-toxic materials that do not discharge dangerous substances into the surrounding environment. This is crucial since hazardous materials harm project participants, animals, and vegetation.

VIII. Invasive Species Prevention

Throughout construction, the teams should take precautions to stop invasive species from becoming introduced and spreading. This is a result of invading species' ability to upend regional ecosystems and displace native plants and animals.

IX. Water Conservation and Protection

To avoid soil runoff and sedimentation in adjacent water bodies, the teams should implement erosion control measures. This preserves water quality and aids in the preservation of aquatic ecosystems.

X. Biodiversity Monitoring

To determine how building is affecting native plants, the teams should monitor and evaluate biodiversity continuously. The results of the monitoring and assessments can be utilised to guide the creation of adaptive management plans and lessen unfavourable consequences.

XI. Compliance with Environmental Regulations

The teams are responsible for making sure that local environmental laws and guidelines about biodiversity and ecosystem preservation are followed. Obtaining permissions and approvals ought to be essential. This would also entail ongoing oversight and adherence to the rules and guidelines.

XII. Collaboration with Conservation Organizations

To create and carry out efficient biodiversity conservation strategies, the teams should work with regional entities such as environmental authorities, conservation organisations, and specialists.

XIII. Long-Term Monitoring and Management

The teams ought to implement extended surveillance initiatives to monitor the post-construction ecological well-being of the construction site. It is imperative to execute strategic management approaches grounded in monitoring outcomes to tackle any unanticipated ecological consequences.

XIV. Certifications and Standards

Follow sustainability standards and certifications, which include requirements for protecting ecosystems and biodiversity. Environmentally related standards are frequently included in certifications.

A built environment that coexists peacefully with the surrounding ecology may be created, local ecosystems can be supported, and natural habitats can be conserved by incorporating these biodiversity and ecosystem protection methods into sustainable construction projects.

3.3.8 Adaptive and Resilient Design

Create construction projects that are resilient to the effects of climate change, natural disasters, and changing environmental conditions. In sustainable building, adaptive and resilient design refers to the use of techniques that let projects adjust to shifting circumstances, absorb shocks, and bounce back swiftly from disruptions. By considering multiple elements like climate change, harsh weather occurrences, and changing social needs, this method seeks to improve the projects' long-term sustainability and performance. Several key practices for resilient and adaptive design in sustainable construction include:

I. Climate-Responsive Design

The projects ought to be planned to adapt to the local weather and environment. Optimising building orientation, implementing natural ventilation techniques, and employing shade components to improve thermal comfort are some actions to take in this endeavour.

II. Flexible Design and Modular Construction

Modular and flexible design elements that facilitate easy adaptation and modification over time should be embraced by the design team. This is so that components can be improved or replaced, giving modular building methods the flexibility to adapt to changing functional requirements.

III. Smart Building Technologies

The teams need to make sure that sensor networks and building management systems are included in the projects. The objective is the real-time monitoring and optimisation of indoor air quality, energy consumption, and other building performance indicators. Being able to respond quickly to changing circumstances is made feasible by these technologies.

IV. Resilient Materials and Construction Techniques

It is recommended that the teams choose construction materials and processes that are resilient to extreme weather, earthquakes, and other natural calamities. This strategy can involve employing resilient materials and cutting-edge building techniques to improve structural integrity.

V. Green Infrastructure and Natural Systems

To increase resistance to climate-related concerns, the teams should put green roofs, rain gardens, and permeable surfaces into practice. These elements enhance biodiversity, lessen the effects of urban heat islands, and manage stormwater.

VI. Energy and Water Efficiency

To improve resource resilience, the teams should prioritise designing with water and energy efficiency. To lessen reliance on outside resources, this can entail using water-saving devices, renewable energy sources, and energy-efficient equipment.

VII. Adaptive Land Use Planning:

It is recommended that the teams incorporate adaptive land use planning principles, which take into consideration anticipated shifts in both land use and climate. This entails considering upcoming developments, the necessity for mobility, and the requirements for infrastructure considering shifting environmental conditions.

VIII. Community Engagement and Social Resilience

To comprehend the vulnerabilities and requirements of the local communities, the teams should interact with them to understand their stake in the projects. The project should support inclusivity, community cohesiveness, and social resilience. The stakeholder should examine how the created surroundings might serve as secure refuges in times of crisis.

IX. Water Management and Flood Protection

To handle potential water-related issues, the teams should implement sustainable water management techniques, such as flood-resistant design. To reduce the risk of flooding, this may entail building permeable materials, raising structures, and creating efficient drainage systems.

X. Lifecycle Assessment and Adaptation Planning

To comprehend the effects of building materials and techniques on the environment, the teams should conduct lifecycle evaluations. In addition, the group should create plans for adaptation that take into consideration prospective upgrades or retrofits as well as the changing requirements of the built environment.

XI. Diverse Planting and Green Spaces

The teams should create landscapes with a variety of plants that enhance biodiversity and the health of the ecosystem. This is because green areas can serve as buffer zones, take up extra water, and improve the built environment's overall resilience.

XII. Collaboration with Local Authorities

To make sure that project environments are resistant to potential risks, the teams should work in conjunction with emergency management organisations and the local authorities to identify the threats and opportunities. In addition, the stakeholders should take part in planning initiatives for community resilience and aid in the creation of robust infrastructure.

XIII. Accessibility and Universal Design

The team should incorporate universal design and accessibility concepts to guarantee that built environments are welcoming and capable of meeting a range of user requirements, such as those of people with impairments or mobility issues.

XIV. Data-Driven Decision-Making

To develop adaptive strategies, the teams should gather and evaluate data on building performance, environmental factors, and occupant well-being. The team should make well-informed decisions about building operations, maintenance, and future renovations by utilising data-driven insights.

Through the integration of these concepts into sustainable construction processes, stakeholders may design built environments that are not only considerate of the environment but also flexible enough to adapt to changing circumstances, strong enough to overcome obstacles, and conducive to community well-being.

3.3.9 Healthy Indoor Environments

Building interior settings are now a crucial factor in determining how well a building performs. Indoor environment is now an objective of sustainability. Thus, interior air quality should be given top priority in sustainable construction. A key component of sustainable construction is the establishment and provision of a healthy indoor environment, as this has a direct impact on the productivity and well-being of building users and occupants. The goal of sustainable construction methods is to create interior environments that maximise human health, illumination, thermal comfort, and air quality. Creating a healthy and comfortable indoor atmosphere can be accomplished using natural lighting, sufficient ventilation, and non-toxic materials. The following are some key factors and procedures that should be followed to guarantee that sustainable building provides healthy interior environments:

I. Indoor Air Quality (IAQ)

The design teams should put plans into action to keep indoor air quality at a high level. This could entail avoiding the use of materials containing volatile organic compounds, installing adequate ventilation systems, and using low-emission construction materials.

II. Ventilation Systems

It is recommended that the teams create and implement efficient ventilation systems to guarantee a steady flow of clean air. Optimal air exchange rates can be maintained by using energy-efficient mechanical ventilation systems and giving special emphasis to natural ventilation solutions like movable windows.

III. Humidity Control

To stop mould from growing indoors, the humidity levels should be managed. This will lower maintenance costs and improve occupant comfort. In addition, appropriate ventilation systems, moisture barriers, and insulation should be installed to help control humidity and stop the accumulation of indoor pollutants.

IV. Low-Emission Materials

Low-emission materials ought to be chosen. These materials have minimal levels of dangerous substance emissions. To enhance interior air quality, the design team should consider low-VOC paints, adhesives, and flooring materials.

V. Biophilic Design

In a similar vein, the design team ought to incorporate biophilic design features that foster occupant-nature interaction. This could be utilising natural materials, having indoor plants, and having access to natural light, all of which have a positive effect on mental health.

VI. Daylighting

In a similar vein, the design team ought to incorporate biophilic design features that foster occupant-nature interaction. This could be utilising natural materials, having indoor plants, and having access to natural light, all of which have a positive effect on mental health.

VII. Thermal Comfort

The teams should plan the areas with the best possible thermal comfort, taking air circulation, humidity, and temperature into account.

VIII. Noise Control

Should implement effective methods to regulate noise levels in interior spaces. This can be accomplished by using sound-absorbing surfaces, acoustic materials, and space design that reduces noise disruptions.

IX. Access to Outdoor Spaces

To promote outdoor activities and exposure to nature, the teams should plan for access to outside areas or include balconies and green roofs. Outdoor areas promote inhabitant well-being and provide leisure possibilities.

X. Non-toxic Cleaning Practices

To maintain cleanliness without adding dangerous chemicals to the interior environment, the project should be planned and executed to employ non-toxic cleaning supplies and techniques. In addition, green cleaning certificates should be considered for building upkeep.

XI. Occupant Engagement

It is recommended that building occupants adopt sustainable and healthful habits. Educational materials about energy efficiency, indoor air quality, and occupant behaviours that support a healthier indoor environment should be given to the inhabitants.

XII. Wellness Certifications

The teams ought to work towards wellness certifications that emphasise designing spaces that enhance the health and well-being of its occupants. Aspects including air quality, lighting, energy, and thermal comfort are covered by these certificates.

XIII. Flexible Space Design

To accommodate a wide range of tenant needs, the teams should design flexible and adaptable environments. To accommodate different work and lifestyle preferences, this involves offering areas for cooperation, privacy, and relaxation.

XIV. Air Purification Systems

The usage of UV-C light or high-efficiency particulate air filters as air purification devices should be taken into consideration by the teams. This will reduce airborne pollutants and improve indoor air quality.

XV. Occupant Feedback Mechanisms

The project must be planned with mechanisms that allow residents to give input on the state of the interior environment. Among the methods used to be sure are sensors, questionnaires, or online platforms where residents can report discomfort or other difficulties.

Stakeholders can prioritise occupant health and well-being by implementing these practices into sustainable building, which will further the objective of sustainable and human-centric design.

3.3.10 Social Equity and Inclusivity

In sustainable construction, social equity and inclusivity refer to the dedication to treating everyone fairly and justly, whether they are directly participating in the project or not. Addressing social inequalities, fostering diversity, and making sure that building projects have a good impact on community well-being are all components of social justice and inclusivity. During the project design and construction phases, the teams and construction team should incorporate social factors such as equitable labour methods, community involvement, and accessibility. Ensuring that building projects positively impact the well-being of all stakeholders is the main goal of this. When it comes to sustainable construction, some of the most crucial factors to consider are:

I. Community Engagement

Strategically, local communities should be involved in the decision-making process for construction projects. The community's opinions should be sought out and issues should be addressed by the project teams. To guarantee that the project is beneficial to all, the stakeholderd should consider the requirements and preferences of the varied community members.

II. Accessibility

It is imperative that design teams create and implement initiatives that are inclusive of those with physical disabilities and impairments. This can be done by implementing accessibility guidelines and universal design principles to create environments that anybody, regardless of ability, can use and enjoy.

III. Job Opportunities and Training

In particular, the initiatives should be designed and built to support local employment in communities that may be experiencing economic difficulties or higher unemployment rates.

IV. Small and Minority-Owned Businesses

The projects should design and built to encourage and assist minority- and small-owned companies by giving them the chance to work on building projects. It is significant to support diverse suppliers and contractors when they submit bids for contracts, as this will help underrepresented populations become economically empowered.

V. Fair Labour Practices

The projects' design and construction teams should consider fair employment practices, such as paying salaries that are reasonable, offering secure working conditions, and adhering to labour laws. Discriminations especially concerning race, gender, status must be prevented. Child labour should be prohibited. The welfare of construction workers should be given top priority by the project teams, and they should make sure that they are treated fairly and with respect.

VI. Inclusive Design Practices

Design teams and clients should embrace inclusive design principles that consider the requirements of all users. This may entail planning public areas, structures, and infrastructure to serve a variety of age groups, genders, cultural backgrounds, and ability levels.

VII. Cultural Sensitivity

The project should be planned and executed with consideration for regional customs and traditions as well as cultural sensitivity. Construction projects should abide by local cultural norms and regulatory regulations, avoiding any actions that could have a harmful influence on cultural heritage.

VIII. Social Impact Assessments

To identify and address potential socioeconomic inequities that may result from building projects, the design team and clients should conduct social impact assessments before the design development process. The details would include the project's potential effects on various demographic groups and the efforts being taken to lessen any negative effects.

IX. Equitable Access to Resources

Ensuring fair access to resources and amenities resulting from construction projects is the responsibility of the project team. Community facilities, parks, and public areas ought to be developed in a way that benefits all locals equally and without prejudice.

X. Affordable Transportation Options

The projects should be designed and built with the transportation needs of building occupants' and users. They should also offer priced, easily accessible transit options both inside and outside of construction sites. This is particularly crucial for areas where having access to dependable transport is essential for both social and economic mobility.

XI. Social Value

Construction inherently generates social value. The projects created are designed to benefit society, often viewed through the lenses of community, aesthetics, accessibility, and prosperity. Public entities such as local authorities and housing associations, along with private enterprises like property developers and contractors, play a central role in communities and are well-positioned to lead social value initiatives and promote local economic and social cohesion. For the construction sector, integrating social value into project delivery goes beyond compliance; it is a way to enhance community well-being, develop workforce skills, and improve environmental health. By committing to social value from the outset of a project, construction activities can make a meaningful, positive impact on society, extending beyond the physical infrastructure to address key issues like social mobility, employment, individual welfare, and environmental sustainability. As part of building projects, some stakeholders-especially the government-should make investments in social infrastructure, such as community centres, schools, and medical facilities. A number of CSR-oriented businesses not only work on building projects but actively support the growth of social values in the neighbourhood. By attending to basic necessities, this enhances the community's general well-being and advances social fairness.

XII. Anti-Discrimination Policies

Anti-discrimination measures should be implemented and upheld by the project team in building projects. Make certain that everyone is treated equally and without bias, regardless of colour, gender, ethnicity, or any other attribute.

XIII. Public Space Design for Inclusivity

Stakeholder should create inclusive and hospitable public areas for everyone, and consider elements like lighting, signage, and seating options to design spaces that accommodate a variety of users.

Through the integration of social equity and inclusion principles into sustainable construction techniques, stakeholders can play a constructive role in fostering the development of built environments that uphold justice, diversity, and the welfare of all. This is in line with the more general objectives of sustainable development, which prioritise social and economic justice in addition to environmental stewardship.

3.3.11 Life Cycle Assessment

Traditionally, the only factor used to determine a construction project's cost was its capital expenditure. However, the operational costs of the project are more and beyond the construction costs only. When deciding whether to purchase, buy, or rent a project-and, if so, how to use and manage it-forward-thinking clients of today want to know what the project's operation and maintenance costs will be in the future. As such, taking the complete life assessment into account is necessary in sustainable construction. The project team should carry out life cycle assessments in order to fully understand the environmental impact of a building or infrastructure project during its entire lifespan, from material extraction and construction to operation and ultimate demolition. Thus, it is fundamentally necessary for sustainable building to take the complete life assessment into account. Life cycle assessments should be carried out by the project team in order to fully understand the environmental impact of a building or infrastructure project from material extraction and construction to operation and final demolition:

I. Goal Definition and Scope

With the help of the design team, the clients should clearly outline the objectives and parameters of the appraisal, outlining the topics that will be included and providing rationale for the project. This entails defining the evaluation's goal, the building project or product under review, and the study's parameters.

II. Inventory Analysis

Each assessment stage's associated inputs and outcomes should be identified and quantified by the team. The extraction of raw materials, transportation, manufacture, building, use, maintenance, and end-of-life procedures may all fall under this category. Information is gathered and examined regarding material usage, emissions, water use, and energy consumption.

III. Impact Assessment

The stakeholders should assess the inventory data's effects on the environment. This entails assessing carbon emissions and resource usage based on factors like land use, water consumption, human toxicity, ozone depletion, and climate change. At each level, the relevant impact evaluation techniques should be used.

IV. Normalization and Weighting

To provide a set of data that is easier to handle and comprehend, the team should normalise and weigh the implications. Weighting is allocating relative importance to various impact categories according to stakeholder or society preferences, whereas normalisation involves comparing the effects to reference values.

V. Interpretation

The evaluation's findings should be interpreted by the team considering the environmental hotspots and areas that need improvement. The team should also discuss how to minimise the total environmental impact and consider trade-offs between various impact categories.

VI. Sensitivity Analysis

To consider, the requirements of sustainable construction, the team should perform sensitivity studies to evaluate the impact of key assumptions or factors on the outcomes. The obtained information will aid in determining the degree of ambiguity and will emphasise the aspects that have the greatest influence on the projects' results.

VII. Data Quality and Completeness

The team should assess the accuracy and comprehensiveness of the data utilised in the evaluation. To incorporate the specifications for sustainable buildings, there must be comprehensive data on the various aspects of the project. For the assessment to be deemed credible, data sources, assumptions, and constraints must be transparently documented.

VIII. Continuous Improvement

The report of the evaluation should be used by the team to guide future developments in environmentally compliant construction techniques and to help make well-informed judgements. The group should find ways to improve in too lessen the effects on the environment and society.

To find environmentally compliant techniques, lessen the ecological impact of construction projects, and improve the built environment's overall sustainability, life cycle assessments are a useful tool for decision-makers in the construction sector.

3.3.12 Land and Sustainable Construction

The term "land" refers to the Earth's surface that is not covered by water. Plains, mountains, valleys, deserts, woods, and urban areas are among the landscape types represented. Land is a central natural resource that sustains a variety of ecosystems, habitats, and human activities. It offers space for agricultural, urban growth, infrastructure, recreation, and conservation.

In the context of sustainable construction, land has a considerable impact on determining the built environment and construction techniques. Followings are techniques on how land practices influence sustainable construction:

- Land Use Planning: Proper land use planning is essential for sustainable construction. To illustrate, the planning authorities either at federal, state or local levels must allocate property efficiently while balancing the requirements of diverse stakeholders, including people, businesses, environmental protection activities and social values. Sustainable land use planning minimise land consumption, preserve natural habitats, safeguard agricultural land, and encourage compact, mixed-use development patterns to limit urban sprawl.

- Site Selection: Choosing the correct location for construction projects is crucial to achieve sustainable construction. This is because sustainable construction management prioritise site selection characteristics such as proximity to public transit, utility accessibility, green space availability, and environmental considerations such as soil quality, slope stability, and flood risk. By selecting suitable sites, construction projects can minimise environmental effect, reduce transportation-related emissions, reduce costs, and boost resistance to climate change.
- Land Degradation: Unsustainable land practices like deforestation, soil erosion, and habitat destruction can degrade land quality and compromise the long-term viability of the construction and the environment. Sustainable construction aims to reduce land degradation by using erosion control measures, protecting natural vegetation, repairing degraded land, and using construction practices that minimise disturbance to the ground.
- Land practices affect biodiversity protection efforts: Sustainable building projects prioritise the preservation and restoration of natural habitats such as wetlands, forests, grasslands, and coastlines. Construction projects that preserve biodiversity-rich places and employ sustainable infrastructure solutions can improve ecosystem services, maintain native wildlife populations, and increase ecological resilience.
- Land reclamation and brownfield redevelopment: In some circumstances, sustainable construction entails repurposing underutilised or contaminated land for future development. Construction projects that reclaim land from the sea or revitalise abandoned industrial sites can minimise greenfield development, reduce pressure on natural ecosystems, and encourage urban revitalization in already built-up areas.

3.3.13 Community Engagement

Construction organisations must consider the social aspect of sustainable development in addition to other numerous benefits that are associated with the projects. The development of sustainable projects that benefit the community hence requires addressing the needs and concerns of local people and involving them in the planning and decision-making processes. Engaging the community is a key component of sustainable construction, involving local people in all stages of a project-planning, design, construction, and operation. This is mostly done to make sure the project meets community needs and goals, resolves any potential problems, and improves the social, economic, and environmental well-being of the locals. Engaging communities effectively in construction projects is vital for building trust, reducing resistance, and ensuring the project meets local needs. Some suitable techniques for community engagement in construction are stakeholder mapping and analysis, public consultations and town hall meetings, workshops and focus groups, community liaison officers, participatory design sessions, surveys and feedback forms, information campaigns, site visits and demonstrations, community advisory committees, and local employment and training opportunities. The primary facets of community involvement in environmentally compliant construction projects encompass:

I. Communication

From the commencement of development of a project, the design team, and the clients—whether private or public-should have open and honest lines of contact with the community. To build trust and understanding, the local populace should be informed succinctly and explicitly about the purpose, goals, and potential effects of the projects.

II. Inclusive Decision-Making

The community should be included in decision-making processes concerning the construction project by the design team and the clients. To ensure that a variety of viewpoints are considered during the decision-making process, they should solicit feedback on design possibilities, site selection, and other crucial issues. The engagement should not be bias. All the segments of the community must be included. Often, only selected members of the community are engaged in such meetings.

III. Needs Assessment

To determine the community's objectives and concerns, the design team and the clients should conduct a needs assessment. The group should also determine which social, economic, and environmental problems the construction project can resolve or lessen, and then modify the project based on those findings.

IV. Stakeholder Mapping

Stakeholders include locals, businesses, government agencies, and community organisations. The design team and the clients should identify and interact with these groups of people. To guide project planning and execution, the team should be aware of the interests, worries, and expectations of various stakeholders.

V. Community Workshops and Meetings

To give details, get input, and encourage conversation, the design team and the clients should plan town hall meetings, community workshops, and other engagement activities. These forums would provide project representatives and community members with avenues for direct communication and participation.

VI. Capacity Building

The projects should fund community capacity-building projects and offer instruction, training, and skill-building opportunities so that the locals can actively engage in and gain from the construction project.

VII. Cultural Sensitivity

In addition to considering the community's cultural background and including culturally relevant design aspects in the building project, the project should exhibit cultural sensitivity and respect for regional customs, traditions, and legacy.

VIII. Job Opportunities and Local Employment

It is anticipated that the project will help the indigenous people find work and employment possibilities. To determine the community's needs for skills and training, the project teams should collaborate with them. They should also make sure that the community has access to job possibilities that the construction project will create.

IX. Health and Safety Awareness

For large projects including dams, heavy bridges, the natives should receive training and information on health and safety precautions connected to construction activities from the design team and the clients. Raising awareness of dangers and mitigating actions would also be necessary to guarantee community members' safety during the construction process.

X. Mitigation of Disruptions

The project management team should devise tactics to alleviate any disturbances created by construction operations to the events of the community cultural festival. To guarantee that everyday life in the community can continue with the least amount of disturbance, efforts should be taken to minimise the effects of traffic, noise, and dust.

XI. Post-Construction Engagement

When the project is finished, the project management should continue to interact with the community. The management team should devise a plan to ask for input on how the project is performing, deal with any problems that arise after construction, and guarantee that the community continues to gain from and feel a part of the initiative.

XII. Local Business Opportunities

The project manager needs to examine ways that nearby companies can get involved in the building project. To support local suppliers and contractors and the community's economic growth, project managers ought to encourage their utilisation.

XIII. Education and Information Dissemination

Information regarding sustainable construction practices and educational resources should be supplied by the project management team. Additionally, to promotes a feeling of pride and ownership within the community by emphasising the project's beneficial social and environmental features.

3.4 Green Building Certifications

The purpose of certification is to prove that certain conditions are fulfilled. The project management team must obtain and follow accredited green building certifications and standards. Third-party verification methods, known as "green building certifications," evaluate and accredit buildings and construction projects according to predetermined sustainability and environmental performance standards. These certifications are intended to encourage resource- and environmentally-conscious building methods, improve energy efficiency, lessen the impact on the environment, and produce built environments that are healthier and more sustainable. There are close to 1000 green building certifications available worldwide, each with its own set of requirements and standards. These certifications play a pivotal role in advancing sustainable construction practices by providing a framework for assessing and improving the environmental, social, and health performance of buildings and construction projects. The certifications help guide developers, architects, engineers, construction managers, facilities managers, quantity surveyors and contractors toward more sustainable design and construction practices while also providing a recognizable standard for environmentally conscious building design.

3.4.1 Innovation and Technology

Most green building ratings place a high value on innovative design. The project management team should use cutting-edge building materials, energy-efficient appliances, and smart building systems, among other cutting-edge technologies and construction techniques that support sustainability. To address environmental concerns, increase efficiency, and produce resilient, resource-efficient, and ecologically responsible buildings, the sustainable construction industry must embrace innovation and technology. Particularly in the construction technology revolution, digital space is crucial. The use of technology is crucial. Technologies can assist in lowering energy usage, lowering on-site accidents, lowering, or avoiding lawsuits or fines, etc. Technology innovation can revolutionise many facets of sustainability when it is incorporated into building methods. The key points that emphasise how crucial it is for sustainable construction to embrace innovation and technology are as follows:

I. Energy Efficiency and Green Building Design

Energy-efficient designs can be precisely planned and optimised with the use of sophisticated building design tools like Building Information Modelling and digital twins. Using energy-efficient technologies, such as smart lighting systems, energy-efficient HVAC systems, and renewable energy sources, can reduce a building's carbon footprint.

II. Materials and Resource Efficiency

Creation of recycled and sustainable building materials in addition to innovative building techniques like 3D printing and modular architecture. Sustainable construction techniques are facilitated by technologies that allow for real-time tracking of material usage, recycling, and waste reduction.

III. Digital Twin Technology

Digital twin technology facilitates real-time monitoring and analysis by creating virtual representations of physical construction. By offering information on energy usage, occupancy trends, and maintenance requirements, digital twins improve the performance of the construction and enable proactive decision-making for increased sustainability.

IV. Smart Building Technologies

Using technologies sensors and the Internet of Things to collect and analyse data. Predictive maintenance, enhanced occupant comfort, and resource optimisation are all made possible by smart building technology, which raises overall operational and energy efficiency.

V. Construction Automation

Construction operations can benefit from automation and robotics. Sustainable construction approaches include increased productivity, shortened construction schedules, and enhanced safety using automated construction procedures.

VI. Water Conservation Technologies

Creation of rainwater harvesting systems, water-efficient fixtures, and intelligent irrigation technology. By lowering water usage in buildings and on construction sites, these technologies help to solve the problem of water scarcity and advance sustainable water management.

VII. Renewable Energy Integration

Developments in renewable energy technology, including wind and solar energy. The construction players should incorporate renewable energy sources, such as wind turbines and solar panels, into buildings can reduce greenhouse gas emissions and dependence on non-renewable energy.

VIII. Blockchain

Application of blockchain technology to the supply chain to improve traceability and transparency. ensures responsibility and openness in the procurement of resources, mitigating the negative effects of unsustainable practices on the environment and society. A major challenge in sustainable construction is the arbitrary shifting of sustainability goals, often by the design team, to accommodate conflicting client priorities. Additionally, blockchain helps combat greenwashing misleading claims about sustainability by enabling the tracking of materials from production to on-site use. This transparency ensures that only verified sustainable materials are used while also tracking and minimizing waste.

IX. Augmented Reality (AR) and Virtual Reality (VR)

Applications of AR and VR include training, virtual tours, and project visualisation for building. improves teamwork, lowers mistakes, and permits stakeholders to evaluate and experience a project before it is physically built, all of which led to more effective and long-lasting results.

X. Waste Technologies

Creation of technology for effective recycling, waste sorting, and building material reuse. Lessens the environmental impact of construction activities, encourages the circular economy, and minimises construction waste.

XI. Sustainable Transportation and Logistics

The use of autonomous and electric vehicles for transportation, as well as creative logistical solutions. contributes to overall sustainability by lowering transportation-related carbon emissions and increasing material delivery efficiency.

XII. Continuous Monitoring and Feedback Systems

Deployment of monitoring systems and sensors to gather data in real-time. The key to adopting innovation and technology in sustainable building is the industry's capacity to change and take advantage of developments to provide more resilient, resource-efficient, and ecologically responsive built environments. By incorporating these technologies, the construction sector is better positioned to fulfil the shifting needs of a world that is changing quickly, in addition to addressing today's sustainability problems. It supports the general shift towards a more resilient and sustainable built environment and cultivates a culture of continual development.

3.4.2 Transportation Planning

Throughout a construction project, transportation planning in sustainable construction entails making strategic, thoughtful decisions about the movement of personnel, equipment, and materials. The goals of sustainable transportation planning are to minimise the effects on the environment, use less energy, increase efficiency, and support the construction process' overall sustainability. Hence, in addition to encouraging access to public transit, creating pedestrian-friendly areas, and reducing the carbon footprint associated with commuting, the construction management team should take sustainability into account while developing transportation systems. The

Table 3.2 List of main tasks in transportation planning in sustainable construction

Task	Activities	Sustainable impact
Site selection and proximity	Should select building locations that are easily connected to the current infrastructure to minimise the requirement for staff and materials to be transported over large distances	Will cut down on travel distances, aid in reducing energy use, and lessen emissions related to moving goods, machinery, and people commuting
Material sourcing	To lessen the amount of energy used for transportation and its impact on the environment, aim to source locally whenever possible	Will lessen the transportation of materials' carbon footprint and contribute to the general sustainability of the local economies
Transportation mode selection	To assess and give preference to eco-friendly means of transportation over conventional fossil fuel-based ones, such as rail transit and cars that run on electricity or hybrids	Greenhouse gas emissions and air pollution will be decreased via cleaner transit options, supporting sustainable building techniques
Logistics and supply chain optimization	To maximise transportation routes and timetables through the application of efficient logistics and supply chain management strategies Optimise transportation schedules and routes by implementing efficient supply chain management and logistics strategies	Simplifying logistics will minimise traffic jams, save fuel use, and lessen the environmental effects of transportation
On-site transportation management	Make plans for effective on-site mobility, such as worker transit and the flow of building materials around the site	An efficient on-site mobility system will reduce traffic, improve safety, and lessen the environmental effect of construction-related activities
Waste transportation and recycling	To put into practice efficient recycling and waste management procedures, including moving construction waste to recycling facilities	By recycling and properly disposing of waste to minimise the quantity that ends up in landfills, construction operations can have a less negative environmental impact
Use of alternative fuels	To equip fleets of construction vehicles with alternative fuels like compressed natural gas or biodiesel	By using alternative fuels instead of conventional fossil fuels, greenhouse gas emissions and air pollutants will be reduced
Telecommuting and remote collaboration	To encourage remote work and telecommuting to lessen the need for regular commutes to construction sites	Options for remote work will improve work-life balance, lessen traffic congestion, and lessen the environmental effect of travel

(continued)

Table 3.2 (continued)

Task	Activities	Sustainable impact
Vehicle fleet efficiency	To maximise the fleet's efficiency through routine maintenance, route planning, and fuel-efficient driving techniques	Vehicles that are fuel-efficient and well-maintained will help cut down on emissions and energy use when travelling
Public transportation access	To design building projects so that employees have convenient access to public transport	Promoting the usage of public transport will help promote environmentally friendly commuting patterns and decrease the number of private vehicles on the road
Transportation demand management	To put in place methods for managing the demand for transport, like carpooling and flexible work hours, to lower the total demand for transport	Practices in transportation demand management will help to lessen traffic jams, cut emissions, and increase the effectiveness of transportation
Monitoring and reporting	Putting in place monitoring tools to track indicators about transport and provide reports on how well transport activities are performing environmentally	Frequent observation makes it possible to pinpoint regions in need of development and guarantees responsibility for environmentally friendly transportation methods
Integration of smart technologies	To maximise transit routes and boost overall efficiency by utilising smart technology, such as traffic management systems and real-time tracking	Smart technology will improve overall sustainability, minimise delays, and lower fuel use in transportation planning

tasks that follow have a sustainable impact on transportation planning when it comes to contracting with sustainable contractors (Table 3.2).

Sustainable construction techniques can lessen the environmental effect of transportation-related activities, increase the overall productivity of construction projects, and be in line with more general sustainability objectives by incorporating these factors into transportation planning. Planning transit well is essential to developing resource- and environmentally-conscious building techniques.

3.5 Some Historical Perspectives on Sustainable Construction

The historical view on sustainable construction charts the development of construction methods and attitudes about resource usage, environmental concerns, and the welfare of society over time. There are historical examples of nations and cultures incorporating sustainable ideas into their built environment, even if the idea of sustainable construction is new. Some quick historical synopses to are presented below:

Ancient Civilizations

Sustainable construction methods were used by many ancient civilizations, frequently because of the necessity to adjust to the local environmental conditions. Using natural materials like adobe in arid areas, passive solar design in classical Greek and Roman architecture, and constructing structures using materials that are readily available locally are a few examples.

Traditional Building Practices

Based on local knowledge and accessible materials, traditional and indigenous communities have historically developed sustainable construction practices. These methods frequently placed a focus on energy-efficient design, the use of renewable resources, and climate adaptation.

Medieval and Renaissance Periods

Incorporating the concepts of longevity and durability, indigenous materials were frequently chosen by contractors during the mediaeval and Renaissance periods. In Europe, for instance, timber-framed buildings were typical, and their design demonstrated a knowledge of the regional ecology.

Industrial Revolution

Construction techniques were more mass-produced and mechanised during the Industrial Revolution. Although this period saw improvements in the building industry, it also brought with its problems including rising resource use and environmental damage.

Early Environmental Movements

Early environmental movements started highlighting the ecological effects of human activity, especially construction, in the middle of the twentieth century. Few in books 60s/70s brought attention to the negative effects that industrial practices have on the environment.

1970s—Rise of Environmental Awareness

The first Earth Day was observed in 1970 because of a notable rise in environmental consciousness during the 1970s. During this time, there was a growing awareness of how construction affects the environment, which had an impact on the development of sustainable design concepts.

1987—Brundtland Report

In 1987, the World Commission on Environment and Development published the Brundtland Report, which introduced the concept of sustainable development. The need of meeting present needs without compromising the ability of future generations to meet their own needs was the report's main focus.

1990s—Green Building Movement:

In the 1990s, the green building movement became formally organized Organisations like the U.S. Green Building Council (USGBC) and the Leadership in Energy and Environmental Design (LEED) certification system were developed to provide a framework for sustainable construction approaches.

Twenty-First Century—Global Focus on Sustainability

The twenty-first century saw an international emphasis on building sustainability. Different countries and organizations created their own green building certification programmes and standards. Life cycle assessment, rainwater harvesting, post-occupancy assessments, energy efficiency, and green building are some of the concepts that have gained popularity.

Advancements in Technology:

Innovations in technology, such as the creation of Building Information Modelling (BIM), smart technologies, and environmentally friendly building materials, have been crucial in advancing sustainable practices. More accurate planning, resource optimisation, and real-time monitoring are made possible by these technologies.

International Agreements:

The significance of sustainable construction in tackling global issues like climate change, resource depletion, and social fairness has been further highlighted by international accords and initiatives like the United Nations Sustainable Development Goals (SDGs).

Sustainable construction's historical viewpoint reflects our growing understanding of the interdependence of natural systems, the built environment, and societal well-being. The construction sector is constantly adapting, innovating, and integrating sustainable techniques to produce a more responsible and resilient built environment as environmental concerns become more pressing.

3.6 The Current Trends and Challenges

As is evident so far, even while the construction is a significant economic sector in every nation, there is a significant risk to the environment. Nonetheless, the construction is evolving and incorporating sustainability into its methods and the way that construction is delivered. The following are some problems and developments in sustainable construction that we can anticipate seeing after 2024.

3.6.1 Current Trends

Net Zero Construction

The desire for Net Zero Energy buildings is one noteworthy trend. In order to generate as much energy as they consume, buildings often integrate energy-efficient design, on-site renewable energy generation, and state-of-the-art technologies.

Circular Economy Practices

A focus on the concepts of the circular economy, encouraging material reuse, recycling, and repurposing to cut down on waste and lessen the environmental effect of construction operations.

Green Building Certifications

Certifications for green buildings become increasingly popular as standards for environmentally friendly building practices. The certificates offer uniform structures for assessing and advocating for ecologically conscious construction methods.

Smart and Connected Buildings

Smart technology integration for improved resource efficiency, occupant comfort, and building performance. Real-time parameter monitoring and control are made possible by smart building technologies.

Health and Well-Being Design

Considering the health and well-being of building occupants during the design process. This covers factors including natural lighting, biophilic design, indoor air quality, and wellness-focused building certifications.

Renewable Energy Integration

Increasing use of renewable energy, such as solar and wind power, in construction projects to reduce greenhouse gas emissions and reduce reliance on fossil fuels.

Prefab and Modular Construction

The use of prefabrication and modular construction techniques is growing in favour due to its effectiveness, low waste production, and capacity for sustainable design. Off-site building can shorten project completion times and reduce its negative effects on the environment.

Water Efficiency Strategies

Water-efficient technology, rainwater collection, and environmentally friendly landscaping techniques are being used in construction projects to reduce water usage and improve water management.

Innovative Sustainable Materials

Searching out and implementing cutting-edge sustainable materials, such as those that are bio-based, contain recycled materials, or have smaller environmental footprints. Using cutting-edge materials in buildings to increase durability and efficiency is one example of this.

Resilient Design

Incorporating resilience considerations into building and design to tackle the obstacles presented by climate change, extreme weather, and additional environmental stressors.

3.6.2 Current Challenges Facing Sustainable Construction Market

Cost Implications

Perceived greater starting costs linked to sustainable construction approaches are one of the persistent challenges. Many clients, contractors, purchasers of real estate, and government agencies continue to hold the view that sustainable development is more costly than traditional construction. This is true, but sustainable construction ends up being significantly less expensive eventually. The cost of operation and maintenance is significantly lower, and residents often save money on transportation and medical expenses. These are due to a shortage of sustainable materials and components in some countries. However, as manufacturers of materials and components make investments in retooling their plants, and the cost of sustainable materials would be lesser or equal to that of conventional materials. Overcoming these challenges requires demonstrating the long-term economic benefits and return on investment associated with sustainable features.

Lack of Standardization

Inconsistencies in the project and regional implementation might arise from different interpretations and applications of sustainable construction techniques due to a lack of worldwide or standardised standards and laws. Many certification bodies only control standard of materials and components in most nations. Also, most of the standards are not suitable outside the countries where they are been established.

Skills and Education Gap

Even professionals in the built environment are in great shortage and are still using outdated methods. They have a challenging time picking up the lasting habit. One difficulty is the shortage of experienced labourers knowledgeable about sustainable construction methods. For the successful implementation of sustainable building, it is imperative to close the skills and education gaps.

Regulatory Compliance

For those working in construction, interpreting, and adhering to changing construction regulations, rules and environmental requirements can be difficult. Keeping up with evolving criteria can also be difficult.

Limited Awareness and Understanding

Widespread adoption of sustainable methods may be hampered by stakeholders' lack of knowledge and comprehension of sustainable construction concepts, including clients, contractors, suppliers, manufacturers, and the public.

Supply Chain Sustainability

It is still difficult to guarantee the sustainability of the whole supply chain, from the extraction of raw materials to manufacturing and shipping. Cooperation between industries is necessary for sustainable sourcing procedures. This will deal with a few of the issues related to Scope 3 emissions. Emissions from activities that businesses produce indirectly along their value chain are included in scope 3. For example, emissions from manufacturers and suppliers related to the commodities that businesses purchased or used.

Overcoming Resistance to Change

The adoption of new, sustainable methods in the construction sector may be impeded by resistance to change. This opposition could stem from ignorance, conventional building techniques, or established conventions.

Life Cycle Assessment Complexity

Undertaking thorough Life Cycle Assessments for construction projects can be difficult and time-consuming. Simplifying and standardising LCA procedures could encourage wider adoption of sustainable construction projects.

Limited Availability of Sustainable Materials

Although the market for sustainable materials is expanding, there are many obstacles to their widespread use in building projects due to their restricted availability and occasionally higher costs.

Global and Local Disparities

Some locations may find it more difficult to implement sustainable construction techniques due to differences in economic growth and resource availability, which could exacerbate social and environmental inequality.

To address these problems, politicians, industry experts, academic institutions, and the public must work together to promote sustainable construction techniques, offer information and training, and foster an atmosphere that is supportive of change and innovation.

3.6.3 *Market Challenges for Sustainable Construction*

A market problem, to put it briefly, is a major impediment or difficulty that companies or sectors encounter in the marketplace. It may result from several aspects, including changes in consumer behaviour, technical advancements, regulatory concerns, economic situations, and competition. Market difficulties may affect a company's capacity to meet objectives, continue to turn a profit, or remain competitive. Resolving market difficulties successfully frequently calls for strategic planning, creativity, and flexibility. To effectively promote sustainable practices in the construction sector, initiatives that take these challenges into account must be developed. The following are the main obstacles in the market for sustainable construction:

Cost Considerations

Initial Investment: Because sustainable construction employs sustainable materials, energy-efficient technologies, and adheres to green construction standards, it often entails greater upfront expenses.

Long-Term Savings versus Perceived Cost: Because short-term costs are prioritised, some stakeholders may be reluctant to invest in energy efficiency and reduced environmental impact, even though these measures have the potential to save money overall.

Market Awareness and Education

Limited Knowledge: It is possible that many stakeholders, such as consumers, contractors, and developers, are only dimly aware of the advantages of sustainable building techniques.

Education Gap: To educate stakeholders on the long-term benefits of sustainable construction on the environment, economy, and society, extensive education and awareness campaigns are required.

Regulatory and Policy Challenges

Inconsistent Regulations: Different nations and areas have different building rules and regulations pertaining to sustainability, which causes variations in how sustainable techniques are implemented.

Lack of Enforcement: In certain situations, current laws and regulations may not be adequately upheld, which reduces the motivation for interested parties to use sustainable building techniques.

Fragmented Industry Collaboration

Construction Industry Silos: Traditionally, the construction industry has been divided into discrete professional sectors that operate in isolation from one another. It is difficult but necessary to achieve cross-disciplinary collaboration and integration for sustainable construction practices.

Supply Chain Complexity: It can be difficult to guarantee that sustainable practices are upheld throughout the whole supply chain while working on sustainable building projects, which frequently involve numerous suppliers and contractors.

Technological Adoption and Innovation

Opposition to Change: Historically, the construction sector has been sluggish to accept modern technologies. One major difficulty is encouraging the integration of innovative, sustainable technology while overcoming opposition to change.

Lack of Standardisation: The broad adoption of sustainable construction is hampered by the lack of standardised technologies and methods.

Market Demand and Consumer Preferences

Limited Market Demand: In certain areas, there might not be as much of a market desire for environmentally friendly buildings, which would discourage developers from making these investments.

Misconceptions: There are a few common misconceptions among consumers regarding the cost, comfort, and design of sustainable buildings.

Risk Perception

Perceived Risks: Owing to unfamiliar technologies, materials, or construction techniques, stakeholders may view sustainable construction practices as riskier.

Decision-making may be impacted by the intricacy of insurance and liability issues when it comes to sustainable building.

Financing Challenges

Limited Access to Financing: It may be difficult to obtain funding for environmentally friendly construction projects, particularly for smaller businesses or initiatives.

Perceived Return on Investment (ROI): Investors and financial institutions could need more convincing proof of the financial benefits of sustainable building.

Skills and Training

Training Gaps: The staff in construction may lack the knowledge and experience needed to successfully adopt sustainable practices.

Professional Development: To close these skill gaps, it is essential to support continuing professional development and training initiatives.

A comprehensive strategy that includes industry stakeholder engagement, benefits communication clearly and concisely, regulatory advocacy, and ongoing developments in sustainable construction technologies and practices are needed to address these market difficulties.

3.6.4 *The Operational Challenges Facing Sustainable Construction*

Operational challenges are, to put it briefly, problems or impediments that come up during a business's or organization's regular course of operations. Aspects of operations like as procedures, supply chain management, technology, logistics, and human resources may face obstacles. Operational difficulties can impair an organization's productivity, efficiency, and general efficacy. To overcome these obstacles, internal procedures must be optimised to raise overall operational performance. Solutions must then be found and put into place. The operational hurdles that sustainable construction faces include a variety of issues and roadblocks that may surface during the regular implementation of environmentally responsible building techniques. Comprehending the difficulties associated with embracing and executing sustainable building techniques is essential. Following are some explanation of the main operational difficulties:

Cost Implications

Higher Initial Expenses: Sustainable construction typically has greater upfront costs since it adheres to green building regulations, uses eco-friendly materials, and energy-efficient technologies.

ROI, or return on investment: It can be difficult to strike a balance between the initial investment and the long-term advantages and return on investment, particularly when stakeholders are preoccupied with immediate expenses.

Supply Chain Complexity

Purchasing Sustainable Products: It might be difficult to locate and acquire ecologically acceptable products, especially if the supply chain is not well-established.

Traceability and Certification: It can be difficult to verify the legitimacy and sustainability credentials of commodities as they move through the supply chain.

Workforce Training and Skills

Skills Gap: When it comes to implementing sustainable construction techniques, the construction sector may be confronted with a skills gap in its workforce.

Qualifications for Training: Although it may provide logistical issues, construction professionals and workers must receive sufficient training on sustainable practices and technologies.

Regulatory Compliance

Changing laws: It might be difficult to follow developing building norms and laws linked to sustainability because they may differ throughout locations and be updated.

Making Sure There's Compliance: It can take a lot of time to navigate complicated regulatory environments and make sure green building standards are followed.

Integration of Technologies

Adoption of Innovative Technologies: Adopting cutting-edge energy-efficient systems or smart building technologies, for example, may encounter opposition or integration difficulties.

Interoperability: It can be challenging to make sure that various technologies are compatible with one another and operate harmoniously with current systems.

Project Management Complexity

Coordination and Collaboration: A variety of stakeholders, including architects, contractors, suppliers, and subcontractors, are frequently involved in sustainable construction. Facilitating efficient teamwork and communication might present difficulties.

Project Schedules: It takes careful planning and coordination to incorporate sustainable practices without pushing back project deadlines.

Waste Management and Recycling

Effective waste Management: It can be difficult to control construction waste and make sure that recycling procedures are followed responsibly, particularly in areas with inadequate recycling infrastructure.

Material Reuse: Careful planning and coordination are needed to promote the reuse of materials from dismantled or demolished buildings.

Energy-Efficient Design

Achieving a balance between energy-efficient features and aesthetically beautiful designs can be difficult since sustainable design aspects occasionally clash with traditional design aesthetics.

Life Cycle Evaluation: It takes knowledge and extra work to conduct life-cycle studies of building materials and systems to determine their total environmental impact.

Certification and Standardization

Getting Certification: Getting certifications for environmentally friendly building projects, like LEED or BREEAM, requires a lot of manoeuvring through complicated procedures and fulfilling strict requirements.

Lack of Standardisation: In certain situations, inconsistent evaluation, and comparison results from the absence of standardised approaches for assessing sustainability.

A comprehensive strategy is needed to address these operational issues in sustainable building, one that includes stakeholder collaboration, continual education and training, adaptability to legislative changes, and integration of cutting-edge technology and procedures.

3.7 Policies and Regulatory Constraints Facing Sustainable Construction

Countries and regions have different laws and regulations governing sustainable construction. Nonetheless, there are several recurring themes and obstacles in the regulatory environment that affect the uptake and application of sustainable building techniques. We have listed some crucial laws and regulations below:

Building Codes and Standards

Sustainable construction norms and codes can be inconsistent and differ between areas. For architects, contractors, and developers to implement consistent sustainable practices, this could provide difficulties.

Zoning Regulations

Certain design features or construction techniques related to sustainable buildings may be restricted by zoning laws. Limitations on the height, materials, or green areas of construction might affect the options for sustainable design.

Permitting Processes

Permitting for environmentally responsive construction projects can be a difficult and drawn-out procedure. The timely completion of projects may be hampered by delays in getting permissions.

Energy Efficiency Standards

The application of cutting-edge designs and technology can be impacted by differences in the strictness of energy efficiency regulations. It might be difficult to strike systemic a balance between innovation and compliance.

Water Conservation Regulations

Water conservation regulations may restrict the application of specific technologies or landscaping techniques that support sustainable water management in building projects.

Waste Management Regulations

Effective waste reduction and recycling strategies may be impacted by regulations controlling the disposal of building waste.

Green Certification Programmes

For certain projects, participating in green certification programmes like LEED, GBI, GBCN, Green Star or BREEAM might be difficult because they require navigating complicated processes and completing requirements.

Tax Incentives and Rebates

The availability and regularity of tax breaks or refunds for environmentally friendly building projects can differ, which affects stakeholders' financial incentives.

Renewable Energy Integration

Regulations about the grid connection of renewable energy systems may have an impact on the viability and economics of integrating sustainable energy sources.

Environmental Impact Assessments

Environmental impact evaluations for sustainable construction projects may be stringent, which could prolong and increase the development process' expenses.

Accessibility Standards

It can be difficult to meet accessibility and sustainability criteria at the same time, particularly when some sustainable elements might contradict those needs.

Local and Regional Policies

Local and regional factors frequently impact policies about sustainable construction. Comprehending and adhering to these discrepancies might provide a challenge for projects involving several sites.

Labour and Safety Regulations

Project budgets and schedules may be impacted by the need for additional training and adherence to safety requirements when incorporating sustainable construction principles.

A thorough awareness of the local regulatory landscape, proactive interaction with regulatory bodies, and lobbying for policy changes that encourage the use of sustainable building techniques are all necessary for navigating these laws and regulations. To create an enabling legislative environment for sustainable construction, policymakers and industry stakeholders must work together.

3.8 Summary

In this chapter, we explore the fundamental requirements of sustainable construction, laying out the essential principles that underpin the practice. We delve into the benefits, prospects, and challenges associated with sustainable construction, offering a detailed discussion that highlights both the opportunities and potential obstacles the industry faces in its pursuit of sustainability. To effectively facilitate the procurement and application of sustainable construction practices, the availability of sustainable or green materials, as well as certification tools, is crucial. Without access to these materials and the proper mechanisms to certify their sustainability, it becomes significantly more challenging to achieve the desired environmental, social, and economic outcomes that sustainable construction promises. We also provide a comprehensive overview of some of the most commonly used sustainable materials, illustrating how they contribute to the overall sustainability of construction projects. These materials not only reduce the environmental footprint of constructions but also enhance their longevity, efficiency, and safety. Moreover, we discuss sustainable practices that go hand-in-hand with the use of green materials, such as energy-efficient design, waste reduction, and water conservation. Throughout the chapter, we emphasize the importance of integrating these sustainable practices and materials into every phase of the construction process, from planning and design to execution and operation. By doing so, stakeholders can ensure that their projects not only meet current sustainability standards but also contribute to the long-term goal of reducing the construction industry's impact on the environment. In addition, we explore the role of certification tools, such as the Green Construction Index (GBI) and LEED, in validating the sustainability of materials and practices. These tools provide a framework for assessing and certifying the environmental performance of constructions, offering a way to measure progress and encourage continuous improvement. Overall, this chapter provides a thorough examination of the essential elements of sustainable construction, offering readers the knowledge and tools they need to implement these practices effectively. By understanding the requirements, benefits, and challenges, as well as the role of sustainable materials and certification tools, stakeholders can better navigate the complexities of sustainable construction and contribute to a more sustainable built environment.

Bibliography

1. Adamson S, Medeiros AS (2023) The greenlight for government buildings: strategies for a low-carbon building portfolio. FACETS 8:1–10
2. Balocco C, Leoncini L (2020) Energy cost for effective ventilation and air quality for healthy buildings: Plant proposals for a historic building school reopening in the covid-19 era. Sustainability 12(20):8737
3. Calkins M (2008) Materials for sustainable sites: a complete guide to the evaluation, selection, and use of sustainable construction materials. Wiley
4. Dhir, Ravindra K., Jorge de Brito, G. S. Ghataora, and Chao Qun Lye. Sustainable construction materials. Elsevier Limited; Woodhead Publishing, 2018.
5. Felgueiras F, Mourão Z, Moreira A, Gabriel MF (2024) Characterizing indoor environmental quality in Portuguese office buildings for designing an intervention program. Build Environ 254:111393
6. Goodhew S (2016) Sustainable construction processes: a resource text. Wiley
7. Halliday S (2008) Sustainable construction. Routledge
8. Hill RC, Bowen PA (1997) Sustainable construction: principles and a framework for attainment. Constr Manag Econ 15(3):223–239
9. Kibert CJ (2007) The next generation of sustainable construction. Build Res Inf 35(6):595–601
10. Kibert CJ (2016) Sustainable construction: green building design and delivery. Wiley
11. Maqbool R, Arul T, Ashfaq S (2023) A mixed-methods study of sustainable construction practices in the UK. J Clean Prod 430:139087
12. Myers D (2022) Construction economics: a new approach. Routledge

13. Raynsford N (2000) Sustainable construction: the Government's role. In: Proceedings of the institution of civil engineers-civil engineering, vol 138, no 6. Thomas Telford Ltd., pp 16–22
14. Dhir OBE RK, de Brito J, Silva RV, Lye CQ (2019) Sustainable construction materials: recycled aggregates. Woodhead Publishing
15. Olanrewaju A, Tan SY (2019) Roles of design and construction teams in sustainable housing delivery. In: Greening affordable housing: an interactive approach, pp 1–17
16. Olanrewaju AA, Chong YS (2021) Post occupancy evaluation of green residential buildings, in the Greater Kuala Lumpur, Malaysia. J Housing Built Environ 36:825–857
17. Olanrewaju A, Shari Z, Gou Z (eds) (2019) Greening affordable housing: an interactive approach. CRC Press
18. Shelbourn MA, Bouchlaghem DM, Anumba CJ, Carrillo PM, Khalfan MM, Glass J (2006) Managing knowledge in the context of sustainable construction. J Inf Technol Constr 11:57–71
19. Ugbaja SC (2023) The implementation of sustainable practices in achieving sustainability goals in Middle Eastern Universities
20. Vallero DA, Brasier C (2008) Sustainable design: the science of sustainability and green engineering. Wiley
21. Zhou L, Lowe DJ (2003) Economic challenges of sustainable construction. In: Proceedings of RICS COBRA foundation construction and building research conference. University of Wolverhampton, Wolverhampton, UK, pp 1–2

Chapter 4
Environmental Considerations in Construction

Abstract This chapter explores the crucial environmental considerations in construction, emphasising the sector's significant impact on climate change and resource depletion. It underscores the urgent need for a shift toward sustainable practices that reduce environmental harm while promoting resource efficiency. The chapter delves into key topics such as the selection of sustainable construction materials, whole-life appraisal methods, and techniques for minimising construction waste. It highlights the role of energy-efficient design and the use of renewable energy sources in reducing the carbon footprint of construction projects. Furthermore, the chapter examines the pressing issue of biodiversity loss due to construction activities, advocating for the integration of comprehensive Environmental Impact Assessments (EIAs) to identify and mitigate environmental concerns early in the planning process. The importance of adopting circular economy principles and recycling practices to address waste generation is also discussed. With increasing regulatory pressure and the adoption of green certifications, sustainable construction practices are now integral to the sector's future. This chapter provides valuable insights into how the construction industry can contribute to a more sustainable and resilient built environment by embracing environmental ethics, innovative materials, and effective waste management strategies. By understanding these environmental challenges and solutions, stakeholders can play a pivotal role in shaping a sustainable future for construction.

Keywords Construction materials · Whole life appraisal · Green certification · Waste management

4.1 Learning Outcomes

The need to include social, environmental, and economic factors into decision-making is one of the current challenges facing efforts to achieve sustainable development. The environmental consideration in the sustainability equation is profound and pivotal. Sustainable development has its roots in environmental economics, which

A. Olanrewaju, *Sustainable Construction Management: Research and Practice Companion*, Management in the Built Environment,
https://doi.org/10.1007/978-981-96-4910-5_4

was first developed in 1960. Environmental economics primarily aims to quantify the costs (i.e., consequences) of economic actions on the external environment and to suggest ways to reduce or eliminate these problems, which are typically significant but frequently unknown at the time the decisions were made. The construction sector is one of the most nature-depleted sectors. The construction sector needs to understand and appreciate the roles of communities in delivering biodiversity net gain. In particular, the communities have valuable knowledge and experience of the local habitat and wildlife. Though nations like the UK have recently (in February 2024) passed legislation establishing a biodiversity net gain policy, which without doubt is critical and encouraging. However, the legislation does not fully address the environmental impact of construction because it does not address the production and processing of construction materials, such as wood, sand, and minerals like rocks, gravel, and iron ore, and how this affects biodiversity. Likewise, the current regulations do not address the embodied biodiversity impacts of a project, which are the effects on biodiversity arising from the processes that occur from the extraction of a material to its disposal at the end of its life. This is a part that will eventually need to be addressed. Resources used in construction include energy, water, and raw materials, the production, processing, and use of which are all influenced by the environment. Waste is produced throughout the construction process, and potentially dangerous materials are released. Without question, each of these is causing the climate changes. The pursuit of sustainable construction has been prompted by worries about resource depletion, pollution, climate change, population growth, and environmental degradation. This chapter's main argument is that construction plays a significant role in the phenomenon of climate change. The implementation of sustainable construction practices will result in notable energy efficiency gains, increased construction durability, waste reduction, enhanced social impacts, a healthy indoor environment, decreased pollution, consideration of the life cycle costs of the construction project, active user-friendliness, user comfort, and overall satisfaction. Following your reading of this chapter, you would be:

- Explain sustainable construction material
- Problems with environmentally friendly construction materials and methods
- Explain whole life appraisal
- Explain construction waste
- Explain techniques to reduce construction waste.

4.2 Introduction

In the rapidly changing world of contemporary construction, client and project users now consider construction sustainably to be a strategic choice. The chapter emphasises how crucial it is for construction projects to use environmental sustainability as a guiding concept. One of the pillars of sustainable development is the environmental dimension. As the world becomes more conscious of climate change and resource depletion, the construction sector is forced to reconsider long-standing methods in favour of more responsible and ecologically sound ones. This chapter examined the significant environmental influences that have shaped the construction sector. This investigation explores the different aspects of environmental concerns in construction and how the participants in the construction sector are adjusting to lessen their environmental impact. The chapter reveals the dynamic interaction between environmental awareness and the construction processes that define our built environment, from sustainable construction materials to cutting-edge construction techniques. The chapter also shows why it is necessary to abandon conventional tactics in favour of ones that put environmental ethics first. One of the most vital instruments for assessing and choosing materials according to their environmental impact is whole-life appraisal. The use of renewable energy sources and energy-efficient design are emphasised as crucial tactics for lowering the carbon footprint of construction projects. The substantial amount of waste generated by the construction sector is examined. The chapter highlights the significance of recycling and waste reduction programmes, highlighting the need for a paradigm change towards circular economy practices.

4.3 Environmental Sustainability

The global nature of sustainable development or sustainability makes it a challenge whose causes, effects, and solutions affect all people. Sustainable development has its roots in environmental economics, which was first developed in 1960. The main goals of environmental economics are to measure the costs and impacts of economic decisions on the external environment, as well as to offer solutions for mitigating or eliminating these effects. These consequences are usually large, but they are often unknown at the time of the economic decisions. As a result, sustainable development objectives are future-oriented. One of the most pressing problem in the quest of sustainable development is the requirement that social, environmental, and economic elements be considered when making decisions. To simplify, the three most commonly referenced indicators of sustainable development are social, environmental, and economic. These three are sometimes called the people, the planet, and profit, respectively.

A recent research shows that Climate change could wipe between \$2.5 trillion \$24.2 trillion off the value of global assets, according to a new set of economic models [4]. 195 countries agreed at the Paris Climate Conference (COP21) to limit global warming to 1.5 °C and to keep average temperatures well below 2 °C. The South African Conventions, the Rio Protocol, and the Kyoto Protocol all address the same problems. Concerns about sustainability (such as pollution, resource depletion, global warming, and climate change) are increasingly included in client decision-making processes, business planning, and corporate. Because of this, clients now check labels for goods and services to determine whether they are beneficial or detrimental to the community and environment. Also, concerned client now verify the equipment and product's water, electricity, and carbon emission rates before making a purchase. Both homebuyers and homeowners are considering the social and financial effects of their decisions. The sustainability credentials of construction companies are coming under scrutiny. As a result, the majority of governments worldwide have established goals to address problems like pollution, resource depletion, climate change, and energy and water scarcity. The GDP is no longer the exclusive indicator of a nation's prosperity. An option that is increasingly being examined is the Social Progressive Index, which gauges social and cultural aspects. The need for a sustainable built environment has grown since sustainable development was recognised. The main danger to the objectives of sustainable development is the built environment. The construction sector has a significant impact on sustainability. For instance, to meet construction demand, a timber processing company can cut down an excessive number of trees. The company and its workers would undoubtedly benefit financially from the enormous number of trees cut down since it would lower the cost of producing lumber and boost earnings. Customers would gain as well because they could buy premium timber for less money. However, the trees and the company will eventually perish if the three cuttings are left unchecked for a long time. In addition to these immediate effects for the firm, the price would go up and other companies that rely on the materials would have to close. For instance, the cost of timber for the construction sector would rise, which will also result in a workforce shortage for other companies as the timber producer is prepared to pay higher salaries and fulfil its need for output.

Long-term environmental damage and social discontent would also result from over-chopping trees. Furthermore, consider the growing need for energy, such as electricity and crude oil, brought on by the growth of the transportation and industrial sectors. While the country's economic growth depends on this, doing so will result in significant emissions of pollutants into the environment, decreasing the quality of the air and water, and ultimately affecting how people live. Medical care costs would rise, population growth might be impacted, and treatment costs for erosion and air and water pollution would rise as well. Cement is another common construction material, the production of which has a significant effect on social aspects of the society and on the environment. The environment and public health will be impacted by the CO_2 emissions from the cement industry.

In particular, building accounts for almost 40% of global energy consumption, 1/3 of CO_2 emissions, 25% of harvested wood use, 50% of released fluorocarbons, 40% of landfill material production, 45% of energy use during operations, and 40% of greenhouse gas emissions. 15% of the world's useable water is used by the construction sector (see [10, 22]). Pollutant levels in building interiors are five times higher than those outside. The impact will be significantly greater when additional greenhouse gas emissions related to construction activities, such as those from the production and transportation of construction and demolition materials, are considered. If we consider additional energy use related to construction, the amount of energy consumed by constructions is very huge no doubt. For instance, the annual energy required to heat and cool a single construction is eight to ten times less than the embodied energy in the construction envelope. Due to the distance between their homes and workplaces, children's schools, and marketplaces, people use cars. Cars are a major contributor to pollution as well. Up to 20% of most households' income is spent on transportation. These figures indicate that one of the threats to sustainable development is construction projects.

4.4 Benefits of Environmentally Sustainable Practices

Utilising renewable or recycled materials lowers the construction sector's overall environmental impact. Recycling is becoming a common component of sustainable practices. Energy-efficient constructions reduce operating costs and their negative environmental effects through the use of eco-friendly materials and techniques. Utilising materials that have been recycled or sourced ethically supports ethical extraction methods and aids in the preservation of natural resources. Low-emission construction materials help to improve indoor air quality, which makes homes and workplaces healthier. Although the initial costs of certain eco-friendly materials may be greater, their longevity and energy efficiency can lead to long-term cost benefits.

4.5 Whole Life Appraisal

Life cycle assessment, or whole life appraisal is a systematic and comprehensive approach called that is employed to investigate the environmental impacts of a product, activity, or process over the course of its whole existence. This criterion has grown increasingly crucial as the desire for sustainable or green constructions has increased. The extraction of raw materials, production, transportation, usage, disposal, and recycling are all included in this assessment. The method is a useful tool for evaluating how well different systems function environmentally and pointing decision-makers in the direction of more sustainable behaviours. Next is an explanation of the on the components of the whole life appraisal.

4.5.1 Primary Components of Whole Life Appraisal

1. Scope and Goal Definition:

Goal: When conducting the appraisal, the construction management team should clearly state its goals. These will involve locating hotspots in the environment, weighing options to meet the client's values system, or providing information for making decisions.

Scope: The stakeholders should specify the parameters of the evaluation, such as the functional unit, system boundaries, and life cycle stages that need to be considered.

2. Whole Cycle Inventory:

The construction management team should thoroughly catalogue all inputs and outputs associated with each stage of the entire cycle, including raw materials, energy consumption, emissions, and waste creation.

3. Whole Cycle Impact Assessment:

The whole cycle catalogue's materials and components should have their possible environmental effects assessed by the construction management team. To evaluate the importance of each impact, this usually entails applying environmental indicators, such as greenhouse gas emissions, water usage, and resource depletion.

4. Interpretation:

At this point, the construction management team's findings and suggestions should be grounded in the impact assessment and life cycle inventory data. This step entails considering risks and uncertainties in the project, spotting areas for improvement, informing stakeholders of the results, and making any required recommendations.

4.5.2 Applications of Whole Life Appraisal

1. Product Design and Development:

Low-impact materials and methods should be identified by the construction management team. This will facilitate the creation of environmentally sustainable products starting with the strategic planning stage of construction projects.

2. Material Selection:

The environmental performance of various materials should be assessed by the construction management team. This will direct the manufacturing and construction sectors in choosing sustainable and resource-efficient solutions.

3. Process Optimization:

The construction management team is responsible for identifying areas of high environmental risk and inefficiencies in the production process. In the end, this will result in optimisation initiatives that lower waste, pollution, and energy use.

4. Policy Development:

By offering a scientific foundation for comprehending the environmental effects of certain actions, materials, or products, the construction management team should assist in the creation of environmental rules and regulations.

5. Consumer Awareness:

The construction management team should raise consumer knowledge by being open and honest about how materials and component perform in terms of their environmental credential. Many materials and products have negative environmental effects; however, some materials are often misrepresented as green or sustainable. Clients should be educated on greenwashing. Increasing consumer knowledge and education will enable them to make more environmentally friendly decisions.

6. Waste Management Strategies:

When it comes to waste management techniques, the construction management team should offer direction. This can be achieved by evaluating how various end-of-life options, such as recycling, burning, or landfill disposal, will affect the environment.

4.6 Benefits of the Whole Life Appraisal

1. Holistic Perspective:

The WLA considers a material's, product's, or process's whole life cycle and provides information on the environmental impact of each stage, from resource extraction to disposal.

2. Informed Decision-Making:

WLA enables decision-makers to make well-informed decisions by taking social, economic, and environmental concerns into account, which will eventually result in more sustainable practices.

3. Identifying Improvement Opportunities:

The method enables targeted improvements and the creation of more environmentally friendly alternatives by assisting construction management in identifying locations with the greatest environmental impact.

4. Environmental Certification:

Sufficient deliberation and implementation of the WLA afford products and procedures the chance and assistance to obtain certifications, like eco-labels or green construction certifications, augmenting their competitive edge in the marketplace.

5. Policy Support:

Governments and regulatory agencies can utilise the data gathered from the techniques to create and carry out efficient environmental policies that establish benchmarks for environmentally friendly business operations.

6. Supply Chain Management:

Businesses can utilise the method to assess how their supply chains affect the environment, promoting sustainability along the whole value chain.

4.6.1 Challenges Facing the Applications of the Whole Life Appraisal Technique

While the technique is useful, its applications face some difficulties, notably the following:

1. Data Availability and Quality:

The precise and thorough data are necessary for the procedure to be applied effectively. On the other hand, getting accurate data for every phase of a construction project's life cycle is extremely tough.

2. System Boundaries:

Determining the bounds of an evaluation is always challenging, particularly when considering the interdependence and interplay of the systems in a construction project. Decision has to be made in the WLA should be applied to materials, components, part of the construction or the entire construction project. The choice is usually not a straightforward one. Nonetheless, choices on which products or processes to include or eliminate have a significant impact on the outcomes.

3. Subjectivity and Interpretation:

Despite being data-driven, the assessment process involves subjective judgements in interpreting the data obtained from the technique's implementations, and various stakeholders may prioritise different environmental implications.

4. Dynamic Nature of Systems:

As industrial processes, energy sources, and waste management technologies change over time, appraisals must also adjust to reflect these changes in systems and technology.

5. Trade-offs and Complexity:

Achieving overall sustainability goals may require making trade-offs to balance environmental, economic, and social factors.

Making trade-offs to balance environmental, economic, and social variables may be necessary to meet overall sustainability goals.

Whole life assessment is a potent and thorough method in environmental sustainability that examines the environmental impact of construction projects from conception to completion. The evaluation goes beyond the conventional parameters of construction assessments. It offers a comprehensive perspective that covers the procurement of raw materials, production procedures, construction projects, operating stages, and the ultimate end-of-life situation.

4.6.2 Understanding the Whole Life Appraisal

1. Cradle-to-Grave Perspective:

The method takes a "cradle-to-grave" viewpoint to material, process, system, component or the construction. This facilitates the evaluation of the environmental impact of construction projects at every stage of their life cycles by interested parties. Purchasing raw materials, shipping, manufacturing, building, operating, and maintaining the system, as well as any eventual demolition or recycling, are all included in this.

2. Quantifying Environmental Impacts:

By using a variety of variables, including carbon emissions, energy consumption, water usage, and resource depletion, whole-life evaluation assists in quantifying environmental impacts. It is a data-driven methodology that makes it possible to fully comprehend the environmental effects of the project.

3. Materials and Resource Considerations:

The evaluation closely examines the choice of construction materials, assessing their procurement, production, and delivery methods. This encourages the use of sustainable alternatives by educating decision-makers about the environmental effects of material choices.

4. Construction Processes and Energy Efficiency:

The assessment of the energy efficiency of construction processes is made possible by the use of the appraisal, which also provides insight into the carbon footprint of construction operations. Construction techniques that minimise energy use and cut greenhouse gas emissions are guided by this realisation.

5. Operational and Maintenance Phases:

The appraisal considers the operational and maintenance stages. It also examines energy use, water consumption, and other continuing environmental impacts. This makes it easier to optimise a construction's performance during its operation.

6. End-of-Life Scenarios:

The evaluation includes end-of-life scenarios and considers recycling or disposal of construction components. Through comprehension of the possible environmental consequences associated with waste management and deconstruction, stakeholders can make well-informed decisions on sustainable practices.

4.7 Benefits of Whole Life Appraisal in Construction

1. Informed Decision-Making:

At every step of a construction project, decision-makers may make well-informed decisions thanks to the appraisal's comprehensive data. This covers the choice of appropriate construction materials, techniques, and long-term project management plans.

2. Environmental Optimization:

Throughout the life cycle, the assessment aids in the identification of environmental hotspots. Moreover, it makes focused interventions for environmental optimisation easier. This could entail giving sustainable materials priority, optimising energy performance, or increasing the efficiency of construction operations.

3. Regulatory Compliance:

Identifying environmental hotspots throughout a life cycle is another benefit of the appraisal. Additionally, it makes focused interventions for optimising the environment possible. This could entail making construction procedures more effective, boosting energy efficiency, or giving sustainable materials priority.

4. Stakeholder Engagement:

Engaging with project stakeholders, including communities, investors, and regulatory agencies, is also beneficial. Additionally, it permits open reporting on environmental issues, which promotes confidence and synchronises construction projects with more general sustainability goals.

5. Innovation and Continuous Improvement:

By exposing areas for improvement throughout a construction project, the assessment also promotes innovation. By encouraging a culture of continuous development, this iterative approach pushes the industry towards ever-more sustainable practices.

4.8 Construction Materials

These are products or substances used in the construction sector to produce infrastructure and construction. These materials have several uses, from supporting structural integrity to improving a construction's appearance and use. Based on their main purposes, construction materials can be generally divided into multiple sorts. Table 4.1 provides examples of typical construction materials:

Table 4.1 Examples of typical construction materials:

Material	Uses	Benefits	Sources
Concrete	Foundations, slabs, walls, columns, and other structural elements	High durability, adaptability, and compressive strength	Cement (produced from limestone), aggregates (sand, gravel, crushed stone), and water
Steel	Structural frames, roofing, concrete reinforcement, and other construction materials	Robustness, malleability, and potential for recycling	Coal, iron ore, and other raw resources used to make steel
Wood	Framing, flooring, decking, interior finishes	Renewable, aesthetic appeal, carbon sequestration	Harvested from trees (softwoods or hardwoods)
Bricks and blocks	Facades, walls, and architectural components	Heat mass, fire resistance, and durability	Clay or concrete
Glass	Windows, doors, curtain walls, and decorative features	Openness, daylighting, and recyclable materials	Silica sand, soda ash, and limestone
Asphalt	Road construction, roofing, waterproofing	Flexible, weather-resistant	Petroleum
Plastics	Pipes, insulation, roofing materials	Lightweight, adaptable, energy-efficient production	Petrochemicals derived from crude oil or natural gas
Insulation materials	Insulating walls, roofs, floors	Energy efficiency, thermal and acoustic insulation	Various materials, including fiberglass, mineral wool, and foam
Adhesives and sealants	Bonding materials, sealing joints	Versatility, durability, efficiency	Petroleum-based or synthetic chemicals
Roofing materials	Covering and protecting roof structures	Weather resistance, durability, aesthetics	Various materials, including asphalt, metal, clay, and concrete
Paints and coatings	Enhancing appearance, protecting surfaces	Decoration, corrosion resistance, weatherproofing	Pigments, binders, solvents, and additives
Aggregate materials	Providing bulk to concrete, road construction	Structural stability, drainage	Crushed stone, gravel, sand
Tile	Flooring, wall coverings, decorative elements	Durability, moisture resistance, aesthetics	Clay and other minerals

(continued)

Table 4.1 (continued)

Material	Uses	Benefits	Sources
Composite materials	Numerous uses where combined qualities are beneficial	Tailored properties, lightweight	Mixtures of materials, like reinforced concrete or fibreglass composites
Reinforcement materials	Reinforcing concrete structures	Improved tensile strength of concrete	Steel (manufactured from steel billets)

These are but a handful of the wide variety of construction materials that are employed in the industry. The selection of materials is influenced by several variables, including financial limitations, climatic circumstances, aesthetic preferences, and structural requirements. As awareness of environmental effects increases, sustainable and eco-friendly options are likewise becoming more and more popular in the construction business. It is significant to remember that the advantages and sources of construction materials can differ depending on things like local availability, project specifications, and the development of new technologies in the sector. Sustainable construction methods are becoming more and more popular as a means of addressing environmental issues and advancing communities' long-term well-being.

4.8.1 Materials Selection and Environmental Sustainability

Sustainable construction, which aims to reduce environmental impact, enhance resource efficiency, and produce healthier built environments, includes environmentally friendly processes and the use of construction materials. The following is an outline of factors to be considered while choosing construction materials and environmentally friendly methods.

Construction Materials Selection:

1. Sustainable Sourcing:

The team responsible for construction management ought to select materials that adhere to environmental ethics when procuring them.

2. Recycled and Recyclable Materials:

Materials with recycled content, like concrete aggregates or recycled steel, should be selected by the construction management team. The group should also give preference to materials that can be readily recycled at the end of their useful lives.

3. Low Embodied Energy:

Low embodied energy materials are those that need less energy overall throughout extraction, production, transportation, and installation. This is something that the construction management team should examine when choosing materials. Recycled steel and glass are examples of materials with lower embodied energy.

4. Local Materials:

The construction management team should prioritise locally sourced materials. This will boost regional economies and lower energy use associated with transportation. Additionally, this will lessen the carbon impact caused by the transit of materials.

5. Durable and Long-lasting Materials:

To minimise the need for frequent maintenance or renovations, the construction management team have to use products and materials with extended lifespans. This will lessen the production of waste and improve resource efficiency overall.

6. Low-impact Materials:

Low-volatile organic compound paints and finishes are examples of products that the construction management team should choose since they have minimal environmental impact. These components enhance the quality of the air inside constructions.

7. Innovative and Sustainable Alternatives:

In addition, the group ought to evaluate and use eco-friendly substitutes like hempcrete for construction and bamboo flooring. These substitutes frequently have distinct qualities and less of an influence on the environment.

8. Green Roof and Wall Systems:

Construction energy efficiency can be improved by integrating green roof and wall systems, according to the construction management team. The integration will support biodiversity and lessen the impact of the urban heat island.

4.8.2 Environmental Compliance Practices

1. Energy-efficient Design:

The construction management team should consider the energy efficiency when conceptualizing the construction projects. To lower the total amount of energy used, the team should include elements like energy-efficient windows, high-performance insulation, and passive solar architecture.

2. Renewable Energy Integration:

In order to provide clean, sustainable energy for the construction projects, the construction management team should take renewable energy sources into account when constructing the projects. Examples of these sources are solar panels and wind turbines.

3. Water Conservation:

Water-efficient techniques, such as drought-tolerant landscaping, rainwater harvesting systems, and low-flow fixtures, should be implemented by the construction management team to save water.

4. Waste Reduction and Recycling:

To reduce the quantity of construction wastes that will be dumped in landfills, the construction management team should implement waste reduction techniques and create a construction waste management plan.

5. Efficient Construction Processes:

To minimise energy consumption and environmental impact, the construction management team should optimise the construction operations. This can be accomplished by implementing modular and prefabrication construction methods to increase productivity and decrease waste generated on-site.

6. Sustainable Transportation:

The team overseeing construction projects should promote the adoption of environmentally ethics transportation methods, like the usage of electric or hybrid vehicles and equipment.

7. Green Construction Certifications:

To guarantee adherence to sustainable construction methods, the construction management team should work towards green construction certifications for environmental compliance design and construction.

8. Stormwater Management:

To lessen runoff and improve water quality, the construction management team should use vegetated stormwater management techniques like bioswales or permeable pavements.

9. Community Engagement:

When working on construction projects, the construction management team should collaborate with the neighbourhood. This will help to address some concerns and advance social sustainability. This can involve making local hires, encouraging regional companies, and designing inclusive environments.

10. Lifecycle Analysis and Maintenance:

To evaluate the environmental impact of construction materials and procedures throughout the construction's lifetime, the construction management team should do lifecycle evaluations. Additionally, the team should give top priority to upkeep procedures that prolong the life of construction elements.

11. Smart Construction Technologies:

For effective energy management, construction system monitoring, and optimisation, the construction management team should take smart construction technology into account.

4.9 Challenges of Environmental-Compliance Construction Materials and Practices

Although using environmentally ethic construction materials and techniques is essential for sustainable growth, doing so presents several difficulties. These difficulties can change between locations and projects. However, some typical problems that it presents include:

1. Cost Considerations:

Challenge: Because environment-compliant materials and technologies have greater upfront prices, some developers and clients may not find them to be financially appealing.

Solution: The government should offer greater tax breaks and incentives for environmentally ethics materials, components and practices. The government should raise public knowledge of the sustainable products' long-term cost reductions as well.

2. Limited Availability and Variety:

Challenge: Most countries have relatively limited availability and range of materials that comply with environmental regulations. Because of these, the design team and clients are finding it difficult to choose sustainable products because of this restriction.

Solution: In addition to encouraging local sourcing, the government should fund and foster the development and manufacturing of a wide variety of sustainable materials.

3. Lack of Awareness:

Challenge: There is a deficiency in knowledge regarding environmental materials and methods among numerous stakeholders, such as clients, suppliers, manufacturers, contractors, and developers.

Solution: To enlighten stakeholders about the advantages and accessibility of sustainable alternatives, governments and developers should conduct awareness campaigns in addition to industry training programmes.

4. Cultural Resistance to Change:

Challenge: Adopting new, sustainable processes and materials can encounter opposition due to the deeply established nature of traditional construction practices.

Solution: Overcoming reluctance can be aided by showcasing projects that highlight the advantages and success of environment ethics construction. In order to solve this issue, construction standards and laws that require sustainable practices can be helpful.

5. Certification and Standardization Problems:

Challenge: There are a lot of environmental compliance certificates and standards, in some countries. For instance, there are 10 of them in Malaysia. This is by no means a problem, since they are designed and established for certain objectives. However, this can lead to misunderstanding in many cases. Furthermore, some areas do not have uniform laws.

Solution: Simplifying certification procedures and creating uniform standards for environmentally ethics construction can help customers and builders make more informed decisions.

6. Misconceptions:

Problem: When compared to standard materials, some environment compliant materials are thought to perform worse. The major reason for this, is greenwashing.

Solution: To improve the performance of environmentally ethic materials, more funding should be allocated to research and development by the government, real estate developers, clients, etc. Additionally, presenting data-backed proof can help clear up misunderstandings and false information.

7. Durability:

Challenge: Adoption of some environmentally friendly materials may be hampered by worries about their longevity.

Solution: Ongoing study and development to increase the robustness of environmentally friendly materials, together with transparent performance disclosures.

8. Technological Limitations:

Challenge: Access to cutting-edge construction technology necessary for the efficient use of sustainable materials may be limited in some countries.

Solution: Businesses, developers, and the government in particular should fund training initiatives, knowledge transfer, and technological infrastructure.

9. Constraints in the Supply Chain:

Challenge: Due to resource shortages, there may be disruptions in the supply chain for environmentally compliant materials, which could have an impact on the prices and availability of sustainable materials and components.

Solution: In addition to encouraging local sourcing, manufacturers, suppliers, and contractors should fortify and diversify their supply chains.

10. Regulatory Barriers:

Problem: The majority of construction codes and laws do not encourage or promote the use of ecologically friendly materials.

One potential solution is to advocate for and adopt legislation that supports sustainable construction techniques. Additionally, construction codes should be updated to incorporate environmental-friendly requirements.

11. Waste Management:

Challenge: At the end of their life cycles, certain environmentally complaint materials and products may provide difficulties for recycling and disposal.

The full life assessment, thorough waste management plans, and increased investment in recycling technology are the solutions. To complete the supply chain, sustainable materials should be manufactured to be recyclable or reusable.

Governments, the research community, and industry stakeholders must work together to address these issues. By removing these barriers, environment complaint construction techniques and materials will become more widely used, fostering the development of a more resilient and sustainable built environment.

4.10 Construction Energy Management

Construction energy management involves strategic planning, monitoring, and optimisation of energy use during the construction process. Ultimately, it aims to improve productivity, lessen environmental effects, and cut overall energy costs. Energy consumption in the construction sector is extensive, and sustainable development and long-term economic gains are enhanced by the application of efficient energy management techniques. The primary components of construction energy management are as follows:

1. Energy Audit and Planning:

Evaluation: The construction management team should conduct an energy audit to find inefficient areas, trends in the project's energy usage, and possible areas for improvement,

Planning: To maximise energy consumption in the projects, the construction management team should create a thorough energy management strategy with precise objectives, targets, and tactics.

2. Design for Energy Efficiency:

Passive Design: the construction management team should implement passive design concepts that optimise natural lighting, ventilation, and thermal efficiency to minimise the need for artificial heating and cooling.

Energy-Efficient Systems: During the design stage, the construction management team should specify and include energy-efficient lighting, HVAC systems, and other construction systems.

3. Material Selection:

Embodied Energy: When choosing construction materials, the construction management team should consider the embodied energy of the material and should ensure that the manufacturing procedures used to produce them use less energy.

Local Sourcing: To cut down on energy expenses associated with transportation, the construction management team should use locally obtained products.

4. Construction Site Practices:

Efficiency of Equipment: Energy-efficient construction machinery and equipment should be used by contractors.

On-site Power Management: To track and maximise energy use during construction, contractors should put in place on-site power management systems.

Temporary Facilities: To conserve energy, the contractors should build temporary constructions like construction offices and worker housing.

5. Renewable Energy Integration:

On-site Generation: The contractors should examine the possibility of on-site renewable energy generation alternative. These alternatives may include solar panels or wind turbines, to supplement the energy requirements of the construction site.

Off-site Renewable Procurement: To reduce energy consumption associated with construction, the contractor should buy renewable energy from off-site sources.

6. Energy Monitoring and Data Analytics:

Instantaneous Monitoring: To monitor energy consumption trends and pinpoint regions in need of enhancement, the contractors had to employ instantaneous energy monitoring equipment.

Data analytics is a tool that contractors should utilise to examine patterns in energy use and make well-informed choices to increase efficiency.

7. Operatives Behaviours:

Training and Awareness: The contractors should impart energy-efficient procedures and the value of energy conservation to the workers and personnel on the construction site. A rewards programme might be established for this purpose.

User Engagement: It is recommended to motivate construction site workers to embrace energy-efficient behaviours and practices.

8. Waste Heat Recovery:

Heat Capture: During construction operations like concrete curing or machine operation, contractors should put in place devices to collect and reuse waste heat data.

9. Energy-Efficient Lighting:

LED Lighting: Energy-efficient lighting should be used by contractors in both temporary and permanent construction sites.

Installing timed controls and sensors can guarantee that lights are only turned on when necessary.

10. Commissioning and Quality Assurance:

Commissioning: To ensure that all energy systems function as efficiently as possible, the commissioning managers and maintenance management team should take care of the commissioning process. drafted a commission schedule. In-depth coverage and plan guidance are provided for a number of crucial commissioning process elements, including roles and responsibilities, list of key equipment and plants integrated testing, safety and security, trade catalogues, emergency preparedness guidance notes, O&M manuals, training, as-built drawings, and warranty administration.

Quality Assurance: To avoid wasting energy on malfunctions or inefficiencies, contractors should routinely inspect and repair their equipment.

11. Certifications and Standards:

Certifications: contractors should make use of certifications related to environmental sustainability. This is to make sure that the procedures, policies, and decisions follow energy efficiency standards and best practices in their projects.

12. Post-Construction Performance Monitoring:

Post-Occupancy Evaluation: To identify any deviations from the projected levels of energy efficiency, the maintenance management group should monitor the completed construction's energy performance.

Modifications: Considering the information gathered from the post-construction assessments, the team should all make the required modifications to the construction systems.

It takes a comprehensive strategy that considers every stage of the project life cycle to manage construction energy effectively. Through the integration of energy-efficient practices from the design phase to construction and beyond, stakeholders can make a positive influence on the environment, save operating costs, and enhance sustainability overall.

4.10.1 Technological Innovation and Collaboration in Construction Environmentally Sustainability Practices

The cores of technological innovation and collaboration in environmental considerations in construction are pivotal to driving the required positive change and fostering sustainability within the construction sector. There are many aspects to consider. The following are the main technological innovations and collaborations:

1. Advanced Materials and Methods:

Advanced construction techniques and materials with less of an adverse effect on the environment should be introduced with the aid of technological innovation. This covers the creation of environmentally compliant, and high-performance materials as well as the application of cutting-edge construction methods.

2. Digital Modeling and Simulation:

Digital modelling and simulation of construction projects are made easier by the use of technologies. This will assist in streamlining designs for maximum energy efficiency, cutting waste, and enhancing the projects' overall environmental performance.

3. Renewable Energy Integration:

On-site energy production is now feasible by the advancement of renewable energy technologies, such as solar and wind power. By incorporating these technologies into construction projects, the dependency on conventional energy sources-which are frequently overly taxing on the environment—is decreased.

4. Smart Construction Systems:

Energy consumption in the projects may be controlled and performance can be monitored in real time with the use of automation and smart construction systems. By doing this, one can make sure that a construction's lifetime is designed with maximum performance and efficiency requirements.

5. Waste Reduction Technologies:

Prefabrication and modular construction are two waste reduction ideas that can be adopted to reduce material waste during construction. In the same way, improvements in recycling technologies will help the construction industry adopt a more circular economy.

6. Environmental Monitoring and Reporting:

Real-time environmental monitoring during construction should be applied by the application of technology. To lessen the negative effects of construction on the environment, this involves noise reduction technologies, air and water quality monitoring, and other actions.

In today's networked workplace, collaboration is essential to leveraging variations in choice, expressions, knowledge and ideas. Expertise and ideas are activated through collaboration. The key collaborations to consider are as follows:

1. Interdisciplinary Collaboration:

Collaboration amongst several disciplines, including architects, engineers, manufacturers, quantity surveyors, construction managers, project managers, maintenance managers, users, suppliers, the haulage sector, and legislators, is necessary to address environmental problems in construction. Multidisciplinary cooperation guarantees a comprehensive strategy for environmentally friendly construction methods.

2. Stakeholder Engagement:

To solve environmental problems in construction, cooperation between multiple disciplines is required, including that of architects, engineers, manufacturers, suppliers, the haulage industry, and lawmakers. interdisciplinary collaboration ensures a holistic approach to environment compliant construction techniques.

3. Knowledge Sharing and Best Practices:

Cooperation makes it easier for professionals in the construction sector to share best practices and information. This involves sharing knowledge on components, goods, materials, procedures, standards, practises. technologies, and construction techniques that adhere to and support environmental sustainability.

4. Public–Private Partnerships:

Public–private collaboration refers to the joint effort between the public and private sectors to complete a project or deliver a service traditionally managed by the public sector. The successful implementation of large-scale sustainable construction projects often depends on strong cooperation between these two sectors. Public–private partnerships enable the pooling of financial resources, technical expertise, and operational capabilities to achieve shared environmental and sustainability goals.

5. Global Collaboration for Innovation:

International innovative ideas and solutions can be exchanged through global collaborative efforts. In the construction industry, international cooperation promotes a common commitment to tackling global environmental concerns.

6. Educational Collaboration:

Collaboration between academic institutions and industrial players fosters lifelong learning and the growth of people with expertise in environmentally responsive construction techniques.

To summarise, pursuing environmental factors in construction requires teamwork and technological innovation. Collaboration ensures that sustainable technologies are successfully implemented and that industry information is shared, while innovation drives the development and implementation of these technologies. When combined, these factors provide the construction sector a chance to reduce its environmental impact and help build a more sustainable built environment.

4.10.2 The Essence of Community Engagement and Social Responsibility in Environmental Considerations in Construction

Environmental considerations in construction include social responsibility and community engagement as essential components. These factors extend beyond how construction projects directly affect the environment. It also discusses the relationships with the communities where construction activities occur and the wider societal ramifications. The following are the fundamentals of social responsibility and community involvement in the context of environmental concerns in construction:

4.10.3 Community Engagement

1. Communication and Transparency:

Transparent communication with locals, companies, and other stakeholders is a key component of community participation. Construction trust and fostering a healthy connection with the indigenous is achieved through keeping the community informed about construction plans, potential impacts, and mitigating methods.

2. Inclusive Decision-Making:

Involving the community in decision-making procedures guarantees that their viewpoints, needs, and wants and concerns are given due consideration. Inclusive decision-making will result in more environmentally friendly project designs and construction techniques that uphold local values.

3. Public Consultations:

Setting up public consultations enables community members to provide direct feedback. It gives locals a forum to voice their worries, make suggestions, and take an active role in determining how the development project will affect the environment and society.

4. Cultural Sensitivity:

It is crucial to acknowledge and honour the community's cultural elements. Thus, ensuring that construction operations are carried out to respect and preserve cultural assets requires appropriate involvement with the local cultures and customs.

5. Education and Awareness Programmes:

To involve the community, educational activities educating locals about the environmental implications of the construction project must be implemented. This raises community members' understanding of sustainable methods and promotes sustainable behaviour.

6. Addressing Local Needs:

It is vital to comprehend and attend to the particular requirements of the nearby neighbourhood. This entails considering elements including the project's potential socioeconomic benefits, infrastructure upgrades, and accessibility to amenities.

4.10.4 Social Responsibility

1. Local Employment and Training Opportunities:

In the construction sector, social responsibility includes giving locals where projects are located job and training opportunities. In addition to having a positive economic impact on the neighbourhood, these kinds of events also help people feel proud and invested in the project.

2. Health and Safety Measures:

The projects should provide measure to improve the health and safety of the community.

3. Community Investment and Philanthropy:

Construction projects should engage and invest in philanthropy and community by supporting neighbourhood infrastructure, educational institutions, or other community development projects. This indicates a dedication to the community's overall well-being.

4. Environmental Justice:

Dealing with problem of environmental justice is part of social responsibility. Construction projects should aim for an equal distribution of environmental benefits and responsibilities, without unduly affecting vulnerable or marginalised groups in the community.

5. Post-Construction Legacy:

Social responsibility is not limited to the construction stage. The project's long-term legacy should be considered to make sure that it will continue to improve the community when it is finished in terms of the economy, society, and environment.

4.11 Regulatory Compliance and Green Construction Certifications in Environmental Considerations in Construction

A key element of environmental considerations in construction is regulatory compliance. It offers a structure for guaranteeing that construction projects follow the accepted environmental norms and encourage sustainable practices. Compliance with regulations entails abiding by rules, regulations, practices, laws, and policies established by governmental authorities, professional bodies, and regulatory agencies to safeguard the general public and organisations. In the context of environmental considerations in construction, these are the essential components of regulatory compliance:

4.11.1 Regulatory Compliance

1. Legal Adherence:

Regulatory compliance involves respecting local, national, and international environmental protection laws and regulations. This includes getting the required licences, adhering to pollution regulations, and following construction activity norms.

2. Environmental Impact Assessments:

Performing Environmental Impact Assessments is a vital component of adhering to regulations. It entails evaluating a construction project's possible environmental effects early in the project's conceptualization phase. The assessment's goal is to locate, lessen, and control the effects to comply with legal and regulatory requirements.

3. Water and Air Quality Standards:

Adherence to water and air quality regulations is crucial is sustainable water management in the construction sector. This entails taking steps to manage stormwater runoff, avoid pollution, and regulate emissions to make sure that construction operations do not worsen the state of natural resources.

4. Habitat Protection and Biodiversity:

Ecological and biodiversity protection strategies are part of regulatory compliance. Regulations protecting natural habitats, endangered species, and ecosystems must be followed during construction projects to ensure that project causes the least amount of disturbance possible to the ecosystem.

5. Waste Management Regulations:

To minimise the environmental impact of material disposal, construction sites must adhere to laws for the disposal, recycling, and reduction of construction waste.

6. Noise and Vibration Controls:

To reduce disturbances to the neighbourhood, regulations about vibration and noise levels must be followed. Construction operations must be kept within allowable bounds by putting controls and mitigating measures in place.

7. Erosion and Sediment Control:

To avoid soil erosion and safeguard water bodies, regulatory compliance entails implementing erosion and sediment management methods. These actions are intended to preserve the quality of the soil and avoid sedimentation of surrounding water sources.

4.11.2 Sustainable Certifications

In the construction sector, certification for sustainable construction is a pivotal example approach toward meeting the sustainability target. The significance of the accreditation for sustainable construction is as follows:

1. Commitment to Sustainability:

A dedication to sustainability is demonstrated by sustainable construction certifications. Construction projects that aim for certification are committed to minimising their overall environmental impact, cutting resource usage, and using environmentally acceptable methods.

2. Energy Efficiency and Renewable Energy:

In certification programmes for sustainable construction, energy efficiency and the use of renewable energy sources are prioritised. This entails utilising energy-efficient equipment, producing renewable energy on-site, and putting an emphasis on lowering total energy usage.

3. Material Selection and Life Cycle Assessment:

Sustainable certifications place a strong emphasis on choosing materials based on sound decision and taking life cycle evaluations into account. This entails selecting materials that have less of an adverse effect on the environment and being aware of the entire life cycle of construction materials.

4. Water Conservation and Efficiency:

Water conservation and efficiency initiatives are encouraged by sustainable certifications. Water-efficient fixtures, rainwater collection, and water-efficient landscape designs are a few examples of this.

5. Indoor Environmental Quality:

Certifications place a high priority on the quality of the indoor environment, addressing elements like lighting, thermal comfort, and air quality. The goal of construction projects is to give users of the projects comfortable and healthful indoor spaces.

6. Innovation and Regional Priority Credits:

Innovation and regional goals are frequently credited towards sustainable construction certifications. This incentivizes construction projects to surpass conventional specifications and take context-specific environmental factors into account.

7. Market Leadership and Recognition:

Obtaining accreditation for green constructions proves market leadership and honours ecologically conscious construction techniques. It is in line with the real estate market's growing desire for environmentally friendly and sustainable constructions.

While green construction certifications go a step further by recognising and supporting proactive, sustainable practices, regulatory compliance makes sure that construction projects adhere to the law's criteria for environmental protection. When combined, these components support the overarching objective of encouraging ecologically conscious construction practices in the built environment.

4.12 Construction Waste

Construction waste encompasses the waste materials and debris produced during the building, renovation, or demolition of infrastructure projects such as roads, bridges, buildings, and other structures. This debris can consist of many different elements, including bricks, concrete, wood, metal, polymers, glass, and insulation. Hazardous elements including asbestos, lead, and chemicals that need to be handled and disposed of carefully can also be found in construction waste. Construction waste disposal may contribute to landfills and other waste management issues, as well as having a substantial negative influence on the environment. Improper disposal of waste can lead to environmental pollution and health hazards, which is why proper waste management and disposal practices are essential. This can include reducing waste through recycling and reuse, properly disposing of hazardous waste, and using environmentally friendly disposal methods. To reduce the impact of construction waste on the environment, it is crucial to manage and dispose of it appropriately. Globally, construction waste is a significant environmental and safety risk. Based on data, the yearly contribution of construction waste is expected to reach 100 billion metric tonnes. Materials left over after excavation, building, refurbishing, maintenance, repair, and destruction are referred to as construction waste. Over 90% of waste coming from construction is inert. Inert materials are substances that do not undergo significant physical, chemical, or biological changes over time. In the context of construction and waste management, inert materials are those that: Do not decompose Do not react chemically Do not contribute to environmental pollution (e.g., leaching harmful substances) are typically non-hazardous. They end up in open-air waste disposal sites. Brick, rubble, soil, concrete, and debris are examples of public fill that can be used for site formation and land reclamation.

4.13 Importance of Construction Waste Management

1. Environmental Protection
   - It helps minimise landfill.
   - Reduces Landfill Use: Proper waste management minimises the volume of construction and demolition waste sent to landfills, preserving natural land and reducing soil and groundwater pollution.
   - Decreases Carbon Footprint: Recycling and reusing materials reduce the need for raw material extraction and processing, which are energy-intensive and contribute to greenhouse gas emissions.
   - Conserves Resources: Reclaiming materials like concrete, wood, metals, and glass helps conserve finite natural resources.
2. Economic Benefits
   - Cost Savings: Efficient material use and recycling reduce the cost of purchasing new materials and minimise disposal fees.
   - Revenue Generation: Recyclable materials such as metals and wood can be sold, turning waste into a source of income.
   - Compliance and Avoidance of Fines: Helps construction firms comply with regulations and avoid penalties related to improper waste disposal.
3. Enhances Project Efficiency
   - Site Cleanliness and Safety: Organised waste disposal reduces hazards on-site, lowering the risk of accidents and improving productivity.
   - Improved Material Management: Encourages careful planning and procurement, which can improve inventory control and reduce waste.

4. Social and Regulatory Impact

   - Corporate Social Responsibility (CSR): Demonstrates environmental responsibility, which improves company image and supports community trust.
   - Supports Circular Economy: Encourages a shift from a take-make-dispose model to one that values reuse and regeneration.

4.14 Major Causes of Construction Waste

The major causes of construction waste can be broadly categorised into design, procurement, materials handling, on-site operations, and human factors.

1. Design-Related Causes

   - Inefficient design
   - Design changes during construction (variation orders, client-initiated changes)
   - Incomplete or unclear drawings/specifications leading to misinterpretation
   - Overdesign or redundant structural elements
   - Lack of standardisation in materials and components
   - Poor coordination between disciplines (e.g., architect vs. M&E services)

2. Procurement and Supply Chain Issues

   - Over-ordering of materials due to bulk purchase incentives or estimation errors
   - Under-ordering that results in urgent orders and increased waste
   - Inappropriate materials procurement (wrong size, quantity, or type)
   - Damage during transportation or delivery
   - Ineffective supplier communication and lack of just-in-time delivery strategies

3. Material Handling and Storage

   - Improper storage conditions (e.g., exposure to rain, sun, pests)
   - Loss due to theft or vandalism
   - Poor logistics planning for on-site storage and movement
   - Multiple handling and double handling increasing the chance of damage
   - Unlabeled materials causing misplacement or misuse

4. On-Site Construction Activities

   - Incorrect construction methods or sequencing
   - Poor workmanship or lack of skilled labor
   - Rework due to errors or inspection failures

- Excessive cutting or trimming of materials without reuse plans
- Use of inappropriate tools and machinery that cause damage
- Trial-and-error construction due to a lack of planning

5. Human and Management Factors

- Lack of worker training in waste minimisation practices
- Low level of supervision and monitoring
- Negligence, carelessness, or poor site culture
- Poor communication between stakeholders
- Lack of incentive programs for waste reduction
- Resistance to innovation in construction techniques or materials

6. External and Environmental Factors

- Weather conditions causing damage or delays (leading to rework)
- Unforeseen ground conditions requiring demolition or redesign
- Legal and regulatory changes during the project
- Limited space on site, leading to inefficient material handling

7. Planning and Scheduling Issues

- Inaccurate demand forecasting
- Last-minute changes in project schedules
- Poor coordination among trades and subcontractors
- Lack of a construction waste management plan

Construction waste often results from a lack of planning, coordination, and quality control, from the design stage all the way to on-site execution. Addressing these causes requires a holistic approach that includes training, better project management, and sustainable design and procurement practices.

4.15 Waste to Value in the Construction Sector

There are various ways in which construction projects could implement the waste-to-value principle. Some of these ways include:

- Design for deconstruction should be the target. This designing structures and infrastructure to simplify their disassembly and material recovery at the end of their useful lives. Instead of having items dumped in landfills, designs for deconstruction allow for their recovery, reuse, or recycling.

- Material reuse: Reclaimed materials from the construction or other projects should be used in construction, maintenance, refurbishment and renovation of another to lower the need for new materials. This can involve working with steel, bricks, or repurposed wood.
- Material recycling can help cut down on the requirement for virgin resources by reusing materials like steel, concrete, and asphalt in new construction projects.
- Waste-to-energy: Construction waste can be utilised to create biomass or wood waste.
- Prefabrication: By constructing components off-site under controlled conditions, prefabricated construction technologies can reduce the amount of waste that will be generated on-site. Prefabrication also allows for more accurate and efficient resource use.
- Concepts of the circular economy: By maximising the lifespan of products and materials, planning for waste, and encouraging resource efficiency, the concepts of a circular economy can be applied to construction projects.

There are numerous ways to cut waste and improve resource efficiency in construction projects by implementing the waste-to-value principle. Construction projects may become more sustainable and support a circular economy by considering the lifecycle of materials and products and figuring out how to reuse and recycle them.

4.16 Waste Reduction and Recycling in Environmental Considerations in Construction

Two essential elements of environmental considerations in construction are waste minimization and recycling. By encouraging material reuse and minimising waste generation, these approaches seek to lessen the negative environmental effects of construction projects. Some of the fundamentals of recycling and waste reduction in the context of environmental concerns in construction are listed below:

Waste Reduction:

1. Minimization at the Source:

Reducing waste at its source is crucial to waste reduction. To minimise waste during the construction process, construction projects should strive to maximise material consumption, minimise overordering, and use efficient construction techniques.

2. Efficient Material Management:

Cost of materials take more than 60% of construction costs of most projects. Therefore, careful planning of the acquisition, storage, and use of materials is necessary for effective cost management. Simplifying the material management procedures is crucial to guaranteeing that resources are used efficiently, minimising surplus, and lowering the possibility of material waste.

3. Prefabrication and Modular Construction:

By reducing on-site construction activities, prefabrication and modular construction approaches help reduce waste. The key is to build construction components off-site, which minimises construction waste and lessens the requirement for customisation on-site.

4. Lean Construction Practices:

Lean construction methods place a strong emphasis on productivity, efficiency, effective and waste minimization. The main purpose of applying the technique is to get rid of tasks that do not improve the construction process to create a more efficient and environmentally ethics workflow.

5. Design for Deconstruction:

A construction's materials can be readily disassembled and used at the end of its life if it is designed with deconstruction features. To minimise the environmental impact of destruction, it is crucial to plan for future disassembly and to promote a circular economy.

6. Recycling:

Reusing materials from dismantled or demolished constructions is known as recycling in the construction sector. The main goals are to increase the lifespan of materials, decrease the need for new resources, and lessen the negative effects that material extraction and production have on the environment.

7. On-Site Material Sorting:

Construction waste is sorted into recyclable and non-recyclable categories by means of on-site material sorting. The key is developing specialised technologies that make it easier to recycle materials like wood, concrete, and metal, thereby lowering the quantity of waste dumped in landfills.

8. Recycled Content in New Materials:

Utilising recycled materials-like recycled aggregates in concrete-is crucial to lowering the need for virgin resources and fostering a more sustainable supply chain.

9. Construction Waste Recycling Facilities:

Construction facilities for recycling construction waste are crucial. The intent is to set up the necessary infrastructure to handle and recycle construction waste so that valuable materials may be recovered and the environmental impact is minimised overall.

10. Circular Economy Principles:

Adopting the concepts of the circular economy entails creating, utilising, and disposing of materials in a closed loop. The main goal is to establish a regenerative system that continuously reuses materials, thereby decreasing the demand for the extraction of new resources and cutting down on waste production.

11. Recycling Construction Demolition Waste:

It is essential to recycle debris from construction and demolition, including wood, asphalt, and concrete. The goal is to keep these items out of landfills, which will increase resource efficiency and lessen the environmental impact of disposing of waste.

To summarise, reducing waste and recycling during construction are crucial methods for reducing the environmental effects of construction operations. To build a more sustainable and environmentally conscious construction sector, the key is to embrace a circular economy perspective, design for reuse and recycling, and employ efficient construction processes.

To surmised, the chapter finished with an explanation of Environmental sustainability-enabled construction management practices.

Environmental sustainability-enabled construction management integrates principles of sustainability into various aspects of construction project management to minimize the environmental impact of construction activities and promote long-term environmental stewardship. By integrating environmental sustainability principles into construction management practices, construction managers can mitigate environmental impacts, conserve natural resources, and create healthier, more resilient built environments for current and future generations. The mechanism for the Environmental sustainability-enabled construction management is:

1. Sustainable Design Integration:

Environmental sustainability starts with the design phase. Construction managers work closely with architects, quantity surveyors, engineers, and other stakeholders to incorporate sustainable design principles into the project. This could entail choosing sustainable materials with low embodied carbon, designing for water efficiency and stormwater management, defining energy-efficient building systems, and optimising building orientation for natural daylighting and passive heating or cooling.

2. Lifecycle Assessment and Environmental Impact Analysis:

Construction managers conduct lifecycle assessments and environmental impact analyses to evaluate the environmental footprint of construction projects. This involves assessing the environmental impacts of materials, construction processes, energy consumption, and waste generation throughout the entire lifecycle of the building. By quantifying and analyzing these impacts, construction managers can identify opportunities to minimize environmental harm and enhance sustainability performance.

3. Green Building Certification and Compliance:

Construction managers ensure that construction projects comply with green building standards and certifications. They coordinate with project teams to meet the requirements for sustainable site development, energy efficiency, indoor environmental quality, materials selection, and water conservation set forth by these certification programs.

4. Material Selection and Waste Reduction:

Construction managers prioritize the selection of sustainable building materials with low environmental impact and high recyclability. They work with suppliers and contractors to source locally produced materials, recycled content materials, and renewable materials whenever possible. Additionally, construction managers implement waste reduction and recycling programs to minimize construction waste and divert materials from landfills.

5. Energy Efficiency and Renewable Energy Integration:

Construction managers oversee the implementation of energy-efficient building systems and renewable energy technologies to reduce energy consumption and carbon emissions. This may include specifying high-efficiency HVAC systems, lighting fixtures, and building envelope components, as well as integrating solar panels, wind turbines, or geothermal systems to generate renewable energy onsite.

6. Water Conservation and Management:

Construction managers implement water conservation measures and sustainable water management practices to minimize water consumption and protect water resources. This may involve installing water-efficient fixtures and appliances, implementing rainwater harvesting systems, and designing green infrastructure for stormwater management and water reuse.

7. Construction Waste Management and Recycling:

Construction managers develop waste management plans and recycling programs to minimize construction waste and maximize material reuse and recycling. They coordinate with contractors and waste management companies to segregate construction waste, recycle materials such as concrete, wood, and metal, and dispose of hazardous materials in an environmentally responsible manner.

8. Environmental Monitoring and Compliance:

Construction managers monitor construction activities to ensure compliance with environmental regulations and permit requirements. They conduct environmental inspections, soil erosion control measures, and habitat protection measures to minimize environmental impacts during construction. Additionally, construction managers implement environmental management systems and best practices to promote environmental stewardship and sustainability performance throughout the project lifecycle.

4.17 Summary

This chapter provides a comprehensive exploration of the environmental challenges and opportunities within the construction sector, emphasizing the urgent need for a paradigm shift toward sustainable practices. Given the industry's substantial impact on the environment, it is crucial to rethink traditional approaches and embrace sustainability at every stage. The chapter covers key aspects such as material selection, life cycle assessments, energy efficiency, waste reduction, and community engagement, all of which are critical in driving the sector's transformative changes. Adherence to green construction rules and certifications is increasingly mandated by law, serving not only as a legal requirement but also as a hallmark of quality. Prioritizing environmental considerations in construction is essential for creating a more sustainable and resilient built environment. This approach addresses the pressing challenges of today while ensuring a positive legacy for future generations. The inherent nature of construction often results in significant negative impacts on biodiversity, leading to the loss and fragmentation of ecosystems. Such activities contribute to the disappearance of natural habitats, impeding the movement and occupation of diverse species across different environments, thereby affecting biodiversity. To mitigate these effects, it is imperative that construction project design includes a comprehensive Environmental Impact Assessment (EIA). The EIA plays a critical role in identifying environmental concerns early in the planning process and determining the extent to which these can be mitigated. By doing so, it ensures that both the public and planning authorities are fully informed before any planning approvals are granted.

Bibliography

1. Allen E, Iano J (2019) Fundamentals of building construction: materials and methods. Wiley
2. Bandler H (1986) Environmental considerations in planning, design and construction of dams. Water Int 11(3):98–106
3. Chan EH, Tse RY (2003) Cultural considerations in international construction contracts. J Constr Eng Manag 129(4):375–381
4. Dietz S, Bowen A, Dixon C, Gradwell P (2016) 'Climate value at risk' of global financial assets. Nat Clim Chang 6(7):676–679
5. Ding GK (2008) Sustainable construction—the role of environmental assessment tools. J Environ Manage 86(3):451–464
6. Glasson J, Therivel R (2013) Introduction to environmental impact assessment. Routledge
7. Goel A, Ganesh LS, Kaur A (2020) Social sustainability considerations in construction project feasibility study: a stakeholder salience perspective. Eng Constr Archit Manag 27(7):1429–1459
8. Hossain MU, Ng ST (2018) Critical consideration of buildings' environmental impact assessment towards adoption of circular economy: an analytical review. J Clean Prod 205:763–780
9. Isaksson A, Linderoth H (2018) Environmental considerations in the Swedish building and construction industry: the role of costs, institutional setting, and information. J Housing Built Environ 33:615–632

10. Killip G (2006) The housing maintenance and sustainability debate, in Proceedings of the Annual Research Conference of The Royal Institution of Chartered Surveyors, London
11. Morel JC, Mesbah A, Oggero M, Walker P (2001) Building houses with local materials: means to drastically reduce the environmental impact of construction. Build Environ 36(10):1119–1126
12. Ohene E, Chan AP, Darko A, Nani G (2023) Navigating toward net zero by 2050: drivers, barriers, and strategies for net zero carbon buildings in an emerging market. Build Environ 242:110472
13. Ortiz O, Castells F, Sonnemann G (2009) Sustainability in the construction industry: a review of recent developments based on LCA. Constr Build Mater 23(1):28–39
14. Papamichael I, Voukkali I, Loizia P, Zorpas AA (2023) Construction and demolition waste framework of circular economy: a mini review. Waste Manage Res 41(12):1728–1740
15. Sterner E (2002) 'Green procurement' of buildings: a study of Swedish clients' considerations. Constr Manag Econ 20(1):21–30
16. Valdes-Vasquez R, Klotz LE (2013) Social sustainability considerations during planning and design: framework of processes for construction projects. J Constr Eng Manag 139(1):80–89
17. Varnäs A, Balfors B, Faith-Ell C (2009) Environmental consideration in procurement of construction contracts: current practice, problems and opportunities in green procurement in the Swedish construction industry. J Clean Prod 17(13):1214–1222
18. El-Adaway I, Abotaleb I, Mohler R, Sheffler J, Fuller H, Orozco C (2018) Contractual considerations for complying with environmental requirements: Lessons learned using a case study approach. In: Construction research congress 2018, pp 235–245
19. Flood C, Gorse C, Scott LM, Smith M (2021) Environmental considerations. In: Building surveyor's pocket book. Routledge, pp 276–309
20. Moshood TD, Rotimi JO, Shahzad W (2024) Enhancing sustainability considerations in construction industry projects. Environ Dev Sustain, 1–27
21. UN (United National Environment Programme Report) (2014) Climate change costs to escalate
22. UNEP-SBCI (United Nations Environment Programme) (2014) Greening the building supply chain. DTI/1753/PA
23. Wong WS, Chan EH (2000) Building Hong Kong: environmental considerations, vol 1. Hong Kong University Press
24. Vachon A (2008) Economic and environmental considerations for construction and demolition (C&D) debris management and policy. University of New Hampshire
25. Yap CK, Leow CS, Ismail AP (2022) Environmental finance under ESG: A literature review and synthesis. Sustain Soc Dev 2(2):2481

Chapter 5
Sustainable Construction Economics

Abstract This chapter explores the economic dimension of sustainable construction, positioning it as a critical pillar alongside environmental and social sustainability. It introduces construction economics as a foundational concept, highlighting its evolution from a cost-control function to a holistic approach addressing the full life-cycle cost of projects. The chapter emphasises the need to assess construction decisions not only by initial capital outlay but also by long-term social and environmental costs and benefits. Through discussions on life cycle costing and cost-benefit analysis, the chapter equips readers with essential tools for evaluating the economic sustainability of construction projects. It also examines financial mechanisms such as tax incentives, grants, low-interest loans, and green certifications that support eco-friendly practices. A key theme is the trade-off between cost efficiency and sustainability, posing questions such as: at what environmental or social cost is a reduction in construction expenditure justified? The chapter ultimately advocates for economically rational and ethically responsible decision-making that aligns with sustainable goals. By integrating client value systems with environmental ethics and social impact, construction economics becomes a powerful tool for achieving resilient, high-quality, and cost-effective built environments. This chapter offers both theoretical insights and practical frameworks to guide construction stakeholders in navigating the complexities of sustainable project delivery.

Keywords Whole life appraisal · Life cycle costing · Feasibility analysis · Client value system

5.1 Chapter Learning Outcomes

This chapter sets out to provide explanations of sustainable construction economics. The integration of the economic dimension into the 'sustainable construction' paradigm signifies the consideration of the costs on the one hand and the emphasis on meeting the environmental and social ethics of the projects on the other hand. The emphasis is placed on achieving the project effectively and efficiently in terms of

A. Olanrewaju, *Sustainable Construction Management: Research and Practice Companion*, Management in the Built Environment,
https://doi.org/10.1007/978-981-96-4910-5_5

cost, social, and environmental ethics. In practical terms, what is the mean construction cost reduction at what environmental equivalent? For example, if a client intends to reduce construction costs, what is the equivalent waste, pollution, or degradation that is accepted? What is the cost of the social value of the projects to the community and users? Costs or benefits need to be allocated to the environmental and social problems associated with the project throughout its whole life span. In this chapter, several techniques applied to the analysis of the cost and benefits of the projects are discussed. After studying this chapter, you would be to:

- Explain construction economics
- Explain construction quality
- Explain the application of life cycle cost analysis for sustainable construction
- Explain the application of cost–benefit analysis in sustainable construction.

5.2 Introduction

Terms including construction economics, building economics, cost engineering, cost management and project economics are used interchangeably in the literature. This book uses construction economics as the umbrella term for these terms. This chapter aims to give a contextual overview of construction economics rather than explain the complex matters in construction economics by simplifying the application of the economics principles to sustainable construction projects. The construction sector is large, complex, and costly and involves a lot of parties, as such that the various aspects of construction economics can be studied separately. It is impossible for one book to cover all the various aspects of construction economics. However, broadly, some books have concentrated on the microeconomics of the construction aspect, while many have also contextualised the construction sector into the macroeconomic realm. Construction economics is difficult to define concisely as it involves various procedures and techniques currently used by the quantity surveyors. Traditionally, construction economics is no more than cost control. The importance of the construction sector as a facilitator or enabler of national developments has given rise to the development of construction economics as a specialism to the extent that it is normally referred to as construction cost consultants, construction cost planners, construction cost engineers, or quantity surveyors more appropriately. In most countries, quantity surveyors perform construction economic duties and responsibilities, but where there are no quantity surveyors, these duties and responsibilities are performed by cost engineers or estimators. Construction economics has many similarities with cost planning, and traditionally, this starts with the cost development bracket for a proposed scheme. Many have studied construction economics from the macro level, focusing on the contributions of the construction sector to gross domestic product and gross national product, as well as its contribution to the employment of labour. Yet, some have considered construction economics as a tool to control the cost of construction projects. This latter consideration receives more attention as it relates to making construction projects more efficient and effective. However, the emphasis has been on the capital costs or the initial costs of the projects. Cost modelling and cost benchmaking are part of the process in construction economics. In this chapter, the intention is to elaborate on the applications of economic principles in the context of the whole-life appraisal of the proposed scheme. More appropriately, this chapter focused on the economic sustainability of construction projects to ensure the long-term viability and resilience of construction projects. Traditionally, clients, the design team, and the construction team emphasized the costs of projects.

Only recently have the operational and use costs of construction taken centre stage in decision-making. Economic decisions on construction projects have a huge corresponding impact on the social and environmental aspects. Traditional construction practice ignores the interconnectivity between the economic value of the projects, the environmental and social dimensions, and the maintenance and operation costs of the projects. The basic questions that are now asked are: environmental sustainability of constructions at what costs? The social sustainability of construction projects at what costs? How do you balance the capital costs of the projects with the maintenance costs? Not only about how much does the project costs, but also what it the cost of the distruptions to environment and societal value system? Sustainable construction involves more expenditure as compared to conventional construction. However, in terms of whole-life costs, sustainable construction is cheap. It also generates less waste or pollution. Sustainable constructions are elegant and attractive. Sustainable construction contributes to the health, comfort, and productivity of its users and plays a crucial role in fostering sustainable, resilient, and vibrant communities. Their positive impacts extend beyond the construction's boundaries, influencing the overall environmental quality and livability of the surrounding area. Therefore, this chapter helps us apply economic concepts and methods to make [rational] decisions on location, design, services, operation, maintenance, methods of procurement, choice of contractors when to construct, and how to construct to lower both capital and operational costs for the clients and revenue and profit for the designers and contractors. The chapter starts with a brief explanation of what construction economics means.

Construction costs are often the main denominator for the success of construction projects. Conventional clients prioritise construction costs over the quality and schedule of the projects. So, it is expected that construction costs will still prioritised over the other dimensions in the sustainability indicators. What has changed lately is that construction clients are now paying more attention to the whole life cost or life cycle cost of the projects. However, this is not exclusive. Cost certainty has remained on the agenda for most clients around the world, irrespective of whether they are private or public. However, geopolitical, and other economic indices are affecting the cost of construction anywhere in the world. Although project costs are measurable, the major barrier to project delivery is the dynamic and unpredictable cost of projects. Clients, developers, contractors, suppliers, and other project stakeholders depend on cost data to make decisions on projects. Construction economics offers stakeholders a more efficient approach to delivering high-quality projects within time and budget and addressing the environmental and social impacts of the projects.

5.3 Definition

It will be appropriate to provide some understanding of what construction economics means. To do this, we will provide insight into economics as a body of knowledge and then tie this to construction. Hence, we pull from and combine two different theoretical perspectives: economics and construction. The literature on construction examines the nature and practice of construction and the various means of delivery, while the economics literature examines the relevancy of the theory to the construction process and practices. Put these two together, the chapter moves further to exploit what frameworks are needed to improve the performance of sustainable construction outputs in terms of meeting the value system of the clients and the profitability of other stakeholders like contractors and consultants. Economics as a body of knowledge is well established. There are many branches of economics, like financial economics, behavioural economics, political economics, and geography economics, just to name a few. However, these and even more are useful for construction economics. Construction, by its nature, has political, social, geographical, financial, and behavioural implications. But the construction projects' nature, clients' size, location, and purposes determine the extent of these implications. The chapter places both the construction economic (sector), professional, and activities into an economic context. Hence, it is pertinent to start with an introduction to the theoretical economic dimensions.

Economics is defined differently, and while some define it broadly, others have taken a narrow path. According to some perspectives, economics is defined as the theories, frameworks, principles, and models that deal with the operations of the market. In this context, it explains creation and distribution and how individuals, businesses, and governments allocate resources. The perspective further stresses that economics concerns how people make decisions about alternative uses to meet their needs and ensure their satisfaction [14]. While the description above encapsulates the context of economics, there are various other definitions, to the extent that today it is agreed that the various definitions need not be a problem due to ontological and epistemological reasons. In other words, the author's worldview about 'reality' or 'truth' guides all the different definitions. Fortunately, or unfortunately, this is not unique to the subject of economics. The fundamental focus of economics concerns scarcity, choice, behaviours, distribution, and coordination [14]. The concerns apply to individual organisations, national governments, and intergovernmental bodies. For instance, parents make choices on suitable schools for their children. Spinster chooses who to marry. Students make a choice on which course to study. Couples make choices on the number of children to have, government make choices on how much to level to tax, on what service to privatize, and what goods and services to subsidize or ban. Businesses choose how much to include as overhead for the bidding /tendering.

Construction contractors choose which contract to tender and, if accepted, whether to go ahead or not. The construction client decides if to prioritise capital costs over the maintenance and operation costs of the buildings. They also decide if to procure conventional construction or sustainable construction. Simply put, we need to evaluate and prioritise different alternatives to arrive at the best decision. Though many think this alternative is in terms of money, it does not have to be. Many decisions we make have no direct link to money. This does not mean that economic decisions do not involve money; what is meant here is that the choices we make could be related to or not related to money. All these choices are necessary because resources are limited. Then, is economics about economizing? That is, avoiding or reducing spending overspending, or, to rephrase, avoiding waste. While being prudent or efficient is quite crucial, economics is not just about being efficient. There is a branch of economics called behavioural economics that explains why people make irrational decisions. While rational decisions in standard economics taught us how to make decisions by preferring the most efficient options, behavioural economics accounts for deliberately biased decisions. On many accounts, we do not choose what we or others believe will economise our investment or efforts. Despite the fact that economics is typically understood to be the study of how humans produce, distribute, and utilise resources, the definition of economics that follows is among the most thorough. It is as follows: Economics is a social science that examines how people, organisations, governments, and countries make decisions in the face of scarcity and the incentives that shape and balance those decisions [2, 13]. From this definition and other modern economics definitions, the main concern of economics is making choices. This is an interesting topic, anyway, especially making choices in an uncertain situation. Everyone makes a choice every time. It is not only humans that make choices; animals also must make choice.

5.4 What Makes Economics a Science?

In most definitions of economics, economics is classified as a science, though some argue it is an art. This has been a long-standing debate, especially in academia. The major yardstick between science and art is the measurement of being able to develop, test, and confirm hypotheses in subjects like physics, chemistry, and biology. In science, experiments are conducted in a controlled environment. In the laboratory, the effects of some variables can be controlled in the experiments. In this case, some definite outcomes can be obtained. However, in economics, the environment cannot be controlled in the same way as in science. This similar to the cases in sustainable construction management, where many of the output's variables are subjective in nature. Hence, the results of science may contain some biases and will always contain some biases. However experimental results in the mainstream of science, like nuclear physics, are not 100% precise. There are a lot of speculations in the formulation of hypotheses in physics, as we have in art or social science. Some do not believe that economics uses positive or quantitative techniques to investigate problems. Many believe it cannot because it involves normative aspects that cannot be objectively measured. However, if any, quantum physics or quantum mechanics made us understand that beyond classical physics, nothing is deterministic. However, economics is not 100% like any of the subjects in mainstream science in many respects. The claim that economics is a science, and an art may, nevertheless, not be that difficult to make. How economic data are analysed, hypotheses are developed, and models are created to demonstrate the scientific side. When applying economic principles to real-world problems, the artistic element of the process frequently involves making subjective decisions, selecting policies, and taking society's values into account. But most people will accept economics as a social science. Social science deals with human behaviours, psychology, anthropology, economics, criminology, and sociology. Unlike the natural sciences of physics and chemistry, which focus on cells, light, chemical structure, and compositions, social science deals with society. Social science examines the features, characteristics, changes, actions, reactions, causes, and consequences of human behaviour and its influence on the environment. It addresses how and why people behave as they do, both as individuals in groups and society.

If one accepts that what constitutes scientific endeavours is the rigorous approach to knowledge towards solving problems, then we have no possibility but to agree that economics is a science but not like natural science. However, since economics cannot use the scientific method as neatly as in physics, economists include assumptions in their models or estimations. The assumption is also in nature science. For instance, no one knows the exact speed of light yet. Even some economists disagreed with making assumptions in their models and called those that did as being unrealistic. In other words, while some economists choose to be positivists, others have remained constructivists or pragmatists. The scientific method requires the isolation of variables and testing to prove causality. In the natural sciences, including chemistry, biology, and physics, in the attempt to conduct investigations on some variables, certain variables are controlled. In such a situation, there is a control variable and an experimental variable. Variables can be specifically allocated to the experimental and control groups without specifying them, and the effect of treatment on the experimental variable can be accurately found, examined, and a conclusion drawn. However, economists cannot correctly isolate many individual variables in a natural setting, so they make some assumptions to create a model with some degree of accuracy. So, the economists creating models are aware of the inherent errors in their models, but efforts are made to reduce them to the extent that they would not have a significant impact on the results. In economics, needs and wants people, family, community, or national desire is scarce, because every time someone must give up something to enjoy what he or she wants or prefers, that thing that they prefer is scarce. Therefore, at the heart of economics is analysing the costs and benefits of alternatives. Even turning a tap on to collect fresh water is scarce, though many may not realise it, but since you must pay bills, the freshwater is considered scarce because you can use the money for the water bills for other purposes. However, while some resources like diamonds and gold are very scarce, at least in terms of money, the air or sun are less scarce comparatively.

The problems that economics aims to solve are applied to individuals, organisations, and national and international governments. For instance, in general, the main fundamental economic problems include insatiable wants, scarcity of resources, choice, trade-offs, and incentives. The basic idea is that individual, group, business, and organisational resources are scarce or limited. For instance, the time available to perform an activity is limited. The construction client's budget is limited. The government has insufficient budget to meet all its promises to the electorate. Individuals only have 24 h in a day. Construction resources are limited. Sustainable materials are very scarce for now. While this is the case, the wants of individuals, businesses, governments, and organisations are many. To interpret, the resources at the disposal of individuals, businesses, and governments are not sufficient to carry out all their needs. If this is then the case, the individual, government, or business will choose by prioritising their needs, thereby using the available resources to achieve the most crucial need. Therefore, economics is concerned with using available resources to meet the most productive requirements or needs.

5.5 What Is Construction Economics?

Construction economics is the branch of economics that studies how individuals, businesses, and governments make choices about acquiring construction projects with scarce resources without sacrificing the criteria within the client value system. For private individuals, this choice might be to buy or build new houses, offices, companies, warehouses, units, fences, space, schools, hospitals, etc. Also, it can be to rent any of these. For corporate organisations or governments, the list may include roads, dams, bridges, and other public facilities. Strictly and in the context of this book, construction economics is the branch of microeconomics that studies the commercial implications of choice on the design, construction, and operation of construction projects.

Construction economics presents a collection of multiple tools and specialisations in finance, engineering, economics, psychology, sociology, and other aspects of other social science disciplines. As will be expected, construction economics principles are useful for new construction, maintenance, refurbishment, retrofitting, alterations, extensions, and repairs for roads, dams, buildings, tunnels, etc. Construction economics involves the application of the principles, techniques, and expertise of economics to the construction sector to increase the efficiency and effectiveness of the construction process and its implementation to meet the requirements of clients and other parties involved in the total supply chain. It is not only about the clients but also about other stakeholders. Construction outputs provide societal values. As a matter of course, any of the parties could use the tools, techniques, and principles of the construction sector to optimise their investments. As explained above, the impact of construction outputs must be examined in terms of how they affect society. The current construction climate and increasing client requirements are intensifying the need for construction economics. Clients now favour systemic analysis of construction projects. Both the construction phase and operation phase of the projects are now examined together for a better-informed decision. The requirements of modern clients are also impacted by sustainability requirements. Clients are now questioning the sustainability profile of their design and construction teams. Many of the clients are concerned about the environmental impact of their construction projects. Many are also conscious of the social implications of their construction projects. At the practical level, many clients are concerned about their future investments in the operation of their built assets.

Construction economics involves extensive areas of research and practice, prominently it comprises methods of economising construction costs, the relationship between space and shape, the relationship between design criteria and client's requirements, the relationship between quality and costs, sustainability requirements and cost, the influence of location and price, the accuracy of cost planning with the final estimate, construction whole cycle, duties and responsibilities in cost modelling, client value system, the relationship between the construction sectors and other sectors, pricing mechanisms, construction organisation and construction team, application of methods of investment appraisal, the relation between value, price and cost, the effect of inflation on the construction cost estimate, property rights, shortage of labour, roles of markets on construction materials and construction projects, property market analysis and taxation. However, this book will not focus on all of these. It would be impossible to address all these in one book. Therefore, this chapter will focus more on the relationship between cost sustainability and environmental and social sustainability. Answers would be provided to the following questions in this chapter and elsewhere in this book.

- What can construction economics help us do?
- It can help understand and evaluate alternative technologies.
- In the selection of alternative designs
- Building models (micro and macro) to predict market penetration of modern designs and products.
- Building a model to assess and evaluate market behaviours
- Enable us to evaluate trade-offs.
- Building an energy cost model
- Determine the social-economic benefits of different proposals.
- Evaluate the safety of alternatives by examining the opportunity cost of preventing hazards.
- Evaluate the opportunity cost of productivity.

The cost of owning and running a construction project is high. It takes a great deal of resources. A lot of money, ranging from a few hundred dollars to many million dollars, would be involved. An enormous amount of time, from one day to many years, may be required, depending on the nature and characteristics of the construction projects. Some projects are abandoned for decades before launching again. Many parties and professionals will be involved. During the design and construction, if not earlier, pertinent decisions will be made. Some questions that required urgent and in-depth answers during the design stage to economise project costs include:

- Is the project feasible and viable?
- Is the information on the cost sufficient and up to date?

- What is the maintenance cost?
- Who are the users?
- Who will finance the project?
- How can the hidden cost be identified?
- What is the recovery period if the project is an investment?
- Is it a sustainable construction or a conventional construction?
- If it is sustainable construction, what is the expected certification level?
- What are the costs associated with sustainable construction?

All these and many more have economic implications for construction projects. As a choice will be made; a trade-off is involved. Construction economics is concerned with all economic aspects of construction projects once the client completes his or her decision to construct. All the decisions from planning, design, production, completion, and operation are considered in making decisions on the proposed projects. The purpose of construction economics is to provide accurate information to the clients so they can make an informed decision as to whether to go ahead with the projects or not, with the clients adequately informed of their decisions. The information provided based on economic analysis would be useful to all members of the design team, construction team, and operation team.

5.6 Sustainable Construction Economics

Ensuring the long-term viability and health of economic systems is the emphasis of economic sustainability, which is an essential component of the overall sustainability requirement. It means managing economic activities to guarantee that current requirements are met without jeopardising the ability of future generations to meet their own needs and desires. The goal of economic sustainability is to build strong, diverse, and prosperous societies that aims to reduce negative effects on the environment and social justice. Economic sustainability is about balancing economic growth with environmental and social considerations, creating a foundation for enduring prosperity and well-being. It requires a shift towards more holistic and inclusive economic practices that benefit both current and future generations. The major components are economic sustainability, financial viability, and balanced consumption and production. Financial viability entails the profitability of an investment or the value of the money invested. For instance, economic sustainability acknowledges the importance of profitability for individuals, companies, and organisations. It emphasises the need for economic activities to generate profits to support ongoing operations, investments, and growth. Economic sustainability encourages resource efficiency to reduce waste and negative environmental effects in balanced production and consumption. Adopting techniques like recycling, material reuse, and production process optimisation are examples of this. Sustainable construction economics is the aspect of economic sustainability that specifically emphasises economic considerations within the context of construction sustainability. It integrates terms that convey the intersection of financial viability and sustainable practices within the sector.

5.7 Construction Cost

Before we examine the concept of construction cost, it is imperative to distinguish between cost and price. There is always a problem in differentiating between price and cost. The owners, buyers, and developers each measure price and cost from an entirely different point of view. According to Ashworth and Perera [2], the distinction between price and cost depends on the stakeholder's concern. In other words, what the supplier considers a cost is what the consumer considers a price. Price includes the cost of production, overhead, charges, and profit. Price is the amount paid in exchange for a final good or service. Hence, it can be surmised that profit is what differentiates a contractor's cost from the client's price. For instance, the developer's cost is the contractor's price, and the developer's price would be the homebuyers' cost for residential buildings. It is often cited that the seller's price is the buyer's cost. Ashworth and Perera [2] put it this way: to the contractor, the cost is her expenditure, while the price is her income. To illustrate, the price of a bag of cement is equal to the cost of production, overhead, charge, and profit. The price of a building is equivalent to the cost of materials, labour, overhead, charge, and profit. The price of an item is normally considered, wrongly though, as the "market value' of the item. However, price and market value are not the same. Price is a relative measurement of market value. Hence, it requires the comparison of the market values of two items. The price of a good is decided by demand and supply forces. In the context of housing, the income of households is a dominant factor in the determination of the housing price. Like any other good or service, a house has a price tag attached to it. Given that construction costs on their own cannot be optimised, the optimisation mechanisms will focus on social and environmental sustainability. By doing so, construction costs will serve as the denominator of the basis for comparison.

5.8 Design Management in Construction

Functionality, construction quality, and construction effect are the components that make up design quality. Design quality can be specifically determined for every facility in the majority of project types. However, a few universal factors-effective spatial utilisation, favourable environmental effects, safe and secure facilities, energy-efficient structures, and accessibility-apply to the majority of projects. A better design strikes a balance between needs, community ideals, aesthetic standards, and technical limitations. Occupiers of high-quality buildings are guaranteed to be very productive. Value for money is delivered by superior design in terms of both overall cost and cost-in-use. In addition to providing outstanding performance, well-designed constructions would sell more quickly and for a greater price than others. All project users are impacted by decisions made at the strategic stage of the project regarding size, form, materials, layout, functionalities, and aesthetics. The requirements that the design team employs as a guide to complete the design are known as the design criteria. Design limitations, buyer requirements, construction codes, and standards all contribute to the creation of design criteria for residential building projects. Developers' earnings and profits are impacted by the decisions made on the design criteria. Finalising the design criteria prior to the building phase is crucial as they are strategic in nature.

5.9 Design Economics

One of the major criteria within the construction clients' value system is the quality of the project. Traditionally, the other two criteria are cost and time. Traditionally, clients that opt for high or premium quality would contend with high cost and delay, or either of the two, depending on the nature of the projects and the client characteristics. But today, this is changing, and not only do clients want all of these; but they want more than these actually. The quality of design is blamed for many of the defects in buildings and infrastructure projects. However, a widespread problem in discussing construction client requirements is the meaning of quality. Because there are numerous persons, the term has multiple meanings. with example, it is frequently mistaken with value. "The totality of features and characteristics of a product or service that bears its ability to satisfy stated or implied needs" is what ISO 8402-1986 defines as quality. It was also described as the level of quality that a good or service possesses. A common definition of quality is "fulfilling the needs of the customer." It is also defined as 'fitness for purpose' (Juran). Achieving quality in projects is, however, a continuous process, as echoed by the total quality management principle. While the delivery quality of the construction projects is the responsibility of all the stakeholders, notably the clients, developers, contractors, specialists, employees, suppliers, governments, policymakers, and many more, depending on the nature of the construction projects. In big construction companies, the quality management functions of a project are dedicated to someone or a department in the company. Specialists in quality management are often referred to as quality managers, quality engineers, quality directors, or assurance managers who have the cognizance, knowledge, tools, and guidance to provide the leading roles. The quality of buildings and infrastructure can be defined in terms of various characteristics, including aesthetics, performance, neighbourhoods, image, and durability.

Who defines the quality of a construction? This question could be easily answered with a statement that says the customer is always right. In other words, the client defined the quality standards for their construction projects. However, granted that the quality of a project depends on the requirements of the client or users, this

Table 5.1 Dimensions of quality in a product

Dimension	Definition
Performance	This is the basic operating characteristic- how well does a building provide shelter?
Features	This is the additional operating characteristic
Reliability	How often does the building require maintenance
Conformance	The extent to which building design meets specifications
Durability	The building's whole life cycle
Serviceability	The extent to is maintained easily
Aesthetics	Appearances and pleasant
Safety	Assurance/confirmation that operators and occupier will not be injured or died

Adapted from Russel and Taylor III [15]

is only achievable if the client's requirements are integrated with the production process. If the production process [the activities of the construction companies] is not considered during the product design and specification, it would be difficult or impossible to deliver the product or services at a competitive price. Hence, the projects would be too expensive to construct and operate. The customer will perceive it as low quality if the cost of acquisition is prohibitive. A product or service will not have an acceptable value if it cannot be provided at a competitive price; this occurs when the price exceeds what consumers are prepared to pay for the specified quality dimensions [15]. Table 5.1 contains the dimensions, or what the client seeks for in projects to assess the level of excellence. The buyer will assess the product's quality by comparing these dimensions to the price.

The dimensions of quality for a product are not the same for services. Services (i.e., maintenance) quality is more related to time and the interaction between providers, clients, and users.

5.10 Dimensions of Quality: Services

Table 5.2 shows that the dimensions of quality for products and service providers are different. Time and user-provider contact are more crucial factors in determining the quality of a service. This is closely related to maintenance services in the construction industry. The client will assess the product's quality by comparing these dimensions to its price.

Table 5.2 Dimensions of quality in services delivery

Dimension	Explanation
Timeliness	How long must the customer to for the maintenance organisation to respond to the complaint, and what is the duration of the maintenance to complete
Completeness	Is the maintenance organisation complete the repair work with the time limit?
Courtesy	How are the customers treated by the maintenance organization/employers?
Consistency	Is the same standard of maintenance provided to all users/building occupiers
Accessibility	How easy to make maintenance complaints
Convenience	How convenient to make a complaint
Accuracy	Is the repair correctly done
Responsiveness	How well does the maintenance organisation respond or react to unusual complaint or request

Adapted from Russell and Taylor III [15]

5.11 Economic Techniques for Sustainable Construction Project Appraisal

5.11.1 Life Cycle Cost Analysis (LCCA)

LCCA is a thorough assessment technique called life used in many industries, including construction, to decide a project's overall cost during its full life Ashworth and Perera [2]. It makes it possible to compare the alternatives economically, considering both the costs of the present and the future. This will enable the final assessment to decide the amount of added capital investment that is necessary now to realise future benefits over the project's full lifespan. LCCA analyses a project's costs over its whole life cycle to decide its sustainability and economic viability. It offers a comprehensive perspective that extends beyond the initial costs of development, assisting stakeholders in making choices that maximise both immediate and long-term financial results. Costs related to planning, designing, building, maintaining, operating, using, disposing of, or decommissioning are considered in the LCCA. The main goal of performing an LCCA is to aid decision-makers in making well-informed decisions by considering both the upfront construction costs and ongoing project expenses. Clients must take the construction projects' operating costs into account. Buildings typically have a useful and economic life of more than 60 years. Nonetheless, the functional and economic lives of construction work are still dynamic. For the building project to fulfil its intended duties, as outlined by the value systems of the clients, maintenance is needed. The materials used to make products and construction components have a limited lifespan.

By offering a thorough and methodical strategy to assess the effects of building projects across their whole life cycle, LCCA plays a crucial role in sustainable construction. To help decision-makers select economically viable, ecologically, and socially conscious solutions that support the long-term viability of construction projects, LCCA for sustainable construction incorporates economic, environmental, and social concerns. There are four main ways that LCCA is applied in the context of sustainable construction:

Cost Assessment

LCCA extends beyond typical cost assessments, which concentrate mostly on construction costs at the outset, in the context of sustainable construction. It considers every stage of the life cycle, including the phases of construction, use, upkeep, and end-of-life. The assessment aids in finding solutions that optimise life cycle costs while minimising initial building expenses. Investing in long-lasting materials, energy-efficient technologies, and other sustainable practices that result in long-term cost savings is the ultimate purpose of the assessment. Over their life cycle, sustainable construction methods that improve energy efficiency can result in lower energy usage and operating costs. Selecting long-lasting and sustainable materials might help the project require less upkeep and operation over time. LCCA evaluates how material choices affect ongoing maintenance requirements and costs for the duration of construction projects.

Environmental and Social Considerations

The integration of sustainability measures into construction projects is facilitated by LCCA for sustainable construction. This entails evaluating the social and environmental impacts of various project alternatives. It encourages a comprehensive method of decision-making that is in line with more general sustainability goals. Decision-makers can strike a balance between social and environmental objectives and economic considerations by using the LCCA. In sustainable construction, LCCA aids in the identification of methods to reduce the negative impacts of construction on the environment and social requirements. The resilience and flexibility of structures to shifting environmental conditions are considered by LCCA. Investing in durable materials and designs can save life cycle costs and promote long-term sustainability. Social cost factors are also considered, such as how decisions will affect nearby communities, the generation of jobs, and the general well-being of society. The integration of sustainability measures into construction projects is facilitated by LCCA for sustainable construction.

Regulatory Compliance

Sustainable projects are less likely to face fines for non-compliance when LCCA makes sure they adhere to certification requirements and regulatory standards. This is in line with sustainable standards and improves the project's overall financial results. The different rules and regulations that apply to construction projects are considered by LCCA. This covers energy efficiency requirements, environmental laws, construction rules, and other pertinent legislation. The LCCA can help ensure that the project follows legal and policy requirements by considering these rules during the strategic stage of building projects. For stakeholders' benefit, LCCA also evaluates the possible expenses of adhering to regulatory requirements. Ensuring compliance with environmental and energy efficiency laws using specified materials, technology, or design characteristics is a crucial aspect of sustainable construction. In many instances, there may not be enough information available to decide the prices, effectiveness, and effects of the new specific materials and products. LCCA aids in the quantification of related expenses throughout the project's life cycle. By assigning funds to compliance-related tasks, LCCA facilitates budget planning. By doing this, it is ensured that the money is accessible to implement sustainable practices that satisfy legal requirements. It also aids in avoiding financial difficulties brought on by compliance issues and budget overruns. Throughout the project's life cycle, continuous compliance monitoring and reporting would be made easier with the use of LCCA. This involves monitoring how well sustainable features, goods, and methods are doing.

Problem Management

Potential hazards are found by LCCA, which supports an open approach to risk management. By considering these problems, decision-makers may create plans to lessen adverse effects and ensure the project is flexible enough to handle unforeseen difficulties. In LCCA, every stage of the life cycle-design, construction, operation, maintenance, and disposal-is thoroughly evaluated in terms of costs. Regulation violations may result in fines from the law, delays in projects, and extra expenses. Problem mitigation techniques can be incorporated into the decision-making process thanks to LCCA. This thorough analysis aids in finding any hazards associated with costs at every stage of the project. The project's financial features, such as the project's initial construction costs, ongoing operating costs, and maintenance costs, are considered by LCCA. Financial problem like budget overruns, funding shortages, or unforeseen expenses that may occur during the life cycle are identified with the use of LCCA by analysing these costs. The LCCA's consideration of regulatory compliance aids in detecting potential problems related to non-compliance. The identification of hazards linked to rigid designs or technologies facilitates the creation of resilient approaches capable of accommodating unanticipated modifications. It is possible to find environmental hazards, such as modifications to laws governing emissions or the disposal of waste. The LCCA techniques involve scenario analysis, in which several future scenarios are considered to evaluate their impact on project costs. This helps in deciding the hazards connected to hazy future circumstances and creating plans to deal with them. Finding various kinds of construction project problems for stainable construction can be facilitated by the LCCA. This kind of knowledge is necessary to adapt construction techniques to evolving environmental restrictions. The information is crucial for modifying construction techniques to comply with evolving environmental laws.

Stakeholder Communication and Transparency

The LCCA encourages transparency by giving interested parties thorough information about the financial, environmental, and social ramifications of building decisions. The effective use of sustainable practices can be facilitated by knowledgeable stakeholders. Throughout the life cycle, LCCA also considers how construction materials, energy use, and waste management affect the environment. To give stakeholders a thorough understanding of the ecological effects of the project, the analysis may consider factors like carbon emissions, resource depletion, and other environmental indicators. With the aid of LCCA, stakeholders may evaluate the life cycle costs of various materials and understand the effects they have on the environment. During the material selection process, this transparency eases informed decision-making. Stakeholders will be able to learn more about how energy-efficient designs and technology lower construction projects' lifetime operating and maintenance costs. This will encourage decision-making about energy to be transparent. The social ramifications of various project options, including their effects on nearby communities, job prospects, and general societal well-being, can be communicated to stakeholders. Information about the project's strategies for reducing risks associated with societal issues, environmental uncertainties, and regulatory changes can be shared with stakeholders. The LCCA offers a transparent view of how sustainable practices used during construction contribute to the project's long-term success and resilience because it encourages stakeholders to adopt a holistic viewpoint by considering the project's complete life cycle.

5.11.2 Cost–Benefit Analysis (CBA)

CBA assesses economic viability of a project by weighing the costs and advantages of a project or decision. Based on the analysis, the advantages of sustainable construction can be classified as either social or environmental. Making educated and defensible decisions is made possible by this approach, which aids in determining whether the advantages of a suggested project or course of action exceed the disadvantages. The cost of sustainability is not capital costs, but rather while-life costs. An outline of the essential elements of a cost–benefit analysis of projects, or even services, is provided below:

(1) Identification of Costs and Benefits

Identification of costs: Every expense related to the project should be noted by the design and construction teams. These would comprise the project's initial construction costs as well as ongoing operating and maintenance costs and any other unforeseen expenses related to its use over time. Since many of the materials and products for a sustainable project are unknown, every piece of information regarding the projects' estimated costs needs to be carefully examined considering future technological advancements.

(2) Identification of Benefits

Identification of benefits: The design and construction teams must ascertain the advantages or satisfactory results that emerge from the construction projects. The social value of the projects as well as their social influence on the construction team should be given special consideration for a sustainable construction. The advantages may include increased revenue creation, cost savings, reduction in transportation costs, increased efficiency, or improved community and site worker quality of life. They may also be direct or indirect, tangible, or intangible.

(3) Monetary Valuation

In contrast, the analysts (such as the sustainable construction team) should assigned monetary figures on the benefits of the projects that have been identified. This is to translate all benefits into a single currency for meaningful comparison so that analysis can be done consistently.

(4) Time Frame

The design team should decide on the duration of the analysis while taking the short- and long-term effects of the costs and benefits into account. The period over which expenses and benefits are anticipated to occur must be taken into consideration. This framework would be determined by the types of clients, the goals of the projects, and any applicable standards, norms, and regulations.

(5) Discounting

Future expenses and earnings are expressed into present equivalents or values using discounting procedures. Comparing expenses and advantages that arise at various times is beneficial. To account for the time value of money, the teams should discount future costs and benefits. To do this, future cash flows' value must be adjusted to reflect their present value.

(6) Calculation of Net Present Value

The initial investment less the net revenue or income inflows that the project generates is the project's net present value, or NPV. Salvation value, if any, is added. The team should be able to determine the Net Present Value (NPV) by deducting the entire discounted cost from the total discounted benefit. If the net present value (NPV), which shows that the benefits outweigh the costs, is positive, the project is considered economically viable.

(7) Benefit–Cost Ratio

The benefit–cost ratio (BCR) (e.g., project or services) is one profitability statistic that may be used in cost–benefit analysis to evaluate the sustainability of cash flows from an investment. It compares the present value of all benefits received from a project or asset against the current value of all costs invested. It can be computed by dividing the total discounted benefits by the total discounted costs. If the BCR is more than 1, an investment in the project might be profitable.

(8) Sensitivity Analysis

The project team should perform sensitivity analysis to evaluate the effects of changes in principal factors on the outcomes. This will assist in determining the degree of ambiguity that may be involved in the analysis. The type of construction project and the sustainability objectives will determine the need for sensitivity analysis in sustainable construction. The project stakeholders will be better able to comprehend the resilience of sustainable construction initiatives and make well-informed decisions by conducting sensitivity analyses on variables and taking probable variations and uncertainties into account. The sensitivity analysis for environmentally compliant project take into accounts:

Construction Costs

Material Costs: And price fluctuations for environmentally ethic materials.

Labour Costs: Variations in labour costs, particularly if specialised knowledge is needed to use sustainable construction techniques.

Energy Efficiency Measures

Energy Performance: Evaluating how sensitive the project is to modifications in energy-saving technologies and architectural elements.

Integrating Renewable Energy: Analysing the effects of various scenarios pertaining to the utilisation of renewable energy sources.

Life Cycle Assessment Factors

Embodied Carbon: Examining how sensitively the project is affected by modifications in the embodied carbon of the materials utilised.

Operational Carbon: Assessing changes in the carbon footprint that occur during an operational period.

Regulatory Compliance and Incentives

Code Compliance: Determining how sensitive a construction code or regulation is to modifications.

Incentives: analysing the effects of differences in the incentives that are offered for sustainable construction.

Project Timeline:

Construction Duration: Examining how sensitive the project is to adjustments to the construction schedule.

Operational Phase Duration: Analysing Differences in Operational Phase Length.

Waste Management

Evaluating the sensitivity to modifications in recycling and waste reduction techniques.

Evaluation of the Effects of Waste Disposal Cost Variations.

Water Conservation Measures

Water-Efficient Systems: Assessing how sensitive water-efficient technology and systems are to changes.

Assessing variances in the sources and availability of water.

Biodiversity and Green Spaces

Green Infrastructure: Evaluating how differences in biodiversity metrics and green infrastructure affect each other.

Analysing the impact of modifications to land use planning to protect biodiversity.

Renewable Materials

Application of Sustainable Materials: Examining Differences in Renewable and Sustainable Building Material Use.

Supply Chain Resilience: Analysing how susceptible the supply chain for sustainable resources is to interruptions.

Resilience Measures

Climate Resilience: Evaluating how sensitive one is to scenarios of climate change and how successful resilience strategies are.

Examining the effects of differences in adaption techniques for environmentally friendly building.

(9) Risk and Uncertainty

Primarily, it is necessary to understand and record the distinctions between risk and uncertainty, especially about sustainable construction. Many of the materials, components, and design ideas have not been used or tasted. They have only been employed in conjecture and suggestions in design, not truly utilising the items all that well yet. In actuality, the items for a few significant projects have not yet been produced. Stakeholders should assess and measure the risks and uncertainties related to project expenses and associated benefits. The majority of which need to be recorded during the analysis phase. It may be possible to account for changes in project outcomes by using machine learning models. There are numerous risks and uncertainties in sustainable construction that the stakeholders should be aware of. However, attention should be given more to the negative risks as compared to the positive risks. Negative risks are the problems, where problem management strategies are required. The positive risks are benefits, and a proper benefit management approach should be deployed to leverage the benefit to the stakeholders. Many impacts on decisions on sustainable construction are indeterminate and unknown.

5.12 Application of Cost–Benefit Analysis in Sustainable Construction

When it comes to sustainable construction, where decision-makers strive to strike a balance between social, environmental, and economic factors, cost–benefit analysis is extremely pertinent. Through a methodical assessment of the advantages and disadvantages of sustainable construction methods, CBA empowers decision-makers to make well-informed decisions that are in line with social, environmental, and economic goals. Highlighting the long-term benefits and financial sustainability of implementing ecologically friendly methods in the construction sector, aids in setting sustainability as a top priority. Table 5.3 lists the uses of cost–benefit analysis in sustainable construction.

Table 5.3 Applications of cost–benefit analysis in sustainable construction

Application	Costs	Benefits
Energy-efficient building design	Add the extra up-front expenses for adding energy-efficient features, like smart building technologies, energy-efficient HVAC systems, and high-performance insulation	Calculate the long-term financial savings from lower energy use. Think about the benefits to the environment from fewer greenhouse gas emissions
Renewable energy integration	Determine how much it will cost to install renewable energy technologies, including solar or wind power	Determine the financial benefits of producing energy and lowering your reliance on traditional energy sources. Think about the advantages to the environment and government incentives
Green building certification	Examine the costs of getting certifications for green buildings, such as certification fees and extra costs for design and construction	Think about the higher market value, increased appeal to tenants, and reduction in operating expenses that come with certified sustainable buildings
Sustainable materials and construction practices	Examine the expenses associated with implementing sustainable construction materials and techniques, such as recycled materials, low-impact building techniques, and waste minimization strategies	Think about the long-term advantages, such as lowered environmental effects, enhanced indoor air quality, and possible financial savings via recycling and reusing materials

(continued)

Table 5.3 (continued)

Application	Costs	Benefits
Water conservation measures	Determine how much it will cost to install water-saving technology including effective irrigation, rainwater harvesting systems, and low-flow fixtures	Calculate the possible cost savings and the reduction in water consumption. Think about how beneficial it is for the ecosystem and the nearby water supplies
Natural habitat preservation and biodiversity	Analyse the costs of maintaining natural habitats and implementing designs that promote biodiversity	Think about the ecological advantages, such as increased biodiversity, better ecosystem services, and increases in property values
Life cycle assessment	Add the expenses for carrying out a life cycle assessment, which examines the construction project's environmental impact throughout its whole existence	Think about the knowledge obtained from the LCA, which enables well-informed choices to reduce environmental effects and maximise resource utilisation
Social equity and well-being	Evaluate the extra expenses associated with social equality initiatives, like accessible features, community involvement initiatives, and fair employment standards	Think about the societal advantages, such as better worker well-being, heightened community ties, and improvements to the project's reputation
Resilience and Climate Adaptation	Examine the expenses related to putting catastrophe resilience and climate adaptation strategies into practice, such as robust building plans and infrastructure	Think about the long-term advantages, such as enhanced project durability, cost savings in disaster recovery, and decreased susceptibility to climate-related hazards
Smart technologies and building automation	Analyse the expenses associated with putting in place smart technology and building automation systems for resource efficiency, occupancy detection, and energy control	Consider the increases in operational efficiency, energy savings, and occupant comfort that result in long-term financial and environmental advantages

5.12.1 Return on Investment

Return on investment is the assessment of an investment's profitability concerning its cost is done using a financial phrase. It provides information about the financial performance of an investment by comparing the gains or losses to the initial investment amount. It is expressed as a percentage. Most of the return on investment is comprised of investment costs and net gain or loss. To calculate the net gain or loss, the initial investment cost is subtracted from the total returns or benefits that the investment generated. A gain is indicated by a good outcome, whereas a loss is indicated by a negative outcome. All costs related to obtaining, putting into practice, and sustaining the investment are included in the cost of the investment. This can include the purchase price at first, ongoing running expenditures, and any other costs spent throughout the investment.

5.13 Application of the Return-on-Investment Technique in Sustainable Construction

Stakeholders can choose projects that correspond with sustainability goals, justify investments in sustainable construction and make financially informed decisions by using return-on-investment analysis to sustainable construction activities. To evaluate the financial sustainability and appeal of investing in green techniques and green building elements, return on investment in sustainable construction is a crucial tool. Quantifying the financial gains from adopting sustainability practices with the related expenses is necessary to assess the return on investment in sustainable construction. Practically speaking, it will be used to evaluate the financial results of investments made in sustainable building, as shown in Table 5.4.

Table 5.4 Application of return on investment in sustainable construction

Features	Application	Justification
Energy efficiency upgrades	Determine the energy-efficient upgrades' return on investment before making any purchases of HVAC systems, high-efficiency lighting, and insulation	Illustrates the financial advantages of lower energy use, lower utility costs, and possible operating cost reductions throughout the course of the investment
Renewable energy installations	Analyse the return on investment of installing solar or wind power systems, among other renewable energy technologies	Calculates the payback duration for the initial expenditure, the possible revenue from selling extra energy, and the financial returns from on-site energy generation
Green building certification	Analyse the financial benefit of earning certifications for green buildings	Evaluates the financial effects of energy savings, rising property values, and the marketability of certified buildings
Sustainable materials and practices	Determine the return on investment of using environmentally friendly building materials and techniques, such as waste minimization, low-impact construction techniques, and recycled content	Demonstrates cost savings, the advantages of waste reduction, and the enhancement of brand reputation and competitiveness in the market

(continued)

Table 5.4 (continued)

Features	Application	Justification
Water conservation measures	Evaluate the return on investment of putting rainwater collection, effective irrigation, and water-efficient technology into practice	Calculates the financial gains from lower water use, cost savings, and favourable environmental effects
Smart technologies and building automation	Analyse the return on investment of installing building automation systems and smart technology	Calculates the financial benefits of energy savings, enhanced occupant comfort, and operational efficiency
Resilient design and climate adaptation	Determine how much it will cost to implement resilient design features to mitigate climate-related hazards	calculates cost savings from a decreased susceptibility to climate effects and the long-term resilience of the built environment

5.14 Essential of Return on Investment in Sustainable Construction

Applying ROI to sustainable construction facilitates the alignment of financial concerns with social and environmental objectives. It facilitates the development of a sustainable building business case. Table 5.5 summarises the key concepts of using return on investment in sustainable construction:

Table 5.5 Essences of applying return on investment on sustainable construction

Essence	Justification
Financial justification	Investments in sustainable building can be financially justified by examine the return on investment. It assists interested parties in appreciating the financial advantages and determining if the expenses of adopting sustainable practices are justified by the profits
Decision-making support	When weighing the pros and disadvantages of different sustainable construction options, decision-makers find that return on investment is a crucial tool. Comparing the financial attractiveness of various projects or initiatives allows for the making of well-informed decisions
Resource optimization	It facilitates the sustainable construction industry's resource optimisation. It directs the distribution of funds to projects with the highest potential for both environmental and economic benefit
Risk mitigation	Give all parties involved the opportunity to evaluate and reduce any hazards related to sustainable building. Making decisions that balance risks and benefits requires an understanding of prospective returns

(continued)

Table 5.5 (continued)

Essence	Justification
Performance monitoring	It is a tool for keeping track of how well sustainable activities are working. It offers insights for ongoing improvement and assists in determining if the anticipated financial gains are realised
Market competitiveness	To prove that sustainable construction projects are economically competitive, return on investment is essential. It demonstrates how social and environmental compliance strategies may support both long-term financial performance and competitiveness in the market
Communication and Stakeholder Engagement	It acts as a communication tool to assist the public, investors, and stakeholders in understanding the financial justification for sustainable building practices. It promotes openness and confidence in environmentally friendly projects

5.15 Financing Sustainable Projects

Financing sustainable projects is securing the funding required to start, carry out, and continue initiatives that adhere to environmental, social, and governance (ESG) standards. These initiatives aim to solve problems related to sustainability, have a beneficial social impact, and encourage ethical business conduct. The primary components of funding initiatives that are sustainable are:

1. Identification of Sustainable Projects

The first step in the process is to identify projects that are distinctly focused on sustainability. Initiatives about renewable energy, energy efficiency, water conservation, social impact, climate resilience, and other socially and environmentally responsible projects may be among them.

2. Project Planning and Design

Upon identification of a sustainable project, careful planning, and design ought to be carried out. This entails establishing the project's aims, objectives, schedule, and necessary resources. Extensive project blueprints facilitate precise financial requirement estimation.

3. Financial Needs Assessment

Establish the project's budget, considering the original capital outlay, operating costs, and any maintenance or monitoring fees that may be required later. Both the direct project expenses and related indirect costs are considered in this evaluation.

4. Sustainable Financing Options

Examine the range of sustainable financing solutions that are offered by the market. Development finance institutions, green bonds, social impact bonds, and sustainable loans are typical choices. Every choice has distinct qualities of its own and fits with various kinds of sustainable initiatives.

5. Green and Sustainable Funds

Consider utilising green and sustainable funds, which are financial instruments created specially to assist socially and ecologically conscious initiatives. Investors in these funds are often driven by the desire to have a beneficial influence in addition to financial returns.

6. Partnerships and Collaborations

To assist sustainable projects, establish alliances and partnerships with development organisations, impact investors, financial institutions, and other interested parties. Collaboration can bring together a variety of resources and experience.

7. Government and Regulatory Support

Examine the grants, subsidies, and incentives provided by the government to sustainably run programmes. Certain governments provide monetary or legal incentives to initiatives that support national sustainability objectives.

8. Corporate Social Responsibility (CSR) Initiatives

Analyse the government's grants, subsidies, and incentives for initiatives that are managed sustainably. Initiatives that fulfil national sustainability objectives may receive financial or legal backing from certain governments.

9. Risk Management and Mitigation

Evaluate and control the project's risks related to sustainability. Projects with strong risk management techniques may attract the interest of investors and financial institutions more than others, especially when it comes to the effects on the environment and society.

10. Impact Measurement and Reporting

Provide a reliable system for calculating and summarising the sustainable project's impact. Providing a clear record of the project's beneficial effects on the environment and society will help attract investors.

11. Innovation and Technology

To improve the effectiveness of financing for sustainable projects, embrace innovation and technology. For example, digital platforms, blockchain, and fintech technologies can boost transparency, lower transaction costs, and streamline fundraising. The steps in financing sustainable projects are contained in Table 5.6.

Table 5.6 Step in financing sustainable projects

Step	Essence
Project identification	Find initiatives with definite sustainability objectives and beneficial effects on the environment or society
Project planning	Create a thorough project plan that outlines the goals, schedule, and resources needed
Financial needs assessment	Calculate the total amount of money needed for the project, considering both one-time and continuing expenses
Sustainable financing options	Establish the project's total budget, taking into consideration both one-time and continuing expenses
Partnerships and collaborations	Determine the overall project budget, accounting for both one-time and ongoing costs
Government and regulatory support	Examine government grants and regulatory assistance that complement the goals of the sustainable project
CSR initiatives	Analyse funding and regulatory support from the government that aligns with the objectives of the sustainable project
Risk management	Examine government financing and regulations that support the goals of the sustainable project
Impact measurement	Provide a mechanism for tracking and presenting the project's positive effects on the environment and society
Innovation and technology	Make use of innovative technologies and solutions to expedite the funding process and improve transparency

5.16 Challenges and Considerations

Table 5.7 contains some of the dominant difficulties in sustainable finance especially for sustainable construction projects.

Collaborating with financial institutions, investors, project developers, and other stakeholders is necessary in the dynamic and ever-evolving field of financing sustainable initiatives. Sustainable project finance is essential for tackling global issues and advancing resilient and responsible development because it synchronises financial goals with favourable social and environmental consequences.

Table 5.7 Difficulties of sustainable finance

Challenges	Essence
Complexity of sustainable finance	Navigating the many possibilities in sustainable finance can be challenging and requires specific knowledge and experience
Balancing financial and impact goals	A crucial factor to consider is finding a balance between positive influence and financial returns. Even though they may take longer to pay for themselves, some sustainable projects have a big positive social impact
Access to capital	Certain sustainable initiatives may have trouble obtaining funding, particularly those that are in developing nations. This problem can be resolved with the help of development finance organisations and initiatives to increase financial inclusion
Regulatory environment	The regulatory landscape may have an impact on sustainable funding solutions. Project success depends on comprehending and navigating regulatory restrictions
Continuous monitoring and reporting	Sustainable financial options may be impacted by the regulatory environment. Regulatory constraints must be understood and navigated for the project to succeed
Integration of ESG criteria	Securing sustainable funding requires increased integration of Environmental, Social, and Governance factors into project planning and reporting

5.17 Financing Sustainable Construction Projects

The process of funding sustainable construction projects includes obtaining the funding required to organise, create, and carry out construction projects that prioritise sustainability. These initiatives seek to develop infrastructure and structures that minimise adverse effects on the environment, advance energy efficiency, and enhance social cohesion. This is a detailed breakdown of the crucial factors and procedures related to funding environmentally friendly building projects:

The process of financing sustainable construction projects is complex and includes several steps, such as strategic planning, working with a range of stakeholders, and making a commitment to social and environmental responsibility. By aligning financial objectives with sustainable development goals, these projects contribute to the creation of a constructed environment that is more resilient, eco-friendly, and socially inclusive.

5.18 Economic Incentives and Policies of Sustainable Construction

Policies and incentives based on the economy are essential for encouraging and supporting sustainable construction techniques. A variety of initiatives are frequently put in place by governments, regulatory agencies, and industry participants to promote the use of sustainable materials, energy-efficient construction techniques, and environmentally compliance construction practices. Listed below are the main financial incentives and regulations related to sustainable construction:

1. Tax Credits and Deductions

Governments may provide tax credits or deductions as an incentive to people or companies who work in the sustainable construction industry. This can include credits for installing renewable energy sources, using sustainable materials, or installing energy-efficient building systems. In principle, these form the highlight of The COP28 UN Climate Change Conference in Dubai 13 December 2023.

2. Grants and Subsidies

Governments should offer tax breaks or credits to individuals or businesses engaged in sustainable building as a means of encouraging their efforts. Credits for utilising sustainable materials, constructing energy-efficient construction systems, or installing renewable energy sources might all fall under this category.

3. Green Building Certification Incentives

Until such a time, the sustainable construction will be the normality, governments should provide incentives for obtaining green construction certifications. These incentives may take the form of expedited permission procedures, waived fees, or other benefits related to regulations. Professional bodies in the built environment should develop, harmonise and promote certifications. Most of the certifications are the moment are quire stringent, which not necessarily be crucial. The may be included as part of universities and polytechnic curricular.

4. Energy Performance Standards

Governments have the authority to impose energy performance standards on buildings, mandating that they fulfil certain energy efficiency requirements. Financial rewards might arise from reaching or surpassing standards, whilst non-compliance may result in penalties.\

5. Low-Interest Loans for Sustainable Construction

Developers and clients should be encouraged to invest in energy-efficient technologies, renewable energy systems, and other environmentally friendly features by offering low-interest financing for sustainable construction projects.

6. Energy Efficiency Rebates

Developers and clients that use energy-efficient technology should be encourage by been eligible for refunds from governments or utility companies. Instances of this include financial incentives for the installation of insulation, HVAC systems, and energy-efficient appliances.

7. Mandatory Green Building Requirements

Government should enforce laws that require specific green building methods guarantees a minimum standard of sustainability. This can involve specifications for sustainable materials, water-saving fixtures, and energy-efficient lighting.

8. Renewable Energy Feed-In Tariffs

Renewable energy put back into the grid is guaranteed a set payment rate under feed-in tariffs. This promotes the use of renewable energy technologies, such solar or wind turbines, in building projects.

9. Public Procurement Policies

Governments can implement sustainable supplying and contracting selection processes by implementing sustainable procurement policies. This generates demand in the market for environmentally friendly building supplies and services.

5.19 The Pros and Downsides of Financial Incentives and Sustainable Building Regulations

1. Public Awareness and Education

The effective implementation of incentives and legislation requires educating the public about the advantages of sustainable construction.

2. Consistency in Regulations

Consistency in regulations and policies across municipal, state, national, and international domains encourages the widespread implementation of sustainable practices and contributes to levelling the playing field.

3. Monitoring and Enforcement

Sustaining the integrity of incentive programmes and guaranteeing adherence to sustainability rules require efficient monitoring and enforcement systems.

4. Adaptability to Local Contexts

Given the distinct difficulties and opportunities that each location presents, policies and incentives should be flexible enough to fit local circumstances.

5. Balancing Economic and Environmental Objectives

Finding a balance between the protection of the environment and economic expansion is essential. Legislation should support sustainable building practices without overburdening the construction sector.

6. Long-Term Planning and Vision

A long-term strategy for infrastructure and urban development should include sustainable construction rules and incentives to promote resilience and flexibility to difficulties in the future.

7. Inclusive and Equitable Implementation

It is imperative to guarantee that laws and incentives are equitable and inclusive to prevent the escalation of social inequality and to facilitate community access to sustainable construction methods.

Policies and financial incentives are crucial in deciding how sustainable construction develops. Governments and stakeholders can match financial rewards with ecologically friendly policies.

5.20 Economic Sustainability Enabled Construction Management Practices

To surmise, the chapter finished with an explanation of economic sustainability enabled construction management practices. Economic sustainability-enabled construction management practices focus on optimising the long-term viability aspects of construction projects to maintain long-term sustainability, profitability, and financial resilience while minimising negative effects on stakeholders and the whole economy, it focuses. By integrating economic sustainability principles into construction management practices, construction managers can enhance project profitability, minimize financial problems, and create value for stakeholders while fostering long-term economic resilience and prosperity of the projects on all the stakeholder. The procedure for the incorporation of economic sustainability into construction management work as follows:

1. Cost Management

Construction managers prioritize effective cost management throughout the project lifecycle. They develop complete and inclusive project budgets, monitor expenses, and implement cost-saving measures to optimize resource allocations, minimize problem-risk, minimize waste, increase profit margins, and maximize profits for all the stakeholders in the projects. By effectively managing costs, construction managers help ensure that projects are completed within budget constraints and deliver a positive return on investment.

2. Value Engineering

Construction managers employ value engineering techniques to optimize project costs while maintaining quality and performance standards of the construction projects. They analyze project designs, materials, and construction methods to identify opportunities for cost reduction without sacrificing functionality of the projects. Value engineering can help streamline construction processes, improve efficiency, and enhance project value by eliminating unnecessary expenses and optimizing resource utilization.

3. Life Cycle Cost Analysis

Construction managers conduct life cycle cost analyses to evaluate the total cost of ownership of construction projects over their entire lifecycle, including capital costs, operation and maintenance expenses, and end-of-life disposal costs. By considering the long-term costs and benefits, the construction managers can make informed decisions about design, materials, components and systems selection to minimize life cycle costs and maximize economic sustainability.

4. Financial Risk Management

Construction managers identify and evaluate the financial problems risks that are associated with construction projects with of aim of protecting the stakeholders and ensure project viability. Some of the likely problem risks include cost overruns, schedule delays, market fluctuations, and regulatory changes, and develop risk mitigation strategies to minimize their impact on project outcomes. Financial risk management practices help the project team and stakeholders to safeguard project investments, maintain financial stability, and enhance overall project resilience.

5. Contractual and Procurement Strategies

Construction managers employ strategic contractual and procurement strategies to optimize project delivery and minimize financial problem-risks. They select appropriate contract and procurement types, negotiate favourable terms and conditions, and establish clear performance criteria to that align with the project objectives with stakeholder expectations. Because it fosters transparency, accountability, and collaboration among project participants, it thus allows the construction managers to promote efficiency, fairness, and trust in project delivery processes.

6. Supply Chain Management

Construction managers implement effective supply chain management practices to safeguard the timely delivery of materials, components, systems equipment, and services while minimizing costs and disruptions. They establish relationships with dependable suppliers, manage inventory levels, and coordinate logistics to optimize supply chain efficiency and resilience. The proactive assessment of the challenges and uncertainties in the supply chain, enables construction managers to mitigate problem-risks and maintain project continuity.

7. Financial Incentives and Financing Mechanisms

Construction managers identify available incentives such as tax credits, grants, subsidies, and discounts for sustainable construction initiatives, energy efficiency improvements, and renewable energy installations. Also, the construction managers explore alternative financing options such as public–private partnerships, green bonds, and performance-based contracts to fund sustainable infrastructure projects and promote economic sustainability.

8. Market Analysis and Strategic Planning

Construction managers conduct market analysis and strategic planning to identify emerging materials, components, problems, trends, opportunities, and challenges in the construction sector. The construction managers also assess market demand, competition, and regulatory requirements to advise project decisions and investment strategies that that would affect the projects. Construction managers by staying knowledgeable and adaptive to market dynamics, can anticipate changes, capitalize on opportunities, and maintain a competitive edge in the construction marketplace.

5.21 Summary

This chapter examined the intricate economic dimensions of sustainable building, highlighting the critical financial factors that influence the adoption of sustainable construction methods. It examines a range of financial incentives and regulations employed by governments, regulators, and sector stakeholders to promote sustainable building practices. Key components such as tax credits, grants, and subsidies designed to offset the costs of sustainable projects are thoroughly explored. Additionally, the chapter investigates the role of low-interest loans, mandated procedures, and green building certifications in advancing environmentally responsible construction. The importance of financial mechanisms, such as feed-in tariffs for renewable energy and reduced insurance premiums, is also underscored, particularly in encouraging the integration of renewable energy systems and risk mitigation strategies. Throughout the chapter, the emphasis is on the need for a delicate balance between environmental stewardship and economic development. It argues for policies that not only foster long-term resilience in the construction sector but also promote inclusivity and innovation, ensuring that the sector can thrive sustainably in the face of evolving environmental challenges.

Bibliography

1. Allen E, Iano J (2019) Fundamentals of building construction: materials and methods. Wiley
2. Ashworth A, Perera S (2015) Cost studies of buildings. Routledge
3. Bon R, Hutchinson K (2000) Sustainable construction: some economic challenges. Build Res Inf 28(5–6):310–314
4. Dobson DW, Sourani A, Sertyesilisik B, Tunstall A (2013) Sustainable construction: analysis of its costs and benefits. Am J Civil Eng Archit 1(2):32–38
5. Du Plessis C (2007) A strategic framework for sustainable construction in developing countries. Constr Manag Econ 25(1):67–76
6. Gruneberg S (1997) Construction economics: an introduction. Bloomsbury Publishing
7. Harris F, McCaffer R, Baldwin A, Edum-Fotwe F (2021) Modern construction management. Wiley
8. Hillebrandt PM (2000) Economic theory and the construction industry. Macmillan, Basingstoke
9. Kibert CJ (2016) Sustainable construction: green building design and delivery. Wiley
10. Liu ZJ, Pypłacz P, Ermakova M, Konev P (2020) Sustainable construction as a competitive advantage. Sustainability 12(15):5946
11. Ofori G (2022) Introduction to the research companion to construction economics. In: Research companion to construction economics. Edward Elgar Publishing, pp 1–17
12. Ofori G (1998) Sustainable construction: principles and a framework for attainment-comment. Constr Manag Econ 16(2):141–145
13. Parkin M (2016) Microeconomics. Pearson Education Limited. Harlow. 12th Edition
14. Parkin M (2022) Microeconomics
15. Russel RS, Taylor III WB (2006) Operations management: quality and competitiveness in a global environment, 5th edn. Wiley
16. Sears SK, Sears GA, Clough RH (2010) Construction project management: a practical guide to field construction management. Wiley
17. Sullivan William G, Wicks Elin M, Luxhoj James T (2009) Engineering economy, fundamentals of engineering economics, 14th edn. Person Education, New Jersey
18. Winch GM (2009) Managing construction projects. Wiley
19. Zhou L, Lowe DJ (2003) Economic challenges of sustainable construction. In: Proceedings of RICS COBRA foundation construction and building research conference. University of Wolverhampton, Wolverhampton, UK, pp 1–2

Chapter 6
Social Sustainability in Construction

Abstract This chapter discusses the social dimension of sustainable construction, highlighting its importance in creating inclusive, resilient, and socially responsible built environments. Social sustainability involves integrating human well-being, community involvement, equitable labour practices, diversity, and safety into construction practices. Moving beyond environmental and economic considerations, this chapter positions construction as a driver of positive social change. It explores how projects can strengthen social cohesion, promote cultural diversity, and support socioeconomic stability by encouraging participatory planning and community empowerment. Key areas such as trust-building, stakeholder engagement, and the creation of social value are examined. The chapter also outlines tools and techniques for assessing social impacts, alongside construction management practices that embrace ethical and inclusive approaches. Health and safety are emphasised not just as compliance requirements but as central values in fostering a culture of care and responsibility within construction settings. It is argued that collaboration among architects, contractors, policymakers, and local communities is essential for embedding social considerations throughout a project's lifecycle. Ultimately, this chapter asserts that construction projects should be viewed as social constructs that shape lives and societies, and must therefore be planned and executed with an emphasis on enhancing the collective human experience.

Keywords Social sustainability · Community engagement · Stakeholder collaboration · Health and safety · Social value

6.1 Chapter Learning Outcomes

The purpose of this chapter is to explain the social aspects of sustainable construction. The integration of the social dimension into 'the sustainable construction' paradigm signifies the consideration of the social aspect of the project on the one hand and the ecological footprint and economic ethics of the project on the other hand. Social sustainability in construction refers to the integration of social factors and effects into

A. Olanrewaju, *Sustainable Construction Management: Research and Practice Companion*, Management in the Built Environment,
https://doi.org/10.1007/978-981-96-4910-5_6

construction projects and practices. The integration relates to human well-being, community engagement, labour practices, diversity and inclusion, and the overall positive social impact of construction activities. In this chapter, several techniques that are applied to the analysis of the cost and benefits of the projects are discussed. After studying this chapter, you will be:

- Explain trust in sustainable construction
- Understand community and social values
- Explain stakeholder engagement
- Explain health and safety management
- Understand social sustainability-enabled construction management practices.

6.2 Introduction

Within the ever-changing field of modern construction, sustainability has come to mean more than just ecological concerns; it is now a comprehensive paradigm that takes social, economic, and environmental aspects into account. It is now more crucial than ever to create resilient, inclusive, and socially conscious built environments as we find ourselves at the nexus of progress and responsibility. The chapter presents the social aspects of sustainable construction which focused on the complex interactions that exist between construction methods and the welfare of local communities, national economies, and global society. A more complete understanding of the built environment's impact on human civilizations has replaced the restricted focus on ecological footprints in the emergence of sustainable construction. In addition to energy-saving innovations and green construction certifications, sustainable construction's social components highlight a dedication to promoting fair, healthy, and lively communities. The outcomes of construction extend beyond physical structures to create lasting societal value. This value is expressed through fostering a sense of community, enhancing the aesthetic quality of environments, improving accessibility for all individuals, and contributing to economic prosperity, ultimately shaping vibrant, inclusive, and sustainable societies for future generations. This exploration of sustainability's social dimensions challenges helps to define construction's role in forming communities both now and in the future. Fundamentally, sustainable construction acknowledges that constructions are essential parts of society, not just physical constructions made of brick and mortar. The social ramifications of construction techniques are becoming clearer as they continue to reshape both urban and rural environments. Every construction project, from housing complexes to infrastructure projects, can affect social dynamics, economic growth, and the general well-being of society. A transformative narrative is introduced by the emphasis on community empowerment, even though the physical infrastructure is still a fundamental part of sustainable development. How do construction initiatives help to create environments that promote inclusivity, social cohesion, and cultural diversity? The answers can be found in collaborative processes that involve communities in the stages of planning, execution, and decision-making as well as in the design and functionality of construction projects. Socioeconomic resilience is under the scope of sustainable construction in its social context. Additionally, it measures how sustainable construction methods might help reduce social inequalities, and meet the various demands of local communities. The chapter emphasises the crucial importance of stakeholder collaboration, acknowledging the intrinsic interconnectedness of the social components of sustainable construction. The collaboration of architects, builders, legislators, and local communities is vital to effectively include social aspects into the construction projects' overall design, construction, operation and deconstruction.

6.3 Stakeholder Engagement in Sustainable Construction

A thorough grasp of the social impact of sustainable construction is vital, even more so than its structural and environmental aspects. Efficient participation of stakeholders is essential for the systemic integration of built environments and societal well-being. Stakeholder engagement is the involvement of persons, groups, or institutions in project with a personal stake or influence in programme or portfolio. It also involves the dynamic and interactive process that involves in collaboration, communication, and active involvement of the members. Stakeholder engagement, within the framework of sustainable construction, is an inclusive and strategic strategy that aims to incorporate the varied viewpoints, requirements, and technical know-how of several stakeholders during a construction project. Table 6.1 contain the primary elements that are involved in stakeholder engagements in sustainable construction management.

Involving stakeholders in sustainable construction is more than just a formality; it is a shared commitment to social responsibility, cooperation, trust, and openness. The key to successful construction projects is understanding how closely the welfare and participation of all parties involved in the project's outcome are related. This understanding is the foundation of stakeholder engagement. Beyond the conventional boundaries of project management, stakeholder engagement appears as a dynamic process including the inclusive participation of various players, all of whom have a stake in the construction project. To create a mosaic of viewpoints that stand for the larger socio-cultural environment, the engagement process brings together a variety of stakeholders, including local community, government stakeholders, architects, contractors, engineers, construction managers, quantity surveyor, and non-governmental organisations. In a deliberate and inclusive approach, stakeholder engagement in sustainable construction actively involves people, organisations, or entities that have a stake in or influence over a construction project. Throughout the whole lifecycle of the conceptualizations, designing, construction and uses process, this dynamic and continuing discourse aims to combine varied viewpoints, requirements, and skills with the overarching goal of producing built environments that are ecologically sustainable, socially responsible, and profitable.

Table 6.1 Elements of stakeholder engagement

Elements	Explanation
Identification and analysis	Determine and evaluate all pertinent parties that could be impacted by or have an impact on the construction project. Local governments, investors, builders, architects, environmental organisations, and others are included in this
Inclusive decision-making	Acknowledges the importance of involving stakeholders—especially local communities-in the process of determining the objectives, layout, and results of the construction project. By being inclusive, a project's success is ensured from a variety of perspectives
Open communication	To give stakeholders timely and accurate information about the project's goals, status, and possible effects, efficient communication channels are set up. This promotes cooperation and fosters trust
Building relationships	Gaining the confidence of stakeholders, responding to their worries, and aggressively seeking their input. Developing trusting connections guarantees a collaborative and encouraging atmosphere for sustainable construction projects
Addressing diverse needs	Acknowledge that different stakeholders have different requirements, interests, and worries; the engagement process identify, find and resolve these differences. This could entail modifying construction layouts to accommodate community requirements, guaranteeing social inclusion, and taking cultural factors into account
Empowerment and capacity building	Enables communities and other relevant parties to actively participate in decision-making by giving them the information, tools, and abilities they require. The ability of stakeholders to meaningfully contribute to the project is improved through capacity-building activities
Feedback mechanisms	Ongoing feedback systems are put in place to give stakeholders a chance to voice their issues, suggestions, and opinions. By using an iterative feedback loop, the construction project is guaranteed to adapt to changing conditions and stakeholder expectations

Community as a Core Stakeholder

The understanding that communities are active participants in the construction process rather than passive beneficiaries is at the core of sustainable construction. Stakeholders' perspectives are heard and appreciated throughout the project lifespan when there is effective stakeholder engagement, which makes local communities become essential stakeholders. Inclusive decision-making procedures provide communities with the power to design their living environments, promoting cultural continuity and a sense of ownership. The community is an essential and crucial stakeholder in the field of sustainable construction. The community has a significant role in defining the success of construction projects in addition to being the recipient of developed structures. The community stands for a paradigm change that is transformative in that it is a core stakeholder in sustainable construction. The key is to go beyond traditional stakeholder models and adopt a cooperative, inclusive, and community-focused strategy that not only addresses the immediate needs of locals but also helps to build long-lasting, sustainable communities for future generations.

Transparency and Trust Building

The development of trust and openness is essential for effective stakeholder involvement. Encouraging stakeholders to be aware of project aims, possible impact, and opportunities for active involvement requires open communication, which is essential to sustainable construction initiatives. Building cooperative relationships that go beyond the construction stage is made possible by the development of trust via honest communication. To effectively engage stakeholders in sustainable construction, transparency and the establishment of trust are essential. These ideas become vital instruments in the complex stand of creating sustainable built environments, encouraging candid dialogue, understanding, and cooperative decision-making. In sustainable construction, collaborative collaborations are sustained by dynamic forces such as transparency and trust-building, rather than static rules. The key is to acknowledge these guidelines as continual procedures that are essential to the achievement of sustainable construction.

Balancing Economic and Social Interests

Engaging stakeholders in sustainable construction needs striking a careful balance between social and economic demands. Aligning financial aims with social responsibility is a necessary part of interacting with corporate stakeholders, like investors and developers. The complex relationship that exists between societal well-being and financial viability needs a balanced strategy that gives priority to both without sacrificing the integrity of either. Sustainable construction practices acknowledge that social cohesion and economic prosperity should not be sacrificed for one another. To create a connection that is mutually reinforcing, integration entails matching company aims with more general society goals. Acknowledging social capital as an crucial economic resource is necessary to strike a balance between economic and social goals. The key is realising the roles that beneficial social effects, community ties, and trust have in long-term economic viability. The strategies for balancing economic and social interests are contained in Table 6.2. In sustainable construction, understanding the mutually beneficial link between these factors is crucial to strike a balance between economic and social goals. It entails recognising their interdependence rather than seeing them as conflicting aims. Sustainable construction paves the way for a harmonious future where social well-being and economic success mutually reinforce each other and have a lasting positive impact on present and future generations. It does this by embracing ethical considerations, fostering community-centric development, and adopting inclusive practices.

Empowering Local Economies and Workforces

One of the most significant players in sustainable construction programmes is the local labour force and companies. Sustainable procurement policies that give preference to local vendors and contractors are the cornerstone of successful engagement programmes, as they support local economic growth and development. Concurrently,

Table 6.2 Strategies for balancing economic and social interests

Strategy	Explanation
Inclusive employment practices	Give preference to recruiting locally, encourage inclusiveness and diversity, and fund skill-building initiatives. This tactic guarantees the equitable distribution of economic gains, having a beneficial social impact
Community-based economic development	Entails finding and assisting neighbourhood businesses to generate employment opportunities that complement the community's distinct qualities and goals
Investment in social infrastructure	Think of community spaces, healthcare facilities, and educational institutions as vital infrastructure for economic growth. This all-encompassing strategy acknowledges that a thriving social infrastructure is a prerequisite for economic sustainability
Ethical procurement and supply chain practices	Prioritise ethical suppliers and materials to make sure that financial advantages do not unintentionally contribute to negative social and environmental effects
Shared value creation with stakeholders	Actively collaborate with stakeholders, especially neighbourhood communities, to co-create solutions that meet social needs and provide economic benefit. This cooperative strategy guarantees that the financial gains correspond with the goals of many stakeholders

the labour force appears as a pivotal part, and endeavours to involve them surpass mere job prospects by incorporating skill enhancement, guaranteeing a durable favourable influence on the nearby community. At the core of sustainable construction management is the empowerment of local workforces and economies, which promotes social resilience in addition to economic development. This revolutionary strategy aims to set up a healthy equilibrium that strengthens regional economies and gives workers more authority by acknowledging the connection between social progress and economic growth. To understand the tactics and ideas that support this catalyst for community resilience, this investigation explores the importance of empowering local economies and workforces in sustainable construction. The Strategies for empowering local economies and workforces are contained in Table 6.3.

In sustainable construction, the idea that trained, engaged workforces and economic empowerment are the cornerstones upon which resilient communities are constructed, forms the core of empowering local economies and workforces. Sustainable construction management is a driving force behind long-term community resilience and prosperity by combining economic development with social well-being. This innovative method guarantees that the advantages of construction projects go much beyond the actual structures, having a beneficial and long-lasting effect on the communities they serve through inclusive practices, skill development, and cooperative decision-making.

Table 6.3 Strategies for empowering local economies and workforces

Strategy	Explanation
Skill development initiatives	Developing skills requires measures to empower local workforces. The key is to fund training initiatives that broaden community members' skill sets, matching them to the demands of the construction sector and generating a labour pool of competent workers
Local hiring practices	Provide local hiring procedures with top priority. By actively seeking to engage and employ members of the local community, sustainable construction creates job possibilities that support the region's economic growth
Support for local businesses	Give local hiring practices first precedence. Sustainable construction generates job opportunities that boost the local economy by actively attempting to engage and employ members of the community
Inclusive decision-making	Actively incorporates local opinions into project planning and execution to make sure that the needs and viewpoints of the community are considered when making crucial decisions
Community engagement and education	Engage the community in the construction process by educating them about sustainable techniques and fostering a common awareness of the long-term advantages for the workforce and the economy
Overcoming skills gaps	Partnering with academic institutions, executing apprenticeship schemes, and customising skill enhancement programmes are ways to close skills gaps and match workforce competencies with business requirements
Balancing economic viability with local sourcing	To find flexible ways to keep projects financially practical while giving priority to neighbourhood companies. Investigating innovative sourcing strategies and creative procurement models may be part of this

Cultural Sensitivity and Contextual Relevance

Sustainable construction acknowledges the significance of cultural heritage and the distinct environments in which projects are implemented. Therefore, maintaining local identities and having a keen understanding of cultural sensitivities are essential components of stakeholder participation. Understanding the diversity of communities and the significant influence that cultural settings have on the built environment, cultural sensitivity and contextual relevance are essential components of sustainable construction management. By designing environments that represent the cultural identities of the communities they serve, this method aims to promote social cohesiveness as well as architectural harmony. Table 6.4 lists the primary methods for cultural sensitivity and contextual relevance.

Table 6.4 Strategies for cultural sensitivity and contextual relevance

Strategy	Explanation
Community engagement and consultation	Engage community people in the planning and design process by actively soliciting their opinions and ideas to make sure the built environment reflects their cultural values
Architectural design informed by culture	Transcends conventional designs and builds structures that complement regional materials, spatial preferences, and aesthetics to promote a feeling of place and cultural continuity
Adaptability to local climate and environment	To create constructions that adapt to the weather by using conventional architectural features that work well in certain situations, assuring comfort and sustainability
Use of locally sourced and sustainable materials	Gives priority to locally accessible materials to reduce the environmental effects, boost local businesses, and respect cultural preferences
Preservation of cultural landscapes	Prioritise materials that are easily obtainable locally to lessen the impact on the environment, support regional companies and honour cultural preferences
Navigating diverse cultural perspectives	Embraces flexible strategies, such as constant communication and interaction, to comprehend and consider different cultural points of view during the construction process
Balancing tradition with modernity	Discover flexible solutions that honour traditional customs and welcome modern innovations, constructing areas that connect the past and the present

Measuring and Communicating Social Impact

Measuring and communicating social impact is a crucial part of engaging stakeholders in sustainable construction. Positive social results and the significance of reporting transparency are highlighted while examining effective communication tactics. To promote accountability, openness, and constructive change in communities, measuring and sharing social effects in sustainable construction management is an essential task for the project success. This procedure entails estimating how construction projects affect societal well-being and communicating these consequences to different stakeholders. A commitment to transparency on the impact of construction projects on the social fabric of communities is shown in the measurement and communication of social impact. It promotes responsible decision-making by holding parties accountable. The methods for measuring and communicating social impact are listed in Table 6.5.

Table 6.5 Strategies for measuring and communicating social impact

Strategy	Explanation
Establishing key performance indicators	For social impact, clearly define your KPIs. Metrics on the creation of jobs, community involvement, affordable housing, and accessibility to facilities are a few examples of these indicators. KPIs offer a measurable framework for assessing influence
Community consultation and involvement	Incorporate community members' opinions into impact reports and actively involve them in the evaluation process by asking them for their thoughts on how construction projects have affected their lives
Utilizing social impact assessment tools	Use instruments for assessing social impact. These resources, which could include socioeconomic research, surveys, and interviews, allow for a thorough assessment of the social effects of construction projects and enable impact communication based on data
Periodic impact reporting	To guarantee that stakeholders are aware of the status and results of construction projects concerning their social aims, regular reporting is necessary
Engaging with local stakeholders	Monitoring and communicating social impact requires constant interaction with regional partners. The key is to set up open lines of contact with local government representatives, community members, and other pertinent stakeholders to make sure that their viewpoints are considered in impact assessments

6.4 Health and Safety in Construction

In the construction sector, health and safety management are crucial factors that demonstrate a dedication to protecting employees' health and safety, reducing hazards, and setting up safe working conditions. Project safety is one of the major indicators in evaluation of construction performance. Safety performance, measures the number of incidents, accidents, and adherence to safety protocols. A safe work environment contributes to project success and team morale. Therefore, strong health and safety protocols must be implemented because construction sites are naturally hazardous environments. Encompassing a diverse approach to protect the well-being of the construction workers, users, and the surrounding environment, is the cornerstone upon which the sector's sustainability and success lie. Construction sites require strict attention to health and safety due to their dynamic character, which includes heavy machinery, complex procedures, and a multitude of duties. It is at this crossroads that safe working environments, robust construction, and vibrant communities are built.

Components of Health and Safety in Construction:

i. Risk Assessment and Management

Despite their inherent dangers to the health and safety of workers and all those engaged, construction sites can be vibrant and transformational environments. The construction sector's health and safety protocols include risk (specifically problems) assessment and management as essential elements that serve as preemptive steps to detect, assess, and reduce any risks. The cornerstone for creating a safe construction environment is this all-encompassing strategy, which also protects the fundamentals of well-being.

Numerous activities, such as operating heavy machines, managing materials, working at heights, and being exposed to a variety of environmental conditions, are what define construction sites. Each of these components has possible dangers that could result in mishaps, injuries, or long-term health problems if they are not rigorously evaluated and controlled. One of the first steps in developing a strong framework for health and safety is to understand the dangers. There are many diverse types of hazards associated with construction work, such as exposure to hazardous materials, falls, electrical hazards, machinery-related risks, and environmental elements like noise or extremely hot temperatures. One essential part of risk assessment is finding these hazards. Because every construction site is different, there are hazards and obstacles that unique to each project. The subtleties of the surroundings are considered when planning safety precautions using a site-specific approach to risk assessment. The foundation of health and safety procedures in the construction sector is risk assessment and management. Not only does a methodical and flexible approach recognise possible risks, but it also gives the construction.

ii. Personal Protective Equipment

The suitable personal protective equipment (PPE) is chosen depending on the unique hazards that are peculiar to the construction site. The foundation of construction safety is personal protective equipment (PPE), which acts as a crucial barrier to shield workers from a variety of possible risks. Examples of PPE include steel-toed boots, gloves, hard hats, and safety glasses. Workers receive training on PPE usage and best practices for maintenance. Safety and worker well-being are non-negotiable concerns in the dynamic and always-changing world of construction, where innovation and structures are the primary indicators of development. To protect workers from the many risks that are common in construction sites, Personal Protective Equipment, or PPE, is essential to this effort. PPE is a crucial element in guaranteeing the workforce's resilience and security, from tall scaffolds to large machinery, chemical exposure, and the difficulties of working at heights. PPE acts as all-encompassing armour, consisting of a variety of specialised tools made to shield workers from certain risks they can face while performing construction-related tasks. PPE is a proactive approach that empowers workers, lowers the chance of injuries, and promotes a safety culture on construction sites, in addition to being a legal necessity. PPE changes along with technology, from the classic hard hat to the ultramodern smart helmet, as our awareness of ergonomic requirements grows. Beyond simple compliance, PPE is crucial because it is a dedication to workers' welfare, making sure that each person working on a construction site goes home safely. The PPE environment will change along with construction methods, strengthening its position as a front-line protector of worker safety in the fast-paced construction industry.

iii. Emergency Preparedness

Plans for emergency response are created and distributed by construction sites, detailing protocols for evacuation, provision of first aid, and coordination with emergency services. All workers on the construction site are guaranteed to be familiar with response procedures through regular emergency drills and training exercises. Construction sites are active places with a multitude of activities going on at once, from complex structural work to heavy mechanical operations. Even with careful preparation and safety precautions, unanticipated events that endanger worker safety and project success can occur. In the construction sector, emergency preparedness is a proactive and essential part of safety procedures that guarantees a prompt and well-coordinated reaction in case of unforeseen incidents. In addition to being a collection of protocols, emergency preparedness is a culture that penetrates every side of the construction. The industry's strategy for guaranteeing the security and welfare of those who advance it must change along with it. Construction sites can strengthen themselves against unforeseen events by including strong emergency response plans, frequent training, and technological developments. By fostering a culture of readiness, the construction sector can remain resilient in the face of disasters and protect its most significant resource-the people who carry out its goals.

iv. Health Monitoring

Regular health examinations are carried out to monitor employees' health, particularly those who are exposed to certain workplace dangers. This covers examinations for ailments linked to noise, dust, and other work-related dangers. The contracting companies should make allowance for the payment, and in fact let design team be aware of this during the tendering stage. Because of their dynamic and demanding settings, construction sites need a proactive strategy to protect worker health and well-being. With a variety of instruments and approaches, health monitoring in construction has become an essential part of identifying and mitigating health problem risks. To provide a safe and resilient workplace, it is critical to check the health of construction workers due to the demands of physical labour and exposure to environmental conditions. Integrating health monitoring tools and personalised care programmes will become more common as the construction sector adopts modern technologies. Creating a health-conscious workplace not only protects individual employees but also fosters a resilient and safe culture on construction sites. The construction sector's dedication to health monitoring reflects its understanding that the people who carry out its projects are its most valuable resource.

v. Site Organization and Signage

Clear signage is used on construction sites to show places with problems, emergency exits, and safety measures. The construction site's plan is designed to minimise traffic, promote effective workflows, and lower the possibility of accidents. Sustainable construction management includes all aspects of the construction process, including the layout and communication of construction sites, and goes beyond using environmentally ethic design and materials. In sustainable construction management, site organisation and signage are essential components that promote effectiveness, security, and environmental responsibility. Through the implementation of deliberate tactics in these domains, construction projects can reduce their environmental impact, improve safety measures, and conform to wider sustainability aims. The site organisation in sustainable construction is shown in Table 6.6.

vi. Compliance with Regulations

Construction projects adhere to national and international health and safety laws, guaranteeing that construction methods meet prescribed criteria. To evaluate regulatory compliance and identify opportunities for improvement, routine audits and inspections are carried out. In addition to emphasising environmental and economics ethics, sustainable construction also prioritises protecting the health and safety of its workforce and the community at large. A key part of sustainable construction is adhering to health and safety laws, which create a framework that protects people, fosters wellbeing, and follows moral and legal requirements. This dedication to health and safety is essential to sustainable construction projects' overall success. Sustainable construction projects place a high priority on following federal, state, and local health and safety laws. This involves making sure that construction operations adhere

Table 6.6 Site organization in sustainable construction

Measure	Explanation
Material storage and management	When material storage areas are well-organized, waste is decreased, recycling is encouraged, and items are easily accessible, which minimises the need for wasteful site movement
Waste segregation and recycling stations	Waste segregation stations that are well-maintained promote proper disposal methods. Recycling sites make it easier to separate recyclables, keeping them out of landfills and promoting a circular economy
Energy-efficient construction offices and facilities	To cut down on overall energy use, make use of renewable energy sources, natural light, and energy-saving technologies
Green spaces and biodiversity considerations	Take biodiversity and green space into account. An ecosystem's equilibrium and a healthier atmosphere are promoted by maintaining and developing natural features
Water management systems	Reducing water use and waste can be achieved by using rainwater harvesting, permeable surfaces, and water-efficient landscaping
Modular and prefabricated construction practices	It takes less time to complete construction on-site and requires less extensive on-site organisation when modular and prefabricated construction technologies are used. This leads to a reduced environmental impact and enhanced resource efficiency

to the legal requirements and meet the criteria for occupational health and safety set forth by regulatory agencies. Design is where sustainable construction management stem from. Therefore, by incorporating health and safety regulations into the design phase, risks during construction and operation are reduced and safety considerations are made an integral part of the project's structure. Achieving certifications or adhering to international standards like ISO 45001 are common prerequisites for sustainable construction. These accreditations demonstrate a dedication to the highest standards of occupational health and safety.

vii. Leadership Commitment

Leadership has a clear dedication to health and safety, promoting an atmosphere where people's welfare is given priority. All phases of the construction project incorporate health and safety concerns into their decision-making processes. The foundation of effective and significant sustainable construction projects is the leadership's dedication to this purpose. Leaders commitment are crucial in the field of sustainable construction because they set the tone, decide priorities, and make sure that social, economic, and environmental factors are incorporated at every stage of a project. Leadership's commitment to sustainability goes beyond compliance to include the development of an innovative and responsible culture that goes beyond conventional construction methods. Sustainability is a core value that leaders in sustainable construction express in an unobstructed vision and mission statement. This entails setting clear, quantifiable, and doable sustainability aims that serve as the project's overall direction. By considering the social, environmental, and economic effects, leaders adhere to the triple bottom line. Because ecological, sociological, and economic factors are interconnected, this integration guarantees that decisions are made by sustainability principles. Some techniques to integrate sustainability into organisational culture is shown in Table 6.7.

A construction project's legacy is shaped by the leadership commitment that goes beyond a simple checkbox on a checklist for sustainable construction. Sustainable construction leaders overcome obstacles, stimulate creativity, and cultivate an environment where ethical building practices are the rule rather than the exception. Every stage of a project is affected by their dedication, which has a long-lasting effect on the neighbourhood, the environment, and the construction sector. Buildings are only one aspect of sustainable construction leadership; another is setting up a future in which environmental stewardship, creativity, and accountability form the foundation of the built environment.

viii. Worker Involvement

In sustainable construction management practice, employees take an active role in health and safety programmes, offering input on possible hazards and helping to create safe work procedures. A proactive attitude to health and safety is encouraged when employees are empowered to report dangers without fear of retaliation. To promote a culture of accountability, creativity, and environmental stewardship, construction workers must be involved in sustainable construction practices.

Table 6.7 Embedding sustainability in organizational culture

Technique	Explanation
Training and awareness programs	Through awareness campaigns and training programmes, leadership cultivates an organisational culture that is centred on sustainability. This guarantees that everyone in the construction team—from the top executives to the labourers on the job site—understands the significance of sustainability
Inclusive decision-making processes	Leaders in sustainable construction embrace participatory, multi-stakeholder decision-making processes. This collaborative approach ensures that sustainability issues are integrated into every area of the project and brings a variety of perspectives to the table, encouraging innovation
Recognition and reward systems	Sustainable practices are celebrated and encouraged through the establishment of reward and recognition systems by leaders. This motivates people to actively support the project's sustainability objectives both individually and in groups
Technology adoption for efficiency	Technology is embraced by experts in sustainable construction to maximise project efficiency and minimise environmental effects. Utilising modern techniques that maximise resource utilisation, such as construction management software and BIM, may fall under this category
Materials and design choices	Leaders prioritise designs and materials with smaller environmental footprints after doing extensive research. This entails taking recycled content, life cycle analyses, and the general sustainability profile of building materials into account
Energy-efficient construction processes	Leaders prioritise designs and materials with smaller environmental footprints after doing extensive research. This entails taking recycled content, life cycle analyses, and the general sustainability profile of building materials into account
Waste reduction and circular economy practices	In order to demonstrate their commitment to sustainability, executives must support initiatives to reduce waste and advance the circular economy. This covers methods for repurposing supplies, recycling construction debris, and reducing the environmental impact of the project overall

Although design and leadership choices often influence sustainable construction practices, frontline workers' empowerment and engagement are crucial to the accomplishment of sustainable projects. As crucial stakeholders, construction workers provide special perspectives, know-how, and hands-on experience that can greatly enhance a project's overall sustainability.

Programmes for raising awareness should be used to introduce sustainability concepts to construction personnel. These courses address sustainability's social, economic, and environmental facets, giving students a solid basis for appreciating the significance of responsible construction methods. Employees must get specialised training on environmentally and social ethic methods. This gives employees the knowledge and skills to apply environmental compliant construction techniques, waste management techniques, and energy-efficient building methods on the job site.

A comprehensive approach to construction techniques is reinforced when workers learn to prioritise not only their safety but also the effects their actions have on the environment and society. Tables 6.8, 6.9 and 6.10 explain how employees can participate in decision-making processes.

Beyond simply putting sustainable principles into effect, construction workers are front-line sustainability advocates. For sustainable construction to be successful, their contributions, empowerment, and involvement are essential. Construction projects can use the collective skills and commitment of workers to produce a constructed environment that is socially and environmentally responsible as well as economically practical by implementing strategies such as education, worker participation in decision-making, and celebration of accomplishments. Equipped with knowledge and an intense sense of responsibility, construction professionals create a legacy of responsible building practices and act as agents of good change.

Table 6.8 Worker involvement in decision-making processes

Technique	Explanation
Participation in planning and design discussions	Planning and design discussions receive help from the practical ideas provided by construction workers. Their practical knowledge allows them to recognise possible obstacles and suggest creative solutions that support sustainability aims
Collaborative problem-solving sessions	Including construction workers in group problem-solving exercises helps them feel more like owners. These meetings give staff members a forum to exchange concepts, breakthroughs, and best practices that improve project sustainability
Feedback mechanisms for continuous improvement	Having feedback systems in place guarantees that project managers and construction personnel can communicate freely. Employees are invited to offer insights, worries, and recommendations about sustainability to support ongoing development

Table 6.9 On-site implementation of sustainable practices

Technique	Explanation
Waste reduction and recycling practices	Building labourers is essential to recycling and waste reduction. They take an active part in reducing waste generated on-site, sorting wastes for recycling, and advancing the circular economy
Energy-efficient construction techniques	Employees put the energy-efficient building methods they learnt in training into practice. This could entail making the most use of the equipment, employing energy-saving devices, and following set procedures to reduce the amount of energy used
Responsibility for material use and selection	The choice and application of materials are the responsibility of construction workers. They adhere to regulations for handling environmentally friendly materials, considering elements like durability, recycled content, and total environmental impact

Table 6.10 Safety and well-being considerations

Technique	Explanation
Promotion of worker well-being	The welfare of employees is given priority in sustainable construction. The physical and emotional well-being of all employees is valued by construction workers, who actively participate in improving health and safety on the job site
Adherence to personal protective equipment	Employees have the authority to follow PPE (personal protective equipment) guidelines. By averting mishaps and injuries, this not only guarantees their safety but also supports sustainability goals
Health monitoring programme	Participants in health monitoring programmes include construction workers. Routine examinations and health evaluations are essential to a proactive approach to well-being and the long-term viability of the labour force

ix. Communication and Training

Clear and effective communication ensures that everyone on the construction site is aware of the health and safety procedures. Employees are educated about the most recent safety precautions through ongoing training programmes, which also emphasise the need to keep a safe workplace. Promoting health and safety in the context of sustainable construction requires two essential pillars: effective communication and thorough training. By placing a strong emphasis on these elements, construction projects foster a culture of responsibility that is consistent with sustainable practices while also ensuring the welfare of their staff. Effective communication channels and focused training initiatives enable employees, supervisors, and other interested parties to actively take part in the creation of a safe and sustainable construction environment, see Tables 6.11 and 6.12.

In the construction sector, communication and training are the cornerstones of a sustainable and safety-conscious workforce. Construction projects can set up a work environment where health, safety, and sustainability are integrated into all aspects of the operation by investing in effective communication channels, focused training programmes, and creative learning methodologies. With the right ability and a strong safety culture, workers actively contribute to the success of sustainable construction projects, making sure that the built environment is socially and environmentally conscientious in addition to being structurally sound.

x. Recognition and Rewards

Acknowledging and praising groups and individuals for following safety procedures produces a positive feedback loop. Establishing safety excellence award programmes promotes a shared commitment to upholding strict health and safety regulations. Encouraging and compensating employees in sustainable construction is essential to developing a culture of quality, drive, and commitment to ethical construction techniques. Employees who actively take part in sustainability efforts receive help from recognition, which not only raises morale but also emphasises the significance of their participation in fostering a built environment that is more socially and environmentally responsible. Establishing a culture of appreciation and success in sustainable

Table 6.11 Communication strategies for health and safety

Essence	Explanation
Clear and accessible policies	Clearly define and make available health and safety policies. All parties involved should be informed of these rules, which define the standards and protocols for preserving a secure and long-lasting workplace
Regular safety meetings	Organise frequent safety meetings to answer concerns, promote open communication, and give updates on health and safety procedures. This encourages the building team to have a sense of shared responsibility and community
Multilingual communication	Adopt multilingual communication techniques in varied construction contexts to guarantee that all workers have access to safety information. This inclusiveness encourages comprehension and adherence among employees
Visual communication tools	To communicate safety information, use visual aids like infographics and signs. These resources can be especially helpful in locations where there is linguistic diversity because they are efficient at conveying crucial messages quickly
Digital platforms for updates	Introduce digital channels for up-to-date health and safety information. By doing this, it is made sure that everyone is aware of any updates to safety measures, emergency protocols, and current information

Table 6.12 Integration of sustainability in training

Essence	Explanation
Linking safety to environmental impact	Makes a connection between safety instruction and the larger environmental effects of construction. Assist employees in realising how their safety procedures minimise environmental damage and advance sustainability
Resource conservation training	Incorporate instruction on waste minimization, energy-efficient techniques, responsible material usage, and resource conservation. This cultivates in the workforce a sense of environmental care
Incorporating eco-friendly material handling	Teach employees how to handle environmentally friendly items properly. This entails being aware of the material life cycle, recycling protocols, and the significance of selecting sustainable substitutes

construction requires both rewards and recognition. Construction projects that value and celebrate the contributions made by their employees to responsible construction practices not only inspire their staff but also fosters a forward-thinking and positive work atmosphere. When workers are recognised for their dedication to sustainability, they become advocates for ethical building, creating a culture that goes beyond personal success and helps create a built environment that is more socially and sustainably responsible.

xi. Feedback Mechanisms

Regular feedback systems are set up to address new health and safety problems quickly, enabling continual improvement. Refining health and safety procedures and averting future problem can be carried out in large part through the analysis of incidents, near misses, and accidents. Building a culture of communication, cooperation, and ongoing development in the framework of sustainable construction requires the establishment of efficient feedback mechanisms. The adoption of sustainable practices is heavily dependent on workers and staff, whose views can be invaluable in streamlining procedures, resolving issues, and achieving sustainability aims. Ensuring that all stakeholders' voices are heard through the implementation of strong feedback mechanisms fosters a dynamic and responsive workplace. Especially when it comes to sustainable construction, feedback systems are essential to creating a cooperative and adaptable construction culture. Construction projects can discover areas for development, use the collective intellect of their workforce, and continuously enhance their approach to sustainable building methods by actively observing for and appreciating the advice of workers and staff. When efficient feedback systems are in place, the construction industry becomes dynamic and responsive, communication is two-way, and employees actively take part in deciding the direction that sustainable construction will take.

xii. Technological Integration

Accepting technological advancements like wearables, sensors, and monitoring systems improves the ability to recognise and respond to health and safety threats at once. By using digital platforms for training and education, employers may make sure that the site operatives are up to date on the newest safety protocols. A key factor in improving health and safety procedures in sustainable construction management is technological integration. Utilising ultramodern technology not only guarantees a safer working environment but also enhances construction projects' overall sustainability and efficiency. Technology integration stimulates creativity and enables construction teams to proactively address health and safety issues through real-time monitoring and sophisticated analytics. The field of health and safety in sustainable construction is changing because of technological integration. These technologies, which range from real-time monitoring to predictive analytics, enable construction teams to improve worker well-being, provide safer work environments, and add to the overall sustainability of building projects. The construction sector is ready to embrace a future where technology actively promotes a culture of safety and well-being in all facets of sustainable construction, in addition to mitigating risks, if advances persist.

Measures to Integrate Diversity and Inclusion in Sustainable Construction

Promoting diversity and inclusion in the construction workforce.

Adopting methods and procedures that support equitable treatment, representation of people from varied backgrounds, and equal chances is necessary to ensure diversity and inclusion in the construction industry. Intentional efforts and policies aimed at fostering a construction sector that appreciates and embraces people with a variety of backgrounds, experiences, and opinions are referred to as diversity and inclusion in the context of sustainable building. This method acknowledges that the long-term viability and success of building projects are influenced by a diverse and inclusive workforce. When discussing sustainable building, the terms "diversity and inclusion" relate to deliberate actions taken to foster a workforce in the construction sector that appreciates and accepts people with a range of experiences, viewpoints, and backgrounds. This method acknowledges that a workforce that is inclusive and diverse is essential to the long-term viability and profitability of construction projects. Inclusion and diversity are crucial elements of sustainable construction methods. See Table 6.13 for instances of the essences.

Table 6.13 Essential of diversity and inclusion sustainable construction

Essence	Explanation
Social equity and representation	To attain social equality, sustainable construction makes sure that everyone, from all backgrounds, has equitable representation and possibilities. A workforce that reflects the variety of the communities where construction projects are found is encouraged by inclusion
Innovative solutions	Diverse teams stimulate creativity and innovation because they bring a variety of viewpoints and ideas to the table. Diverse perspectives in sustainable building can result in the creation of creative solutions that tackle difficult problems, boost project effectiveness, and increase sustainability overall
Community engagement	Local communities are regularly affected by construction projects. Engaging with a varied range of community people to understand their needs and concerns is a key part of an inclusive strategy. Trust, cooperation and the development of building projects that are in line with the goals and values of the communities they serve are all eased by inclusive community participation
Cultural sensitivity	The cultural variety seen in construction teams and communities is recognised and respected in sustainable construction practices By preventing detrimental effects on regional customs, cultural heritage, and traditions, cultural sensitivity in building methods promotes social sustainability
Workplace well-being	Positive and encouraging work environments are fostered by inclusive practices, which enhance the well-being of every employee. Making sure the construction sector is inclusive contributes to drawing in and keeping a varied talent pool, which in turn improves the workforce's overall sustainability
Ethical supply chain practices	Sustainable building involves the whole supply chain in addition to the labour force. Working with a variety of partners, suppliers, and subcontractors helps to ensure inclusive supply chain standards, which uphold morality and equity throughout the building process

(continued)

Table 6.13 (continued)

Essence	Explanation
Local employment opportunities	Sustainable building initiatives make a concerted effort to give residents job possibilities. An inclusive strategy makes sure that people from diverse backgrounds may take advantage of these possibilities, which supports the local economy
Equitable access to benefits	All workers, regardless of background, will have equal access to benefits, training, and opportunities for professional growth thanks to inclusive policies. This helps ensure that the benefits of construction projects are distributed more fairly

The larger aims of social, economic, and environmental sustainability are in line with the incorporation of diversity and inclusion into sustainable construction techniques. It encourages the development of a creative, resilient, socially conscious, and adaptable construction sector that meets the needs of all stakeholders. The steps and factors to improve diversity and inclusion in the construction industry are detailed in Table 6.14.

Table 6.14 Steps to enhance diversity and inclusion in the construction sector

Step	Explanation
Create a diverse workforce	Use proactive recruitment techniques to draw in people with diverse backgrounds. Encourage a diverse and inclusive work environment that accepts staff members with a range of backgrounds, abilities, and viewpoints
Equal employment opportunities	Create and give policies that prioritise giving equal opportunity to all workers, irrespective of their gender, race, age, handicap, or other attributes. Make sure that recruiting, promotion, and pay policies are impartial and fair
Diversity training	All staff members should receive diversity and inclusion training, which will increase team member comprehension and increase awareness of the value of diversity. Include training courses on unconscious prejudice to help reduce stereotypes and prejudices
Inclusive leadership	Encourage inclusive leadership that recognises and makes use of a varied team's advantages. Motivate team leaders to proactively ask for feedback from every member of the team and set up a setting where a range of perspectives are acknowledged and valued
Employee resource groups	Provide a forum for workers with similar hobbies or histories to interact, exchange stories, and foster an inclusive work environment by setting up affinity networks or employee resource groups
Flexible work arrangements	Provide flexible work schedules to meet a range of needs, including those related to various family configurations, childcare obligations, or health issues. Make sure that all employees' work-life balance is supported by the policies

(continued)

Table 6.14 (continued)

Step	Explanation
Mentorship and sponsorship programs	Put in place mentorship programmes that pair staff members with mentors who can offer advice and aid. Sponsorship initiatives can take it a step further by actively promoting underrepresented people's professional progress
Supplier diversity	Promote diversity among suppliers, subcontractors, and contractors by actively searching for companies run by members of underrepresented groups. Set aims for supplier diversity and check your success with metrics
Diversity metrics and reporting	Establish mechanisms for monitoring diversity measures, such as staff composition and representation across organisational levels Report progress and activities related to diversity and inclusion regularly to stakeholders
Promote a culture of inclusion	Encourage a culture of inclusion at all levels and celebrate diversity in the workplace. Promote transparent communication and provide avenues for staff members to offer their opinions and suggestions
Zero-tolerance for discrimination	Make it clear that harassment and discrimination will not be tolerated. Make sure that all concerns are at once and completely investigated and provide clear reporting methods
Continuous improvement	To find areas for improvement, evaluate and reevaluate diversity and inclusion activities regularly. Get input from staff members and other interested parties to guide future initiatives

Construction companies should actively support a more diverse and inclusive business by putting these strategies into practice. This will encourage innovation, creativity, and a healthy work environment for all team members.

Measures to Integrate Fair Labour Practices in Sustainable Construction

The project team should ensure that construction workers receive fair compensation and working conditions. Adherence to labour laws and rules to safeguard employees' rights. In the context of sustainable construction management, fair labour practices are morally and practically sound employment policies and procedures that put workers' rights, welfare, and dignity first. To guarantee that construction g projects have a good impact on the social, economic, and environmental sides, this part of sustainability is essential. The measures to include labour practices in sustainable construction are listed in Table 6.15.

Beyond merely adhering to the law, fair labour practices in sustainable construction represent moral principles that put workers' well-being first and support the social sustainability of construction projects as well as the communities they affect.

Public Health and Safety in the Context of Sustainable Construction

In the context of sustainable construction management, public health and safety refers to the extensive efforts and actions made to safeguard and advance the welfare of the public and employees throughout the design, carrying out, and upkeep of construction projects. This part of sustainability emphasises the significance of reducing risks,

Table 6.15 Measure to integrate labour practices into sustainable construction

Measure	Explanation
Compliance with labour laws	Sustainable building initiatives closely follow national and international labour rules and regulations Ensuring compliance with legal standards for salaries, working hours, safety, and other fundamental labour rights is crucial in safeguarding workers' rights
Safe and healthy working conditions	Construction workers' health and safety are given priority under fair labour norms To reduce accidents and occupational health problems, sustainable construction projects offer secure working environments, proper training on safety procedures, and access to personal protective equipment
Equitable wages and benefits	Fair compensation for labour is provided to workers in sustainable construction Fair pay and benefits help communities escape poverty by ensuring that workers and their families can support themselves financially
Elimination of discrimination and harassment	Zero-tolerance rules against harassment and discrimination are in place for sustainable construction projects Employees receive the same respect and decency regardless of their gender, race, religion, or any other attribute. Equal opportunities are promoted by inclusive behaviours for all
Labour union rights	Respecting employees' freedom to form unions and organise is a key part of sustainable construction Workers are guaranteed a say in negotiating just terms and conditions of employment through collective bargaining
Prevention of child and forced labour	Child labour and forced work are prohibited in sustainable building To make sure they do not engage in exploitative labour practices, contractors and subcontractors are screened, and procedures are in place to spot and address such acts
Training and skill development	Sustainable construction initiatives put money into their employees' education and skill-building Training initiatives improve employees' abilities, provide them with useful skills, and increase their employability eventually

(continued)

Table 6.15 (continued)

Measure	Explanation
Social responsibility in the supply chain	Fair employment practices are extended throughout the supply chain via sustainable building It is expected of suppliers and subcontractors to uphold moral labour standards and encourage just and accountable procedures on every side of the construction project
Transparent employment practices	Fair labour practices are characterised by open communication about job terms, conditions, and expectations Employees receive information about their rights, obligations, and prospects for growth within the company
Employee well-being programmes	Beyond the terms of basic employment, sustainable construction projects may incorporate well-being programmes The comprehensive well-being of construction workers can be guaranteed through initiatives including mental health aid, access to healthcare services, and health and wellness programmes

averting mishaps, and creating a healthy environment by acknowledging the connection between construction operations and public health and safety. The benefits of incorporating public health and safety into sustainable construction management practices are listed in Table 6.16.

In sustainable construction management, public health and safety go beyond merely following the law to represent a proactive, all-encompassing strategy that considers the welfare of employees, nearby communities, and the environment.

Table 6.16 Usefulness integrating public health and safety in sustainable construction

Measure	Explanation
Occupational health and safety	The health and safety of construction workers is given top importance in sustainable building Strict safety procedures, proper instruction, and the availability of personal protective equipment all help to lower workplace risks and guarantee employee well-being
Community safety	Sustainable building initiatives take the communities' safety into account To reduce disturbances and potential risks to nearby people and businesses, strategies like dust control, noise reduction, and traffic management are put into place

(continued)

Table 6.16 (continued)

Measure	Explanation
Emergency preparedness and response	Plans for sustainable construction incorporate emergency response and preparation techniques To safeguard employees and the public in case of an accident, natural disaster, or other unanticipated incident, protocols are set up to handle these situations
Air and water quality management	The goal of sustainable construction methods is to preserve or enhance the area's water and air quality Controls over dust and emissions are put in place to reduce pollution and protect the health of nearby residents as well as construction personnel
Hazardous material management	In sustainable construction projects, hazardous materials are managed carefully and disposed of appropriately Proper handling and disposal of materials such as asbestos, lead, and other hazardous substances is necessary to prevent environmental contamination and public health risks
Noise pollution controls	Sustainable building practices reduce noise pollution by considering the possible effects on the businesses and residents in the area It may be possible to change construction schedules to reduce noisy disturbances during sensitive hours and to safeguard public areas by erecting sound barriers
Infection control (in healthcare construction	Infection prevention is given considerable consideration in construction projects related to healthcare To protect patients', employees', and guests' health, precautions are taken to stop the spread of pollutants
Accessibility and universal design	Universal design principles are included in sustainable architecture to guarantee accessibility for all people, including those with disabilities Infrastructural accessibility and safety support inclusivity and public health
Green spaces and recreational areas	The establishment or preservation of green spaces and recreational areas may be incorporated into sustainable construction Having access to natural settings is beneficial for both community health and mental wellness

(continued)

Table 6.16 (continued)

Measure	Explanation
Public communication and awareness	Public communication is a key part of sustainable construction initiatives Educating the public on building activities, possible effects, and safety precautions increases awareness among the public and builds trust
Post-construction maintenance and safety	Plans for continuous infrastructure maintenance are part of sustainable construction, as they guarantee continuous safety Regular maintenance and inspections help to ensure the public's long-term safety and health

Sustainable construction projects prioritise these elements, which improves the general well-being and adaptability of the people they serve.

Zero-Tolerance for Discrimination in the Context of Sustainable Construction

Incorporating Zero-Tolerance for discrimination into sustainable construction management practice entails putting in place procedures, and regulations that categorically opposes discrimination of any kind and fosters an inclusive, varied workplace. This dedication goes beyond only following the law, stressing moral issues and acknowledging the societal responsibility of building projects. The steps to incorporate a zero-tolerance approach for discrimination in sustainable construction are listed in Table 6.17.

By incorporating a zero-tolerance policy for discrimination into every facet of the construction project, sustainable practices can help to promote a fair and inclusive work environment, which will ultimately improve the project's social sustainability and its influence on the larger community.

Table 6.17 Measures to integrate a zero-tolerance approach for discrimination in sustainable construction

Measure	
Develop and communicate inclusive policies	Develop and give a thorough set of policies that expressly declare your commitment to a workplace free from discrimination All sides of work, such as recruiting, training, promotions, and daily contacts, should be covered by policies
Provide anti-discrimination training	Organise frequent training sessions to impart the value of diversity and inclusion to staff members, subcontractors, and stakeholders Training ought to involve developing cultural awareness, finding and correcting unconscious prejudices, and encouraging polite communication

(continued)

Table 6.17 (continued)

Measure	
Establish equal opportunity employment practices	Establish equitable and open hiring and advancement procedures to guarantee that everyone has the same chances, irrespective of gender, colour, ethnicity, age, religion, handicap, or other attributes Make sure your candidate pools are diverse and avoid biased wording in job descriptions
Diverse representation in leadership	Encourage a diverse mix of candidates for leadership roles in the construction project team Diverse viewpoints at the table promote an inclusive atmosphere and serve as an example for the entire workforce
Implement a reporting system	Provide a private, easily available reporting system so that workers can report instances of bias, harassment, or discrimination Make certain that people feel comfortable reporting issues and that there are well-defined protocols in place for handling and resolving concerns
Conduct regular audits and assessments	Conduct surveys and audits regularly to evaluate the working environment and find any potential problem areas Utilise the results to rectify any discriminatory patterns and keep improving
Cultivate an inclusive culture	Encourage a culture of inclusion and diversity in the workplace Promote an atmosphere where everyone feels appreciated and respected, celebrate cultural awareness, and encourage candid communication
Engage in community outreach	Take the zero-tolerance policy for discrimination outside of the workplace Involve stakeholders and the community to advance diversity and inclusivity in the larger construction sector
Collaborate with diverse suppliers and contractors	Seek out and work with a variety of vendors, contractors, and subcontractors Diversifying the supply chain promotes economic inclusion and gives marginalised groups more opportunities
Promote work-life balance and inclusivity	Adopt policies that support family-friendly practices and flexible work schedules in order to foster inclusivity and work-life balance Recognise and satisfy the workforce's diverse needs

(continued)

Table 6.17 (continued)

Measure	
Monitor and address unconscious bias	Increase knowledge of unconscious prejudice and offer training to lessen its effects Promote self-awareness and lifelong learning to foster an inclusive workplace
Celebrate diversity	Plan activities and events that honour the diversity of your workforce Acknowledge cultural customs and foster an environment of respect for one another

6.5 Social Sustainability Enabled Construction Management Practices

To surmise, the chapter finished with an explanation of social sustainability enabled construction management practices. The social sustainability-enabled construction management emphases the integration of social considerations into the management of the construction projects to promote social equity, well-being, and inclusivity. The integration of social sustainability principles into construction management practices, provide strategic opportunities for the construction managers to create positive social impacts, reinforce community relationships, and contribute to the overall well-being and resilience of society. The mechanisms for this involve:

1. Stakeholder Engagement and Community Involvement

Construction managers actively engage with all those that have stakes in the projects, including local communities, residents, businesses, and other relevant parties, throughout the entire span of the project. The construction managers seek inputs, feedback, and participation from stakeholders in project planning, conceptualization, design, construction and decision-making processes to ensure that the project meets the needs and wants and priorities of the community and fosters social inclusion and empowerment.

2. Labour Practices and Workforce Development

Construction managers prioritize fair and ethical labour practices to ensure the health, safety, and well-being of the construction workers and others. The construction managers adhere to all applicable labour standards and regulations, provide training and professional development opportunities for workers, promote diversity and equal employment opportunities, and establish mechanisms for addressing grievances and resolving disputes, claims, conflicts that are associated with the to create a supportive and inclusive work environment. Special arrangements are made to recruit females and people with disabilities. The construction managers have measures to social discrimination all forms and shape.

3. Local Hiring and Economic Development

Construction managers should empower the local economics by supporting local hiring and economic development by prioritizing the recruitment and training of local workers, contractors, and suppliers and buying materials from the local markets. The construction manager should collaborate with local workforce development agencies, educational institutions, and community organizations to create job opportunities, build local capacity, provide training, and promote economic empowerment within the communities where construction projects are located.

4. Community Benefits and Social Impact Mitigation

The construction managers should identify, evaluate and implement community benefits schemes and social impact mitigation measures to address the potential adverse effects of the construction activities on the local communities. This can involve reducing disturbance, noise, and pollution to safeguard the health and well-being of locals and companies close to building projects, as well as offering affordable housing, public facilities, infrastructure upgrades, and job training programmes.

5. Health and Safety

The construction managers are required to prioritize the health and safety of workers, community, and the general populaces through the implementation of robust health and safety protocols, procedures, and training programmes. The construction managers are required to ensure compliance with occupational health and safety regulations, conduct risk assessments and hazard analyses, provide PPE to workers and visitors, and promote a culture of safety awareness and accountability to prevent accidents and injuries on construction sites.

6. Accessibility and Universal Design

The construction managers are required to integrate principles of accessibility and universal design into construction projects to ensure that constructions are accessible and inclusive for people of all ages, abilities, classes, and backgrounds. The construction managers are required to design and construct facilities with features including ramps, elevators, wide doorways, signage, and sensory accommodations to enable equal access and participation for individuals with disabilities and diverse needs.

7. Cultural Heritage Preservation and Cultural Sensitivity

The construction managers are required to respect and preserve cultural heritage sites, landmarks, and traditions within the construction project area. In order to identify and preserve culturally significant objects, landscapes, and assets, construction managers should also confer with indigenous groups, local communities, and cultural heritage specialists. And, they should also incorporate culturally sensitive design elements and narratives into construction projects in order to celebrate and honour local heritage and identity.

8. Resilience and Social Cohesion

The construction managers are required to design and construct resilient, adaptive, and inclusive built environments that enhance social cohesion, community resilience, and quality of life. Furthermore, the construction managers should assimilate features including green spaces, community centers, public art, and gathering places into construction projects to foster social interaction, connectivity, and a sense of belonging among residents and stakeholders.

6.6 Summary

In the construction sector, health and safety are fundamental values that need a proactive and comprehensive approach. Construction sites can be made safer places for all parties involved by including risk assessment, training, emergency planning, and setting up a culture that values well-being. In the construction sector, health and safety are not merely compliance measures but rather essential elements of a culture that values and safeguards every individual involved in the building process. This is shaped by continuous improvement, adaptability to appearing technologies, and a shared commitment to safety excellence. To sum up, in the construction sector, construction health and safety are unavoidable standards. A seamless fusion of procedural rigour, cultural reform, and a never-ending quest for improvement is needed to achieve safer construction sites. The sector is dedicated to making sure that every building block placed, and every structure built reflects a culture that appreciates and safeguards the lives and well-being of everyone engaged, even as it continues to sculpt skylines and communities. In this endeavour, construction health and safety are not only requirements; they serve as the cornerstones around which a future that is resilient and safe is built. It is essential to involve community in decision making. The values, rights, norms should be identified, respected, projected and appreciated.

Bibliography

1. Almahmoud E, Doloi HK (2015) Assessment of social sustainability in construction projects using social network analysis. Facilities 33(3/4):152–176
2. Atanda JO (2019) Developing a social sustainability assessment framework. Sustain Cities Soc 44:237–252
3. Baldwin C, King R (2018) Social sustainability, climate resilience and community-based urban development: what about the people? Routledge
4. Bamgbade JA, Nawi MNM, Kamaruddeen AM, Adeleke AQ, Salimon MG (2022) Building sustainability in the construction industry through firm capabilities, technology and business innovativeness: empirical evidence from Malaysia. Int J Constr Manag 22(3):473–488
5. Brinkmann R (2016) Introduction to sustainability. Wiley
6. Dujon V, Dillard J, Brennan EM (2013) Social sustainability. Taylor & Francis
7. Edum-Fotwe FT, Price AD (2009) A social ontology for appraising sustainability of construction projects and developments. Int J Project Manage 27(4):313–322
8. Fatourehchi D, Zarghami E (2020) Social sustainability assessment framework for managing sustainable construction in residential buildings. J Build Eng 32:101761

9. Goel A, Ganesh LS, Kaur A (2020) Social sustainability considerations in construction project feasibility study: a stakeholder salience perspective. Eng Constr Archit Manag 27(7):1429–1459
10. Hossain MU, Poon CS, Dong YH, Lo IM, Cheng JC (2018) Development of social sustainability assessment method and a comparative case study on assessing recycled construction materials. Int J Life Cycle Assess 23:1654–1674
11. ISO 45001:2018 Occupational health and safety management systems
12. Kiani Mavi R, Gengatharen D, Kiani Mavi N, Hughes R, Campbell A, Yates R (2021) Sustainability in construction projects: a systematic literature review. Sustainability 13(4):1932
13. Lami IM, Mecca B (2020) Assessing social sustainability for achieving sustainable architecture. Sustainability 13(1):142
14. Mahdavinejad M, Sadraie A, Sadraie G (2014) Social sustainability of high-rise buildings. Am J Civil Eng Archit 2(1):34–41
15. Manzi T, Lucas K, Jones TL, Allen J (eds) (2010) Social sustainability in urban areas: communities, connectivity and the urban fabric. Routledge
16. Ng E (ed) (2009) Designing high-density cities: for social and environmental sustainability. Routledge
17. Montalbán-Domingo L, García-Segura T, Amalia Sanz M, Pellicer E (2019) Social sustainability in delivery and procurement of public construction contracts. J Manag Eng 35(2):04018065
18. Montalbán-Domingo L, García-Segura T, Sanz MA, Pellicer E (2018) Social sustainability criteria in public-work procurement: an international perspective. J Clean Prod 198:1355–1371
19. Ortiz O, Castells F, Sonnemann G (2009) Sustainability in the construction industry: a review of recent developments based on LCA. Constr Build Mater 23(1):28–39
20. Shirazi MR, Keivani R (eds) (2019) Urban social sustainability: theory, policy and practice. Routledge
21. Sierra LA, Pellicer E, Yepes V (2017) Method for estimating the social sustainability of infrastructure projects. Environ Impact Assess Rev 65:41–53
22. Sierra LA, Yepes V, García-Segura T, Pellicer E (2018) Bayesian network method for decision-making about the social sustainability of infrastructure projects. J Clean Prod 176:521–534
23. Smith D, Beeck S, Lommerse M, Metcalfe P (2014) An introduction to social sustainability and interior architecture. In: Perspectives on social sustainability and interior architecture: life from the inside, pp 1–11
24. Sodangi M (2018) Social sustainability efficacy of construction projects in the pre-construction phase. Proc Inst Civil Eng-Eng Sustain 172(2):57–67
25. Toole TM, Carpenter G (2013) Prevention through design as a path toward social sustainability. J Archit Eng 19(3):168–173
26. Valdes-Vasquez R, Klotz LE (2013) Social sustainability considerations during planning and design: framework of processes for construction projects. J Constr Eng Manag 139(1):80–89
27. Vallance S, Perkins HC, Dixon JE (2011) What is social sustainability? A clarification of concepts. Geoforum 42(3):342–348
28. Wang H, Zhang X, Lu W (2018) Improving social sustainability in construction: conceptual framework based on social network analysis. J Manag Eng 34(6):05018012
29. Yates JK, Castro-Lacouture D (2018) Sustainability in engineering design and construction. CRC Press
30. Van der Leeuw S (2020) Social sustainability, past and future: undoing unintended consequences for the Earth's survival. Cambridge University Press
31. Yıldız S, Kıvrak S, Gültekin AB, Arslan G (2020) Built environment design-social sustainability relation in urban renewal. Sustain Cities Soc 60:102173
32. Zuo J, Jin XH, Flynn L (2012) Social sustainability in construction—an explorative study. Int J Constr Manag 12(2):51–63

Chapter 7
Net Zero Within the Built Environment

Abstract This chapter explores the imperative of achieving net-zero carbon emissions within the built environment, a sector that significantly contributes to global greenhouse gas emissions. As the impacts of climate change intensify, the construction industry is under pressure to transition toward sustainable practices that balance economic development with environmental responsibility. The chapter unpacks the complexities of decarbonising a carbon-intensive sector reliant on materials like steel and cement, and energy sources such as fossil fuels. It critically examines the concept of net zero, capital cost implications, carbon benchmarking, and the systemic challenges in offsetting emissions across construction lifecycles. Recognizing that the path to net zero is neither linear nor intuitive, it highlights the need for a collaborative, multi-stakeholder approach that includes innovation, energy efficiency, and behavioural change. The chapter also discusses the role of carbon markets, offsets, and the ethical dimensions of carbon management. While acknowledging debates surrounding CO_2's role in climate change, it underscores the importance of reducing excess emissions and enhancing natural carbon sinks. Ultimately, the chapter advocates for a paradigm shift in construction practices one that sees the built environment not just as a consumer of resources but as a vital contributor to global climate action.

Keywords Decarbonization · Carbon market · Construction carbon offsetting · Net zero · Sustainable construction

7.1 Learning Outcomes

The construction sector is a carbon intensive sector, no doubt. The impact of climate change on construction is complex because the impact is nonlinear, and to fully address the problem, there is a cogent need to transition to a net-zero economy, which must be addressed to adapt to a changing climate in the meantime. The carbon dioxide intensity of the construction sector is high with each construction GDP. The values are even higher when the CO_2 multiplier is considered. It is even high when the intensity of the real estate is considered and the emission from electricity

A. Olanrewaju, *Sustainable Construction Management: Research and Practice Companion*, Management in the Built Environment,
https://doi.org/10.1007/978-981-96-4910-5_7

and gas, and others are considered. The built environment commits to accelerating climate action moving forward. The activities of the built environment had left the net-zero equation unsolved. Because the greenhouse gas emissions from the built environment increasing continuously unabated. There are no measures in place to counterbalance the generation through an effective removal approach. This is not by any means implying that the built environment alone can solve the problems but could contribute significantly towards reducing the impact. Addressing the net-zero carbon conundrum is not always intuitive. The main source of the emissions needs to be addressed together, not fragmented. A significant paradigm shift in carbon management is fundamentally necessary. Many construction companies and clients now have a net zero plan. Meanwhile, there is a need for the project stakeholders to understand the carbon impact of their thought process and decisions of the finished construction projects. It is obvious that multiple answers are needed to the number of questions; indeed, many of the questions are not closely related to the activities of the built environment. In this chapter, attempts were made to answer some of these questions about the built environment. After studying this chapter, you will be:

- Understand the zero-carbon concept
- Understand the up-front capital costs for sustainable construction
- Explain Net Zero in the sustainable Built Environment
- Benchmarking the net zero carbon within the built environment in sustainable construction.

7.2 Introduction

"Net Zero" has become increasingly crucial in the unrelenting quest for a sustainable, conscious future, serving as a beacon to direct the built environment's change. Today, innovation and accountability meet at a critical crossroads in the fields of architecture, construction, engineering, surveying, and urban planning. This chapter examine how the net zero is changing the construction landscapes, including its size, impact, and adjustments. The built environment is under increasing strain as environmental awareness and climate concerns grow, drawing harsh criticism from both inside and beyond the sector on the account of sector carbon generation and poor carbon management. A dramatic paradigm change in the commitment to sustainable development is represented by the Net Zero principle. According to this paradigm, the built environment is seen as a responsible contributor to the maintenance and restoration of our planet, in addition to being a resource consumer. Thus, this chapter will investigate the comprehensive ideas that characterise a Net Zero approach in the built environment, including waste management, materials, energy, and water. In the built environment, achieving Net Zero needs teamwork rather than being a solitary undertaking. By using energy-efficient measures like insulation and swapping out fossil fuel-powered heating and cooking equipment with low-emission alternatives, the sector aims to reduce its carbon emissions. Between 2020 and 2050, the construction sector would invest, on average, $1.7 trillion annually in the physical assets needed to enable the net-zero transition [16]. Through the retrofitting of construction with insulation, decarbonisation of buildings might lead to a net gain of around 500,000 direct jobs by 2050 assuming a net-zero transition. Managing the upfront capital costs for end users to retrofit equipment and coordinating incentives amongst multiple stakeholders-including construction clients who invest capital and tenants who might receive help from lower operating costs-will be the construction sector's largest adjustment during this transition. Over time, consumers may see a reduction in other operational costs because of the first capital expenditures needed for the net-zero shift. Even if the Net Zero idea is aspirational, there are some problems that the stakeholders in the built environment need to consider when making decisions. The road to a sustainable built environment is paved with obstacles, both technological and policy-related, requiring careful considerations and thoughtful solutions. Unexpectedly low capital costs for net-zero technologies could be achieved through technological innovation. The chapter examines how energy-efficient construction materials, techniques, and initiatives, along with innovation, are influencing the built environment.

There is debate, though, about whether CO_2 is a problem in and of itself or what causes believe that the sun, not CO_2, is what alters the climate. However, by intensifying the greenhouse effect, which traps heat in the Earth's atmosphere and causes global warming, CO_2 adds to rising temperatures. CO_2 emits and absorbs infrared light in the thermal infrared spectrum since it is a greenhouse gas. It appears that photosynthesis, the process by which plants turn CO_2 and sunlight into energy and oxygen, depends on CO_2. Through photosynthesis, plants are essential in eliminating CO_2 from the atmosphere. This procedure aids in controlling CO_2 levels and preserving the atmospheric gas balance. While CO_2 is required for plant growth, it's crucial to understand that high atmospheric CO_2 concentrations can have negative consequences on ecosystems and the climate. Global warming and climate change are caused by human actions that have resulted in a major increase in CO_2 emissions, such as the burning of fossil fuels and deforestation. While there are situations in which higher CO_2 levels are beneficial to plants, the overall balance of CO_2 in the atmosphere and addressing the underlying causes of CO2 emissions are crucial to minimising the effects of climate change on plants and the planet as a whole. These broader impacts of climate change include rising temperatures, altered weather patterns, and habitat loss. Although CO_2 is an crucial part of the Earth's atmosphere and is necessary for life, high CO_2 levels can be harmful to the environment and to people's health. Responsible CO_2 emission management is crucial, as is striving for a long-term equilibrium in the carbon cycle on Earth. If there is an excessive amount of CO_2, that is the issue. If and to what degree the quantity is higher is the question that is outside the scope of this book.

7.3 Definition of Zero Carbon

Carbon is a chemical element and one of the most abundant elements in the universe. It plays a crucial role in the chemistry of life and forms the basis of all known life on Earth. It is a major component of organic compounds. The element exists in several forms, such as graphite, diamond, nanobuds, nanotubes, nanoribbons, and amorphous carbon. Carbon dioxide (CO_2) is a chemical compound that comprises of one carbon atom bonded to two atoms of oxygen. It is a colourless, odourless gas and is naturally present in Earth's atmosphere as a gas. It is produced through various processes, including respiration and volcanic activity, as well as human activities like burning fossil fuels and deforestation. Net zero carbon is created when CO_2 emissions are offset or removed to the same extent as they are emitted into the atmosphere, the situation. The aim is to bring the carbon emissions produced and the carbon sequestration or offsetting methods into a state of harmonious equilibrium such that there is no net balance. Reducing carbon emissions is the first step towards achieving net zero carbon. This entails putting plans into actions and acquiring technology to drastically cut down on the amount of greenhouse gases-primarily carbon dioxide-that are released into the atmosphere. The primary segments whose emissions must be decreased through the use of renewable energy sources, energy-efficient technologies, and sustainable practices are buildings, transportation, and industries. Reaching net zero carbon emissions requires switching to renewable energy. Reducing the carbon footprint linked to energy production requires a shift away from fossil fuels and towards renewable energy sources like solar, wind, and hydropower. Since these sources generate electricity without releasing carbon into the environment, they contribute to an overall reduction in net emissions.

Although reducing emissions is crucial, reaching zero emissions is often difficult. Sequestration and carbon offsetting are essential elements of the net zero carbon approach. This entails funding initiatives that offset or stop an equal quantity of carbon emissions in other places. Initiatives to offset emissions include reforestation, afforestation, and carbon capture and storage (CCS) programmes. Using ultra-modern technologies is crucial to achieving net zero carbon emissions. Innovations in carbon capture, energy storage, and environmentally complaints transportation help to lower emissions and improve the general sustainability of several industries. Energy-efficient technology, smart grid systems, and circular economy methods are a few of the technological advancements easing the shift to net zero. Achieving net zero carbon requires an understanding of the environmental impact of processes and products over their whole lifecycle. Focus of lifecycle assessments on the evaluation of the environmental impact of resource extraction, production, transportation, and use, as well as disposal or recycling. This all-encompassing strategy directs efforts to reduce emissions at every turn in the life cycle of the project. Adopting sustainable practices by companies and individuals is encouraged by the implementation of supportive laws, legislation, and incentives. The establishment of strict emission regulations and carbon pricing mechanisms helps to foster an atmosphere that supports net zero aims. It will take a team effort from communities, companies, and individuals to get net zero carbon. Raising awareness, promoting sustainable practices at the local level, and stimulating community involvement all help to move society in the direction of sustainability. Energy-saving techniques, sustainable lifestyles, and local projects all support the main aim.

7.4 Overview of the Global CO_2 Emissions

The analysis of the world's CO_2 emissions statistics presents a multifaceted understand of industrialization, economic growth, and changing energy conditions. Although there may have been a trend in favour of sustainability in recent years, significant worldwide efforts must be made to mitigate climate change and promote the development of a future that is more ecologically resilient. The global CO_2 emission (kt) for the previous 30 years is shown in Fig. 7.1. The following observation and conclusions can be made on emissions in the recorded periods.

Era of Accelerated Growth (1990s to Early 2000s)

The data points to a noteworthy period between the early 1990s and the early 2000s when global carbon emissions grew at an increased rate. Rapid economic growth and industrialization throughout this time, especially in emerging nations, resulted in a significant rise in energy consumption and carbon emissions. The plateaus seen in the middle of the 1990s and the 2000s point to brief intervals of stability that may have been brought about by changes in energy sources, efficiency gains, or economic downturns.

Peak Emissions and Transition (2010s)

Concerns over the effects of growing industrial activity and energy demand on the environment are raised by the 2013 peak in carbon emissions. This timeline aligns with global efforts to mitigate climate change, including the United Nations Framework Convention on Climate Change (UNFCCC) conferences and the Kyoto Protocol. The following three years, with emissions slightly declining in 2018, 2019, and 2020. Although the global economic slump and the COVID-19 epidemic may have had some influence on this decline, it may also show a move towards more environmentally friendly methods. The 2020 data may reflect the impact of the Covid-19 pandemic.

Global Efforts and Agreements

International accords, such as the Paris Agreement of 2015, aim to limit greenhouse gas emissions in order to mitigate global warming. The influence of these international measures and growing public awareness of climate change may be partially reflected in the observed drop in emissions in recent years. The information emphasises how crucial it is to stick to these agreements going forward and how cooperative worldwide efforts are needed to achieve substantial and long-term reductions in carbon emissions.

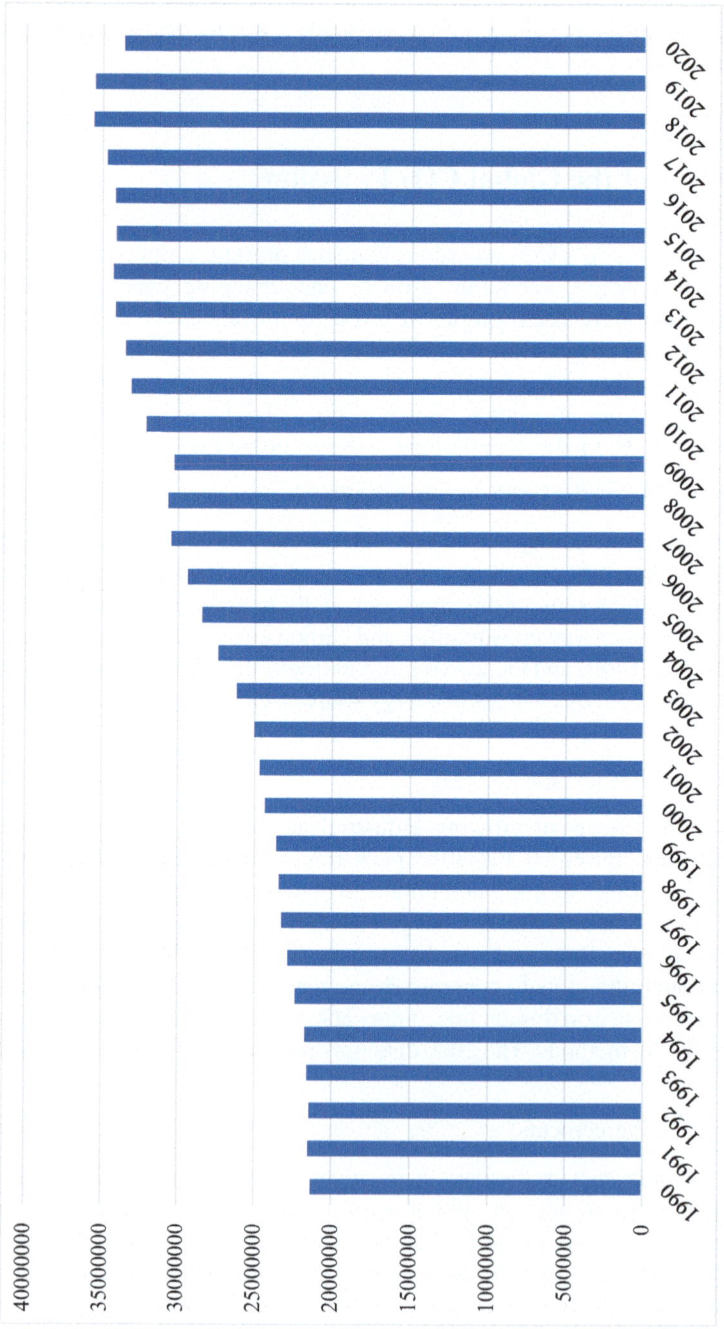

Fig. 7.1 CO_2 emissions (kt)

Technological Advancements and Renewable Energy

Technological advancements have the ability to alter the trajectory of global carbon emissions, particularly in the area of renewable energy. The growing use of renewable energy sources, improved energy efficiency, and a move away from fossil fuels might all have an impact on the drop-in emissions. Maintaining this good trend and reducing the negative effects of human activity on the environment will need ongoing investment and awareness and implementation of clean energy technology.

Challenges and Future Considerations

The global transition to low-carbon and sustainable practices is nevertheless fraught with difficulties, even with recent advancements. As the data revealed, continuous efforts are needed to establish policies that support environmental sustainability and disentangle economic growth from carbon emissions. A comprehensive approach that considers energy generation, as well as circular economy techniques, sustainable urban design, and a reassessment of consumption patterns, should be the focus of future concerns.

Resilience and Adaptation

The COVID-19 pandemic response on a global scale illustrated the possibility of swift behavioural adjustments and a reduction in carbon emissions. Though mainly unintentional, this calls into doubt the viability of sustainable practices and the ability of society to adjust to new, low-carbon norms. Long-term sustainability goals will mostly depend on promoting and maintaining beneficial behavioural changes even in the absence of outside shocks.

As may be seen, the analysis of the world's carbon emissions highlights the relationship between economic endeavours, scientific discoveries, government initiatives, and natural consequences. Even though there have been positive signs of a decline in emissions recently, ongoing global collaboration efforts are necessary to address the complex issues raised by climate change and move towards a more resilient and sustainable future.

7.5 CO_2 Emissions (Metric Tons per Capita)

The difference in emission is proven in terms of capital. The emission per capita for the previous 30 years is shown in Fig. 7.2. The examination of the patterns throughout this period might be clarified as:

1990–2000: Stability and Early Indicators

Global CO_2 emissions per person saw a period of relative stability during the 1990s. With the end of the Cold War, the world community began to move on, and developing economies began to take centre stage in international affairs. Emissions remained stable during this time, which can be explained by a balance between environmental

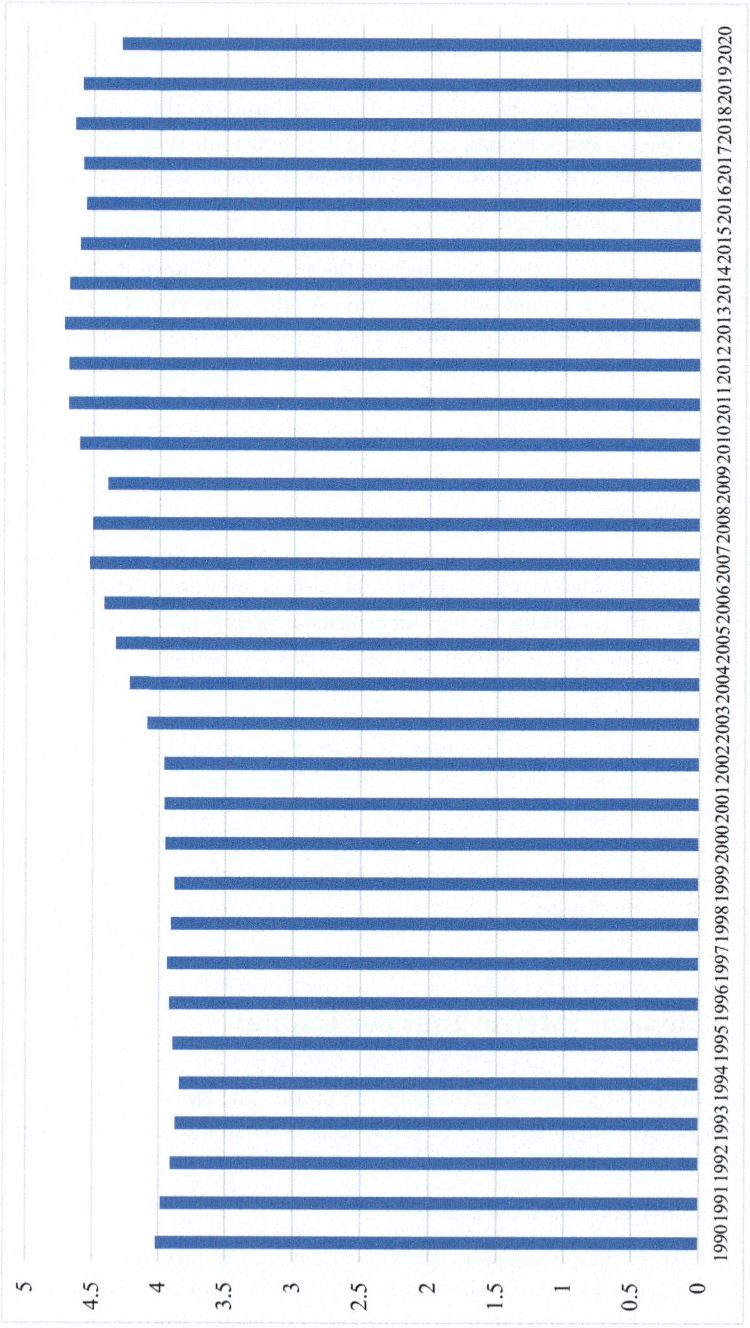

Fig. 7.2 CO_2 emissions (metric tons per capita)

consciousness and economic growth. Numerous nations started to realise how crucial is the sustainable development practices.

Early 2000s: Economic Growth Impact

Between 2000 and 2004, there was a possible correlation between the increase in global economic activity and the rise in CO_2 emissions per person. Rapid industrialization may have played a role in the reported growth, especially in emerging economies. The increase in emissions over this period possibly highlights the convoluted connection between the effects of economic expansion on the environment. Fossil fuels remained a major source of energy for many nations.

Mid-2000s to Late 2010s: Continued Growth and Awareness

The CO_2 emissions per capita show a general rising trajectory from the mid-2000s to the late 2010s. Globalisation, urbanisation, and industrialization have all expanded over this time. Because the 2013 peak suggests that carbon emissions may be accelerating, it raises questions about how difficult it will be to strike a balance between environmental sustainability and economic growth. It might also point to a slowdown in the adoption of renewable energy sources and greener technology during this time all over the world.

2018 and Beyond: Signs of Improvement

The modest decline in emissions from the 2013 high to 2018 may be the first indication of a move towards more sustainable behaviours. This decline may be caused by elements including improvements in energy-efficient practices, increase in the global initiatives to mitigate climate change, and developments in renewable energy technologies.

Global Factors and Challenges

It is imperative to acknowledge that many elements, such as governmental regulations, technical breakthroughs, economic conditions, and international partnerships, impact global CO_2 emissions. The need for greater global collaboration to address climate change, greater adoption of renewable energy, and increased energy efficiency measures are ongoing problems in reducing emissions.

Overtly, the data points to a complicated relationship between the effects of economic development and the environment consideration. Even while there are encouraging signals, combating the effects of climate change calls for persistent work, creative solutions, and international cooperation. Future data will offer crucial insights into whether the trends that have been seen continue and whether global efforts to create a future that is low-carbon and more sustainable are being made.

7.5.1 A Carbon Markets

A carbon market is a mechanism created to lower greenhouse gas emissions by offering financial rewards to businesses who cut back on their emissions of CO_2. It works on the basis of the cap and trade theory, in which the government sets a maximum limit on the total quantity of emissions that are permitted from specific sectors or enterprises. Permits are given to companies so they can emit a specific amount of CO_2. A company may sell its extra licences to other enterprises that exceed their limitations if it emits less than the quantity allotted. As a result, businesses will have financial motivation to cut emissions and spend money on greener technology. Carbon markets put a price on carbon emissions and incentivize companies to switch to low-carbon alternatives in an effort to enhance environmental sustainability. Operating within the cap-and-trade framework, a carbon market aims to lower greenhouse gas emissions into the atmosphere, specifically CO_2. Within their authority, governments or regulatory agencies impose a cap or restriction on the overall quantity of emissions permitted from particular businesses or sectors. This cap is frequently progressively reduced over time to promote additional emission reductions. Companies are granted a set number of emissions allowances, also known as permits, under this system, each of which denotes the ability to emit a given volume of CO_2. A corporation can sell its extra allowances on the carbon market to other enterprises that might exceed their emission restrictions if it emits less than the permits it has been granted. On the other hand, if a business uses more permits than it has been granted, it has to buy more allowances to cover its emissions. Hence, the carbon market offers businesses a financial incentive to cut their emissions through the purchase of greener technology, increased energy efficiency, or the adoption of sustainable business practices. Additionally, it gives businesses the freedom to select the emission reduction tactics that will save them the most money while still keeping overall emissions below the set cap. Different levels of operation, such as national, regional, or global ones, are possible for carbon markets. By placing a cost on carbon emissions and promoting the shift to a low-carbon economy, they play a critical role in reducing climate change.

7.5.2 Construction Carbon Market

The construction carbon market functions within the framework of commodity trading in carbon, which regulates and markets emissions of greenhouse gases like CO_2. Within the construction sector, the main causes of carbon emissions include energy use in building operations, material and equipment transportation, construction material manufacturing, and the actual construction processes. The construction carbon market creates systems for trading carbon allowances or credits in an effort to encourage reductions in these emissions. Developers, consultants, construction companies, and other building process players are potential participants in the construction carbon market. By adopting sustainable construction methods, they can earn carbon credits for reducing emissions. Companies can then monetize their emission reductions or offset their own emissions by acquiring credits from others. These carbon credits can be exchanged on carbon exchanges or through bilateral agreements. The construction carbon market helps achieve the overarching objective of mitigating climate change and shifting to a low-carbon economy by providing financial incentives for emissions reductions in the construction sector.

7.5.3 Challenges of Construction Carbon Market

The construction carbon market faces several challenges that hinder its effective operation and growth. To overcome these obstacles, industry players, legislators, and regulatory agencies must work together to establish uniform procedures, enhance data gathering and reporting systems, and establish financial incentives and laws that promote emissions reductions in the building industry: following are key challenges of construction carbon market:

- Measuring and Verification: Because construction is a complicated process involving many stakeholders, it can be difficult to accurately calculate carbon emissions related to construction activities. The validity of carbon credits depends on accurate measurement and verification of emissions reductions.
- Data Transparency and Availability: It may be difficult to obtain trustworthy data on emissions connected to construction, such as energy use, material usage, and transportation. Robust carbon accounting approaches may be hampered by project participants' opaque reporting and data sharing practices.
- Market Fragmentation: Due to differing standards, procedures, and reporting requirements among various countries and regions, the construction carbon market is quite fragmented. In particular, trade barriers and inefficiencies can result from a lack of uniformity and harmonisation.
- Regulatory Uncertainty: Investor confidence and long-term planning in the building industry may be impacted by uncertainties around future climate policies and regulations. Regulations that are inconsistent or change frequently could make it difficult for businesses to fund carbon reduction projects.
- Project Lifecycles: From planning and design to completion and operation, construction projects usually have lengthy lifecycles. Construction activities are long-term in nature, which makes it challenging to precisely anticipate and quantify emissions over time. This creates issues for carbon accounting and offsetting.
- Lack of Knowledge and Capacity: It is possible that many stakeholders in the construction sector are unaware of the carbon market and its possible advantages. Increasing participation and engagement in the construction carbon market requires capacity building and training in carbon management strategies and methodology.

7.5.4 Construction Carbon Credit

Construction carbon credits are tradable certificates that signify a reduction in greenhouse gas emissions related with construction activities. Implementing carbon reduction strategies, such as utilising low-carbon construction materials, implementing energy-efficient construction techniques, or making investments in renewable energy sources, results in the creation of these credits. Quantifying the quantity of carbon dioxide or other greenhouse gases that the building project avoids, reduces, or sequesters in relation to a baseline scenario is usually the first step in the process. A third-party organisation then confirms this reduction to guarantee correctness and legitimacy. The construction carbon credits can be used to offset emissions from other sources or bought and sold on carbon markets once they have been certified. Companies can buy these credits as part of their sustainability plans to fulfil legal obligations or carbon reduction goals. Incentives for emissions reductions in the construction sector and the promotion of sustainable construction practices are greatly aided by construction carbon credits. They help to reduce climate change by offering financial incentives to developers, contractors, and other stakeholders to invest in environmentally friendly construction processes and technologies.

7.5.5 Mechanisms of Construction of Carbon Credits

Construction projects can produce carbon credits as a concrete demonstration of their efforts to lower greenhouse gas emissions and mitigate the effects of climate change by adhering to these processes:

- Baseline Establishment: Setting up a baseline, or reference point, is the first step in monitoring carbon emissions. This baseline shows the anticipated amount of emissions in the event that no mitigation strategies are put in place.
- Opportunities for Carbon Reduction: Following that, construction projects examine for ways to cut carbon emissions. Using energy-efficient materials, green building techniques, renewable energy sources, and carbon capture and storage technology are a few examples of how to do this.
- Quantification of Emissions Reduction: After carbon reduction strategies are put into place, the next stage is to calculate the precise amount of emissions that have been reduced in relation to the predetermined baseline. This entails computing the difference in emissions between the project's actual emissions and those from the baseline scenario.
- Verification by a Third Party: To guarantee accuracy and legitimacy, the quantified emissions reductions need to be confirmed by an independent entity. As part of this verification procedure, the project's methodology, data gathering techniques, and emissions computations are evaluated.
- Emission of Carbon Credits: Carbon credits are based on the certified emissions reductions and are issued following successful verification. One metric tonne of

averted, reduced, or sequestered carbon dioxide equivalent (CO_2e) is represented by each carbon credit.

- Trading and Registration: The issued carbon credits can be exchanged on carbon markets and are listed on carbon registries. Building firms can utilise these credits to meet legal requirements or achieve sustainability goals, or they can sell them to other organisations need to offset their emissions.

7.5.6 Construction Carbon Offset

The technique of offsetting greenhouse gas emissions from construction operations by funding initiatives that lower or eliminate CO_2 emissions elsewhere is known as construction carbon offset. The objective is to reduce the net emissions related to construction projects or reach carbon neutrality.

The process typically involves the following steps:

- Emissions Assessment: The construction project measures the quantity of CO_2 emissions produced by a range of activities, including energy use, transportation, material manufacturing, and construction operations. This helps to determine the project's carbon footprint.
- Finding Offset Projects: Following the quantification of emissions, the project finds appropriate carbon offset projects to assist reduce or offset the estimated emissions. These offset projects can be energy efficiency projects, afforestation or reforestation programmes, renewable energy projects (like wind or solar farms), or investments in carbon capture and storage technologies.
- Investment in Offset Projects: By acquiring carbon credits or offsets, the construction project makes an investment in the chosen offset projects. One metric tonne of CO_2 equivalent (CO_2e) has been reduced, avoided, or removed from the atmosphere as a result of these credits.
- Verification and Certification: To guarantee that the emissions reductions are actual, quantifiable, long-lasting, extra, and confirmed by independent auditors, the offset projects go through verification and certification procedures. This supports the offset market's integrity and credibility.
- Offsetting Emissions: The carbon offsets are used to offset the emissions of the construction project after they are certified and purchased. This lessens the project's negative environmental effects by offsetting the emissions with comparable reductions made elsewhere.

7.5.7 Construction Carbon Pricing

The term "construction carbon pricing" describes the application of a financial system that places a monetary value on the carbon emissions produced by building operations. It seeks to encourage the building industry to cut greenhouse gas emissions by internalising the cost of carbon emissions to the environment. By incentivizing emissions reductions, encouraging innovation in sustainable construction methods, and stimulating the development of low-carbon technologies and infrastructure, construction carbon pricing seeks to accelerate the shift to a low-carbon economy. The process of construction carbon pricing typically involves the following steps:

- Setting a Price: Using a variety of tools, like carbon taxes, cap-and-trade programmes, or carbon offset markets, governments or regulatory agencies can set a price for carbon emissions. The value of reducing the negative effects of climate change on the environment and society, or the social cost of carbon, is frequently used to determine the price.
- Carbon Accounting: Construction firms evaluate and measure the carbon emissions resulting from their operations, such as energy consumption, transportation, material manufacture, and construction. This entails figuring out how much CO_2 is released for each activity or output unit.
- Cost Internalisation: When creating project budgets and financial plans, construction companies include in the cost of carbon emissions. Throughout the course of a project, the carbon price incentivizes businesses to implement greener technologies, increase energy efficiency, and lower emissions.
- Carbon Offsetting: As previously mentioned, construction companies may choose to invest in carbon offset projects in order to offset their carbon emissions. The entire carbon pricing strategy takes the cost of buying carbon offsets into account.
- Compliance and Reporting: Construction companies adhere to legal obligations pertaining to carbon pricing, including disclosing their emissions and making required payments of carbon taxes or purchases of carbon credits. Accountability and transparency are crucial components of carbon pricing schemes.

7.5.8 Challenges of Construction Carbon Pricing

The application of carbon pricing in construction is fraught with difficulties. A coordinated strategy combining industry cooperation, government regulation, technological innovation, and stakeholder involvement is needed to address these problems. The imperative for bold climate action must be weighed against the interests of social justice, economic competitiveness, and sustainable development. Following are some major challenges of carbon pricing of construction projects:

- Complexity: There are many parties involved in the construction sector, and each has distinct emission profiles and difficulties. It can be difficult to create a carbon pricing mechanism that takes these complexities into account and appropriately represents the carbon footprint of construction activities.
- Data Availability: There may not be enough or sufficient data on carbon emissions from construction operations, particularly in developing countries. It is challenging to evaluate emissions, determine suitable carbon pricing, and monitor the advancement of emission reduction targets in the absence of reliable data.

- Cost Implications: Project prices may rise as a result of construction carbon pricing, especially for carbon-intensive activities. This may affect a project's viability, profitability, and affordability—particularly for smaller businesses or initiatives with constrained funding.
- Competitiveness: When construction firms in jurisdictions with carbon pricing mechanisms compete with those in areas with less stringent environmental laws or without carbon pricing, concerns regarding competitiveness surface. For instance, when emissions-intensive businesses shift to places with laxer rules, it may result in carbon leakage.
- Equity and Social Impacts: Certain groups, such as low-income households or employees in carbon-intensive industries, may be disproportionately impacted by carbon pricing policies. Therefore, it is imperative to guarantee that policies pertaining to carbon pricing are socially just and do not worsen pre-existing disparities.
- Enforcement and Compliance: To guarantee adherence to carbon pricing laws, efficient enforcement methods are required. The efficacy of carbon pricing initiatives could be compromised by non-compliance and evasion in the absence of strong monitoring, reporting, and verification mechanisms.
- Transition Challenges: Making the shift to a low-carbon construction industry will involve large expenditures in infrastructure, technology, and worker development. It may be difficult for construction companies to adopt sustainable practices, comply with new requirements, and incorporate low-carbon technologies into their daily operations.

7.5.9 How Can a Construction Company Trade in Carbon?

By taking part in carbon trading, construction companies can meet regulatory requirements, offset their carbon emissions, and possibly even make more money by selling carbon credits. There are several ways for a construction company to get involved in carbon trading:

- Carbon Offsetting: Investments in reforestation, renewable energy, and energy efficiency projects are examples of projects that construction businesses can engage in to mitigate or offset carbon emissions. The generated carbon credits can subsequently be sold on the carbon market.
- Energy Efficiency: Construction companies can lessen their carbon footprint by integrating energy-efficient technologies and practices into their operations. Any extra carbon credits they obtain as a result of their actions can subsequently be sold.
- Renewable Energy: By producing renewable energy on-site or investing in off-site renewable energy projects, construction businesses can earn carbon credits. On the carbon market, these credits are exchangeable.

- Carbon Capture and Storage (CCS): Businesses are able to store or utilise carbon emissions from their activities underground or for other industrial uses. After that, they can sell the carbon credits these actions produced.
- Carbon Trading Platforms: Through brokers or directly on carbon trading platforms, construction companies can purchase and sell carbon credits. In the carbon market, these platforms help buyers and sellers transact.

7.5.9.1 Business Case for Construction Carbon Offsetting

Construction carbon offsetting aligns with broader sustainability goals, fosters innovation and efficiency, and strengthens business resilience in a rapidly changing regulatory and market landscape. The business case for construction carbon offsetting involves several key factors:

- Environmental responsibility: Energy use, transportation, and material production are the main ways that construction activities add to carbon emissions. Construction companies can exhibit their dedication to environmental sustainability and aid in global efforts to prevent climate change by investing in carbon offsetting.
- Compliance and regulation: Construction companies are under pressure to adhere to environmental standards and emissions targets as a result of the growing number of laws and mandates aimed at lowering carbon emissions. By using carbon offsetting, construction company can comply with regulations and avoid fines for going over emissions limitations.
- Reputation and branding: Implementing carbon offsetting programmes helps a construction company project a more socially conscious image. Because, it shows a dedication to environmental care and can draw partners, investors, and clients who value sustainability and the environment.
- Advantage over competitors: In a cutthroat industry, adopting carbon offsetting can set construction companies apart. Businesses can obtain a competitive advantage and secure contracts from clients needing environmentally friendly building options by providing carbon–neutral projects and sustainable construction solutions.
- Risk management: Construction organisations can reduce possible financial, regulatory, and reputational risks connected with climate change consequences by addressing carbon emissions and climate-related hazards through offsetting programmes. Resists upcoming changes in the market, regulations, and natural disasters.
- Efficiency and cost savings: Carbon offsetting can result in long-term cost savings through increased waste reduction, energy efficiency, and operational optimisation, even though it may require an initial investment. Over time, construction organisations can increase profitability by reducing operating costs and investing in renewable energy sources through the implementation of sustainable practices.

7.5.9.2 How Sustainable Construction Management Help Toward Carbon Offsetting

Sustainable construction management plays a crucial role in contributing to carbon offsetting through various strategies and practices aimed at reducing carbon emissions and promoting environmental sustainability. Below are techniques how sustainable construction management help toward carbon offsetting:

- Energy-efficient design and construction: The use of energy-efficient building designs and construction techniques is emphasised by sustainable construction management. This entails utilising high-performance construction- materials, implementing passive design techniques, and installing energy-efficient appliances including insulation, lighting, and HVAC. Over the course of their lifetime, construction projects can reduce carbon emissions by using less energy during the operating period.
- Renewable energy integration: The use of renewable energy sources in construction designs, such as solar, wind, and geothermal energy, is encouraged by sustainable construction management. Construction projects can offset the carbon emissions associated with the generation of electricity based on fossil fuels by producing renewable energy on-site or obtaining it from off-site renewable energy initiatives.
- Sustainable material selection and management: Sustainable construction management involves selecting sustainable materials with lower embodied carbon and conducting lifecycle assessments to minimize environmental impacts. This includes prioritizing recycled, locally sourced, and low-carbon materials, as well as implementing efficient material management practices to reduce waste generation and landfill contributions.
- Carbon accounting and monitoring: Robust carbon accounting and monitoring procedures are part of sustainable construction management, and they allow for the tracking and measurement of carbon emissions during the course of a project. This entails calculating carbon footprints, keeping an eye on energy use, tracking emissions from building projects, and seeing chances for emission offsets and reductions.
- Carbon offset projects and initiatives: To make up for inevitable carbon emissions from construction operations, sustainable construction management may entail funding carbon offset initiatives. This can involve buying carbon credits from reputable offset suppliers, participating in energy efficiency initiatives, forestry projects, and renewable energy projects. Construction companies can lessen their carbon impact and help reduce emissions globally by sponsoring these projects.
- Stakeholder engagement and awareness: Engaging stakeholders, such as project teams, customers, suppliers, and communities, is essential to sustainable construction management because it promotes sustainable practices and increases public awareness of carbon emissions. Building projects have the power to motivate positive behavioural shifts and group action towards carbon offsetting targets by promoting an innovative and environmentally conscious culture.

7.5.9.3 Sustainable Construction Management Climate

Sustainable construction management climate is the overall atmosphere or environment within a construction organization or industry that supports and promotes sustainability principles and practices. It encompasses the attitudes, values, policies, and practices that influence decision-making and actions related to sustainability in construction projects. A positive sustainable construction management climate fosters a culture of environmental responsibility, social equity, and economic viability. It encourages the adoption of sustainable construction methods, materials, and technologies, as well as the implementation of environmental ethic throughout the project lifecycle. Primarily, a sustainable construction management climate sets the tone for how sustainability is prioritized, integrated, and implemented within construction organizations and projects. It plays a crucial role in driving the transition towards more environmentally and socially responsible construction practices. The sustainable construction management climate is need because:

- Environmental Impact: Traditional construction practices contribute significantly to environmental degradation through resource depletion, pollution, and greenhouse gas emissions. Sustainable construction management aims to minimize these impacts by implementing sustainable building materials, energy-efficient designs, and waste reduction strategies.
- Regulatory Compliance: Governments worldwide are increasingly imposing stricter regulations on construction projects to mitigate environmental impacts and promote sustainable development. Adhering to these regulations is essential for avoiding penalties and legal liabilities, making sustainable construction management a necessity.
- Social Responsibility: Sustainable construction management prioritizes the well-being of communities by minimizing disruptions, ensuring worker safety, and promoting social equity. This approach enhances the reputation of construction companies, fosters positive community relations, and contributes to long-term business success.
- Economic Benefits: While initial investments in sustainable construction management practices may be higher, they often result in long-term cost savings through reduced energy consumption, lower operating costs, and enhanced asset value. Additionally, sustainable construction projects may qualify for financial incentives, grants, or tax breaks, further incentivizing their adoption.
- Future-Proofing: Climate change and resource scarcity pose significant challenges to the construction industry. Embracing sustainable construction management practices allows companies to future-proof their operations, remain competitive in a rapidly evolving market, and adapt to changing regulatory requirements and consumer preferences.

7.5.9.4 Sustainable Construction Management Culture

Sustainable construction management culture refers to the shared values, beliefs, attitudes, and behaviors within a construction organization that prioritize and promote sustainability principles and practices. It encompasses the collective requirements and approach toward integrating environmental, social, and economic considerations into all aspects of construction project planning, implementation, and management. Economic viability, social responsibility, and environmental care are the hallmarks of a sustainable construction management culture. In order to accomplish sustainability goals and objectives, it places a strong emphasis on cooperation, innovation, and continual development. Sustainable practices are welcomed, acknowledged, and deeply embedded in the organisational DNA in this culture. Driving significant and long-lasting change towards more sustainable building methods and outcomes requires a sustainable construction management culture. It gives people and organisations the authority to decide and do things that lead to a built environment that is more socially, economically, and environmentally sustainable. The justification for fostering a sustainable construction management culture is rooted in several critical factors:

- Environmental Preservation: Traditional construction practices often result in significant environmental degradation, including resource depletion, pollution, and habitat destruction. Adopting a sustainable construction management culture is imperative for minimizing these negative impacts, preserving natural resources, and mitigating climate change.
- Regulatory Compliance: Governments worldwide are enacting increasingly stringent regulations to promote sustainable construction practices and reduce environmental harm. A sustainable construction management culture ensures compliance with these regulations, helping companies avoid fines, legal liabilities, and reputational damage.
- Market Demand: Consumers are increasingly demanding environmentally responsible products and services, including sustainable buildings. Embracing a sustainable construction management culture allows companies to meet these market demands, enhance their reputation, and attract environmentally conscious clients and investors.
- Cost Savings: While initial investments in sustainable construction practices may be higher, they often result in long-term cost savings. Energy-efficient designs, waste reduction strategies, and sustainable materials can reduce operational costs, improve resource efficiency, and increase the lifespan of buildings, ultimately enhancing profitability.
- Innovation and Competitiveness: Embracing sustainability fosters innovation and drives competitiveness in the construction industry. Companies that prioritize sustainable construction management are more likely to develop innovative solutions, stay ahead of regulatory changes, and outperform competitors in a rapidly evolving market.

- Social Responsibility: A sustainable construction management culture promotes social responsibility by prioritizing worker safety, fair labor practices, and community engagement. It contributes to positive workplace environments, fosters employee morale and loyalty, and strengthens relationships with stakeholders and local communities.

7.6 Type of Greenhouse Gas Emissions

The information shows the yearly air emissions of greenhouse gases (GHGs) for various industries between 2010 and 2021. The information emphasises how sector-specific greenhouse gas emissions are. Different variables affect different sectors, such as changes in consumption patterns, technical improvements, and economic activities. The peaks noted in a few industries in 2021 demand more research on the occurrences or modifications that led to these increases. The oscillations and patterns highlight how difficult it is to manage emissions reduction in a variety of industries. Sector-specific policies and actions may be necessary for effective mitigation efforts. The combination of wider policy measures and sector-specific policies is necessary to achieve significant and long-lasting reductions in overall emissions. As shown in Fig. 7.3, the gases that are released are carbon dioxide (43%), fluorinated gases (1.1%), greenhouse gases (50%), methane (6%) and nitrous oxide (0.80%). The breakdown of emissions by source is provided in Table 7.1.

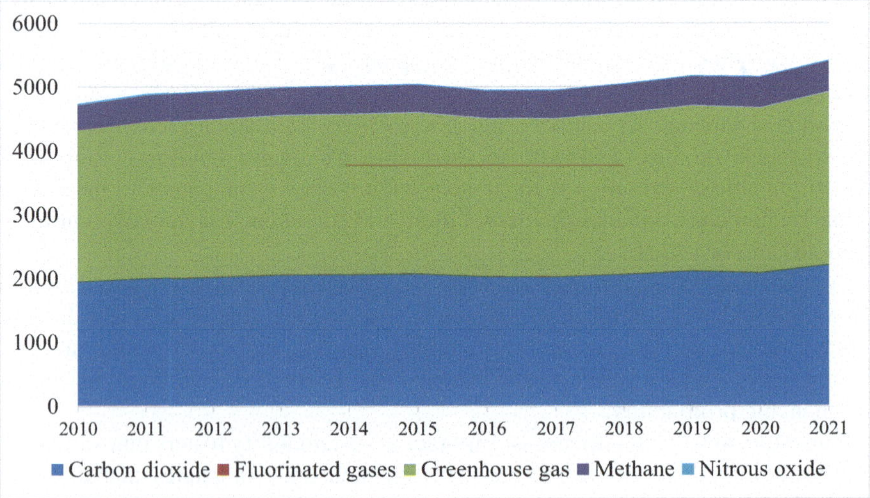

Fig. 7.3 Million metric tons of CO_2 equivalent

Table 7.1 Source greenhouse gas emissions (GHG); air emissions accounts

Year	Construction	Electricity, gas, steam and air conditioning supply	Total households	Water supply; sewerage, waste management and remediation activities
2010	4735.75	24,753.02	9868.84	4722.75
2011	4890.73	25,835.49	9927.15	4823.55
2012	4945.79	26,468.27	10,028.38	4930.14
2013	5001.68	26,897.75	10,293.87	4989.07
2014	5028.12	27,131.91	10,343.20	5080.30
2015	5054.12	26,502.99	10,443.72	5151.33
2016	4958.81	26,643.94	10,564.57	5210.35
2017	4958.94	27,232.88	10,695.57	5306.79
2018	5057.36	28,084.66	10,793.64	5419.24
2019	5181.29	27,844.76	10,748.89	5514.35
2020	5167.66	26,898.40	9825.54	5621.79
2021	5416.53	28,376.95	10,389.16	5682.36

Reference IMF (2024) https://climatedata.imf.org/pages/access-data. [Accessed on 12 February 2024]

7.6.1 Construction

The data about the construction sector shows a general upward trend in greenhouse gas emissions over time. There is a steady increase from 2010 to 2021, with a pronounced peak in 2021. Higher levels of construction activities, greater use of energy-intensive materials, and maybe an increase in total energy consumption related to construction are all probable causes of the growing emissions from the construction sector. The construction-based trends in emissions are associated with multiple factors.

Increased Construction Activity: There is a correlation between higher global construction activity levels and the reported increase in greenhouse gas emissions from the construction sector. The need for added commercial, residential, and infrastructure developments increases as economies expand and become more urbanised. Due to the growth in construction activities, there are more heavy machines, vehicles, and energy-intensive procedures used, all of which raise emissions. The growing usage of energy-intensive construction materials is a major contributing element to the growth in emissions. A significant carbon footprint is associated with certain materials, such steel and cement, because of the energy-intensive methods used in their manufacturing. Increased emissions from the construction sector are a result of the increased demand for these materials, which is often linked to rising construction activities. An increase in the total energy consumed in construction sector associated with the energy-intensive operations, such as resource extraction, transportation, on-site construction, and equipment operation, are crucial to the construction sector toward carbon management. These processes' overall energy consumption has probably increased over time, which has contributed to the trend of rising greenhouse gas emissions that has been seen. The energy requirements of heating, cooling, and powering construction sites, along with the use of large machinery, equipment, and vehicles, all contribute to the construction industry's increased carbon footprint.

Urbanisation and Infrastructure Development: The population in the world is expanding very fast, which has led to rapid urbanisation and increased demand for infrastructure development to meet the rapid growth. This cover construction roads, bridges, structures, and other essential infrastructure. Large-scale projects have the potential to increase emissions significantly from the construction sector due to their size and speed, especially if sustainable construction methods and energy-efficient technology are not extensively used. Reducing emissions from the construction sector calls for an all-encompassing strategy that considers the use of low-carbon materials, energy-efficient technology, and sustainable construction techniques. Emissions can be reduced by pushing the construction sector to use more sustainable designs, make investments in renewable energy sources, and employ more effective construction techniques. The creation and use of innovative construction materials with lower carbon footprints can also significantly lessen the sector's negative environmental effects. The environmental performance of the construction sector can be significantly influenced by governments and regulatory agencies. Active and deliberate encouragement in the use of sustainable materials, encouraging green building certifications, and implementing and enforcing energy-efficient construction rules are ways to promote good change. The construction sector's increasing trend in greenhouse gas emissions highlights the environmental problems brought on by growing urbanisation and increased construction activity. Reducing the construction sector's carbon footprint can be achieved in part by adopting technical breakthroughs and striking a balance between the need for infrastructure development and sustainable construction methods. Academics, industry stakeholders, and policymakers may work together to create and put into practice solutions that will make the building industry more ecologically sensitive and sustainable.

7.6.2 Electricity, Gas, Steam, and Air Conditioning Supply

Over time, there are variations in this industry's greenhouse gas emissions. There is a mix of gains and dips between 2010 and 2021, with 2021 marking a significant peak. Variations in this field are frequently associated with modifications in the ways that energy is consumed, modifications in the sources of energy, and more general advancements in the energy sector. Changes in the energy mix or higher energy consumption could have an impact on the peak in 2021. The variations in the emission in this industry's demonstrate how susceptible this industry is to shifts in the patterns of energy demand. Emissions can fluctuate because of changes in energy consumption brought on by changes in the economy, the seasons, and technological improvements. Variations in the energy mix utilised to produce power are often linked to variations in emissions. The amount of fossil fuels, renewable energy, and nuclear power among other energy sources can affect how much carbon is produced in generating electricity. These changes are influenced by developments in technology, government policy, and infrastructural investments in renewable energy. Variations in emissions could be a sign of more significant achievements in the energy industry, like the use of greener technology, increases in energy efficiency, and developments in energy storage. There are concerns about the causes affecting the significant peak in emissions in 2021. A likely reason would be a rise in the total amount of energy consumed during this time. An increase in greenhouse gas emissions could be caused by industrial activity, economic growth, extreme weather, or other causes that boost energy consumption.

Modifications to the energy mix could be another factor leading to the peak in 2021. The energy generation mix may be affected, affecting overall emissions, if there were variations in the price or availability of specific energy sources, or if there were problems in the supply chain. This sector adoption of renewable energy sources and technical advancements may have an effect on changes in emissions. Innovations in energy storage and grid management, along with a greater reliance on renewable energy sources, can help reduce the carbon footprint of producing power. The emissions profile of the energy industry is significantly shaped by government policies and regulations. Emissions criteria modifications, backing for renewable energy projects, and carbon price schemes can all affect how environmentally responsible the industry is. To control variations and reduce overall emissions from the sector of electricity, gas, steam and air conditioning supply, targeted mitigation techniques are essential. These can include increasing energy efficiency, increasing the use of renewable energy in the energy mix, and implementing smart grid technologies. The intricacy of the energy landscape is clear from the examination of variations in greenhouse gas emissions from the industry that supplies electricity, gas, steam and air conditioning supplies. It is essential to understand the energy consumption dynamics, changes in the energy mix, and industrial advancements to create policies and plans that will effectively move the energy sector towards a low-carbon and more sustainable one.

7.6.3 Total Households

Between 2010 and 2016, home greenhouse gas emissions have shown a comparatively constant trend. But starting from 2017, emissions have been noticeably rising, peaking in 2019, then declining in 2020 and rising once more in 2021. Changes in energy consumption patterns needed for heating and cooling, and the adoption of more energy-efficient technologies in homes can all have an impact on variations in household emissions. Relatively Stable Trend (2010–2016): The stability in home greenhouse gas emissions found between 2010 and 2016 points to a degree of consistency in emissions. During this period, housing technology, set-up energy consumption habits, and general lifestyle choices might all have an impact on this. Notable Increase (2017–2019): There appears to have been a change in the dynamics of home-level carbon footprints, as shown by the noticeable increase in household emissions that began in 2017 and peaked in 2019. This increase could be caused by several things, such as changes in home energy consumption patterns, population expansion, or an upsurge in economic activity. The peak in 2019 might be linked to higher demands for electronic gadgets, heating, and cooling, which would show higher home energy requirements.

Decline in 2020 and Increase Once More in 2021: The COVID-19 epidemic and other extraordinary events may have contributed to the 2020 emissions decline. During the epidemic, lockdowns, decreased economic activity, and lifestyle changes may have caused a brief drop in home emissions. The ensuing increase in emissions in 2021 may be attributed to the relaxation of limitations connected to the epidemic and a return to more normal patterns of energy consumption. Factors Affecting Household Emissions: Modifications in household energy usage habits have a significant impact on trends in emissions. Energy use can be affected by modifications to daily activities, adoption of innovative technology, and lifestyle changes. Weather-related variations in the demand for heating and cooling systems also add to variations in household emissions. Unusual weather patterns could cause climate control systems to use more energy. Reducing household emissions can be achieved by using energy-efficient technologies including smart home appliances, LED lights, and energy-efficient appliances. The widespread adoption of such technology may be influenced by government incentives and awareness efforts.

Urbanisation and population growth have the potential to raise the number of households and, in turn, total emissions. The environmental effect of households is influenced by the development and growth of residential areas as well as by the rise in the demand for energy services. The variations seen emphasise how crucial energy-saving measures are for homes. Long-term emission reduction depends on encouraging the use of energy-efficient appliances, supporting sustainable construction techniques, and teaching households about energy conservation. Positive trends in emissions reduction can be attributed to policy interventions, which are measures taken by the government to encourage sustainable housing, incentivize energy-efficient technologies, and set emission reduction aims for households. Raising awareness of environmental issues can bring about behavioural changes in households that result in more energy-conscious consumption. Promoting sustainable behaviours in homes can be aided by community projects and education efforts. To sum up, the examination of greenhouse gas emissions from the Total Households data shows that the environmental effects at the home level are dynamic. To encourage sustainable practices in homes and aid in the overall reduction of emissions, specific strategies, regulations, and initiatives must consider the complex interaction of factors affecting emissions.

7.6.4 Water Supply; Sewerage, Waste Management, and Remediation Activities

Over time, this industry's emissions have generally increased with some volatility. There has been a steady rise between 2010 and 2021. Increased urbanisation, changes in waste management techniques, and population growth may all be contributing factors to the rise in emissions from waste management and water supply activities. It is imperative to consider the emissions in this industry that are associated with waste as well as energy. Between 2010 and 2021, there was a consistent upward trend in greenhouse gas emissions from the wastewater, remediation, waste management, and water supply sectors. This indicates that these essential services had a greater environmental impact over that period. Although there may be some variability, the upward trends in variations point to an overall consistent increase in emissions.

There is a strong correlation between the increase in emissions from this sector and both factors. The need for waste management and water supply services rises with population growth and urbanisation, leaving a larger environmental impact. Increased waste production, infrastructure development, and construction activities can all be consequences of urbanisation and raise emissions. Modifications to waste management procedures, such as the adoption of alternative disposal techniques or improvements in waste treatment technology, may have an impact on emissions. For instance, the industry's overall emissions profile may be impacted by the switch from conventional waste disposal to more sophisticated waste-to-energy technology. Assessing the environmental efficacy of waste management techniques and pinpointing areas in need of enhancement is crucial. Emissions from this industry can be divided into two categories: emissions linked to energy and emissions related to waste. Emissions related to energy can potentially originate from energy-demanding procedures associated with waste management and water treatment. Methane released from landfills, the breakdown of organic waste, and other procedures involving the handling and disposal of waste materials are all considered waste-related emissions.

The significance of implementing sustainable practices is highlighted by the industry's influence on the environment. Crucial strategies for reducing emissions include using energy-efficient technologies in water treatment plants, encouraging recycling and waste reduction programmes, and investigating the concepts of the circular economy. The industry stands to gain from these developments in recycling and waste treatment technology. Emissions can be decreased by using creative strategies such as enhanced waste sorting technology, more effective waste-to-energy operations, and the creation of sustainable packaging materials. Government policies and regulatory frameworks have a significant impact on how this industry performs in terms of the environment. Enforcing strict environmental standards, encouraging sustainable practices, and setting emission reduction aims can all lead to satisfactory results.

Three key elements of lowering emissions from this industry include encouraging water conservation, raising knowledge of proper waste disposal, and involving communities in sustainable activities. For environmental programmes to be successful, community involvement is essential. An extensive and comprehensive strategy is required, as demonstrated by the analysis of greenhouse gas emissions from the water supply, wastewater, waste management, and remediation activities sectors. Mitigating the industry's influence on climate change and promoting a more resilient and sustainable future depends on striking a balance between the growing demand for these services and ecologically friendly, sustainable practices.

7.6.5 The Up-Front Capital Spending for the Net-Zero for Construction Projects

The construction sector is attempting to reduce its carbon emissions through energy efficiency measures, such as the use of insulation, and the replacement of fossil fuel-powered heating and cooking equipment with low-emission models. Between 2020 and 2050, the construction sector would invest, on average, $1.7 trillion annually in the physical assets needed to enable the net-zero transition [16]. By 2050, net-zero transitions driven by construction insulation retrofits might result in a net increase of roughly 500,000 direct jobs due to the decarbonisation of building projects. The biggest change the construction sector will have to make is managing the initial capital expenditures for end users to retrofit equipment and arranging incentives for various parties (such as clients who make capital investments and tenants who may benefit from lower operating costs). Consumers may eventually be able to save money on additional recurring operating expenses due to the net-zero transition's initial capital expenditures.

The market would need to evolve structurally when the comparative advantages of construction enterprises change, as the demand for low-emission construction products rises and the demand for high-emission construction goods declines. For example, enterprises or the market that develops low-emission projects would receive help from changes in their clientele. Businesses that engaged in sustainable construction practice would be more in demand. Furthermore, while the economics of operating coal plants are becoming more difficult to justify due to the declining costs of solar and wind power, the installation of new renewable generating assets still needs significant capital expenditures. When it comes to sustainable construction, "up-front capital costs" refers to the one-time expenses required during the design, construction, and operation of the project. Although implementing sustainable practices can result in long-term economic savings through reduced environmental impact and increased energy efficiency, there are significant upfront expenses involved. The following are typical initial capital expenses for sustainable buildings:

Energy-Efficient Design: Expenses related to hiring competent design teams in sustainable design methods, such as integrating energy-efficient systems, materials, and building layouts.

Green building certifications: The cost of getting certifications for green buildings, which may involve extra testing and documentation or other regional certifications.

High-Performance Building Envelope: Investing in windows, insulation, and other components of the building envelope of superior quality to increase energy efficiency and lower loads on heating and cooling systems.

Renewable Energy Systems: Installing renewable energy sources to provide on-site power, such as solar panels, wind turbines, or geothermal systems, will incur costs.

Energy-Efficient HVAC Systems: The costs associated with installing energy-efficient heating systems that are intended to use as little energy as possible.

Water Conservation Features: Expenses related to installing rainwater harvesting systems, water-efficient fixtures, and other water-saving techniques.

Sustainable Materials: Added costs for the selection and acquisition of sustainable materials, such as those with low environmental effects, quickly renewable resources, or recycled content.

Waste Management: investments made both during the construction and its whole lifecycle in waste reduction and recycling systems.

Smart Building Technologies: The expenses associated with integrating smart building technology, like advanced building automation and energy management systems, to maximise energy efficiency and boost operational effectiveness.

Green Infrastructure: The costs associated with integrating green areas, porous surfaces, and added elements that support sustainable urban planning and landscape architecture.

The long-term advantages of sustainable building include fewer environmental impacts, better occupant comfort, and cheaper operating costs, even though the initial outlay of money may be more than that of typical construction. When making decisions concerning upfront capital investments, stakeholders should consider the lifespan costs and return on investment related to sustainable construction techniques.

Some sectors may experience operating benefits from the upfront capital expenditures associated with a net-zero transition due to decreased fuel usage, enhanced material and energy efficiency, and lower maintenance costs. A substantial part of this capital investment is already profitable and cost-effective. For instance, studies examining alternative net-zero scenarios have discovered that between 40 and 50% of expenditures have a favourable investment rationale. Furthermore, technological advancement may cause net-zero technology capital expenditures to drop sooner than expected.

7.6.6 Carbon Storage Potential of Construction Projects

Construction projects can both emit and store carbon, depending on various factors such as the materials used, construction processes, and project design. Examples are provided as:

- Carbon Emissions: Construction activities frequently involve the use of heavy machinery and plants, transportation of materials and components, lighting

construction sites, and energy consumption, all of which will lead to the emission of CO2 and other greenhouse gases. To illustrate, the manufacturing of cement, one of the primary materials used in construction, is energy-intensive and releases significant CO2 emissions. The manufacturing of steel, and glass often involves processes that release carbon dioxide into the atmosphere.
- Carbon Sequestration: However, many construction materials have the capacity to store CO_2, effectively acting as carbon sinks. For instance, timber used in construction retains the carbon absorbed during the tree's growth, thereby reducing net carbon emissions. Additionally, innovative materials like carbon capture and storage (CCS) concrete can actively sequester CO_2 during the curing process. Furthermore, the incorporation of green infrastructure elements such as green roofs and living walls can enhance carbon sequestration by promoting the growth of vegetation.
- Green Building Practices: Sustainable construction practices aim to minimize carbon emissions and enhance carbon sequestration. Strategies such as energy-efficient design, use of renewable materials, and incorporation of green spaces can mitigate the carbon footprint of construction projects while promoting carbon storage.
- Regulatory Frameworks and Certifications: Governments and regulatory bodies increasingly enforce regulations aimed at reducing carbon emissions in construction projects.

While construction projects typically contribute to carbon emissions, adopting sustainable construction management practices can help mitigate environmental impacts and promote carbon neutrality in the built environment. Understanding the dynamics of carbon emissions and storage in construction is crucial for promoting sustainability and mitigating the environmental impact of built infrastructure. By adopting sustainable practices, utilizing carbon-storing materials, and adhering to regulatory standards, construction projects can contribute to carbon neutrality and environmental stewardship.

7.7 Contextualising the Sources of Emission into Sustainable Construction

7.7.1 Construction Sector

The emissions from the construction sector show a steady increase over time. This implies that the construction sector has been a contributing factor to the total emissions of greenhouse gases. The construction sector's emissions data show a consistent rising trend between 2010 and 2021. This steady rise shows that construction operations are having an increasing negative influence on the environment.

Factors Contributing to Emissions—The construction sector's emissions, which are shown in the table above, have been trending upward over time. This shows that a part of the greenhouse gas emissions may have come from the construction sector. By encouraging energy-efficient construction designs, the use of environmentally compliant materials, and the use of construction methods that reduce carbon footprints, sustainable construction practices seek to improve the environmental impact of the construction industry. From 2010 to 2021, there was a consistent increase trend in the emissions data for the building industry. This steady rise denotes an increasing environmental cost linked to building operations. To effectively decrease the sector's contribution to overall greenhouse gas emissions, it is imperative to understand the underlying drivers that are driving this trend.

Construction Sector's Role in Overall Greenhouse Gas Emissions—The construction sector has a noteworthy impact on the total amount of greenhouse gas emissions released into the atmosphere. The sector plays a significant role in driving economic expansion and meeting the demand for new buildings, infrastructure, and residences. Acknowledging this contribution highlights the need for proactive measures to reduce the environmental impact of construction.

Sustainable construction procedures are designed to mitigate the adverse environmental impacts that are commonly linked with conventional construction processes. These methods emphasize cutting emissions, conserving resources, and building social and environmental resilience over the long run. Table 7.2 holds the main aspects of sustainable construction.

Table 7.2 Key elements of sustainable construction

Key	Explanation
Energy-efficient building designs	Giving top priority to designs that minimise energy use, make use of natural lighting, and integrate energy-saving HVAC systems
Use of sustainable materials	Reducing the embodied carbon in building projects can be achieved by selecting materials with smaller carbon footprints, such as recycled or sustainably sourced materials
Carbon footprint reduction	implementing methods and strategies for building that reduce carbon footprints, like modular construction, prefabrication, and effective waste management
Life cycle assessment	LCAs are used in sustainable construction techniques among other purposes to assess a building's environmental impact throughout its whole life cycle. This involves estimating the environmental impact of extracting raw materials, manufacturing, building, operating, maintaining, and finally recycling or demolishing a structure
Green building certification systems	Sustainable construction standards are set by the green building certification programmes that have been developed in several regions. Meeting standards for indoor air quality, material choice, water conservation, energy efficiency, and overall environmental performance is necessary to obtain certification

(continued)

Table 7.2 (continued)

Key	Explanation
Technological innovations	Technological improvements that improve efficiency and lessen environmental effects are embraced by sustainable construction. This covers the application of smart construction technology, building information modelling, and innovative building materials with enhanced sustainability profiles
Regulatory compliance and policy support	Governments and oversight organisations are essential in advancing environmentally friendly building practices. It is essential to implement and enforce building codes that prioritise environmental performance, waste reduction, and energy efficiency. The adoption of sustainable techniques is further accelerated by incentives and subsidies for sustainable construction methods
Stakeholder collaboration	A variety of stakeholders, including communities, legislators, builders, developers, and architects, work together to implement sustainable construction. A shared commitment to minimising the environmental impact of building projects is fostered when stakeholders are involved in the adoption of sustainable methods
Education and awareness	It is crucial to inform experts and the public about the advantages and methods of sustainable building. Growing public awareness stimulates the market for sustainable buildings and pushes the construction sector to adopt sustainable practices

Acknowledging how sustainable construction affects greenhouse gas emissions is an essential first step in guiding the sector towards sustainability. The use of sustainable construction techniques can help reduce environmental effects while satisfying the increasing need for infrastructure and development. The construction sector may contribute to the creation of a more resilient and sustainable built environment by using sustainable materials, energy-efficient designs, and creative construction techniques.

7.7.2 Electricity, Gas, Steam, and Air Conditioning Supply

The data about the of electricity, gas, steam and air conditioning supply industry shows variations in emissions, with a noteworthy maximum in 2021. Since this industry provides the energy for construction activities, its emissions are significant when it comes to sustainable construction. Using smart grid technology, switching to renewable energy sources, and enhancing building energy efficiency are all ways that the construction sector may adopt more sustainable practices. Table 7.3 shows the aspects of the gas, steam, electricity, and air conditioning supply industry to sustainable construction.

Table 7.3 Aspects of the gas, steam, electricity, and air conditioning supply industry to the sustainable construction

Aspect	Explanation
Fluctuations and notable peak	The information shows variations in emissions from the industry that supplies gas, steam, electricity and air conditioning, with a significant peak in 2021. Variations can be ascribed to several things, including modifications to patterns of energy consumption, changes in the sources of energy, and more general advancements in the energy sector
Critical role in sustainable construction	A key part of sustainable construction is the supply of electricity, gas, steam and air conditioning. It serves as the main energy source for structures, construction-related operations, and other industry-related processes. Promoting sustainability in the built environment requires an understanding of and commitment to resolving the environmental effect of this sector
Energy supply for buildings and construction	The construction sector is strongly dependent on electricity and other energy sources, and buildings are major energy consumers. The goal of sustainable construction methods is to minimise a building's carbon footprint through energy-efficient design, increased energy consumption, and the switch to cleaner energy sources
Transitioning to renewable energy sources	Using renewable energy sources is a crucial tactic in sustainable construction to reduce the environmental impact of gas, steam, electricity and air conditioning supplies. This means increasing the share of renewable energy sources like wind, solar, and hydropower in the energy supply mix. Renewable energy sources contribute to a reduction in greenhouse gas emissions from electricity production because they are inherently cleaner
Improving energy efficiency in buildings	Energy efficiency in buildings is prioritised in sustainable construction techniques. This entails integrating improved ventilation and insulation systems, utilising energy-efficient technologies, and designing buildings to maximise natural lighting. Buildings' reliance on energy supplied by the relevant industry is minimised by lowering their energy demand
Smart grid technologies	The implementation of smart grid technology is crucial in optimising the effectiveness of energy distribution. Real-time energy use optimisation, control, and monitoring are made possible by these technologies. Smart grids support a more responsive and robust energy infrastructure, which eases greater integration of renewable energy sources and enhanced building energy management in the context of sustainable architecture
Demand-side management	Demand-side management, sometimes known as active management of energy demand, is a part of sustainable construction. This covers tactics including demand response plans, time-of-use pricing, and peak load management. These actions support efficient energy use, lessen grid strain, and balance energy demand

(continued)

Table 7.3 (continued)

Aspect	Explanation
Community microgrids	Community microgrids are becoming a practical possibility, particularly for self-sufficient building projects. These microgrids, which are often fueled by renewable energy sources, offer a dependable and localised energy supply for towns or building sites
Energy storage solutions	The construction sector can store and use renewable energy more effectively by including energy storage technologies, such as batteries. This keeps the energy supply stable and offers backup power during times of high demand or grid failures
Policy support for clean energy	The shift to greener energy sources is mostly driven by government regulations and incentives. Subsidies, legal frameworks that promote clean energy investments, and support for renewable energy projects all help to create a sustainable energy sector that complements the aims of the building sector

Sustainable construction needs to address the emissions and volatility from the Gas, Steam, Electricity, and Air Conditioning Supply industry. By adopting smart grid technology, increasing building energy efficiency, and embracing renewable energy, the construction sector can lessen its environmental footprint and help create a more robust and sustainable energy infrastructure. This shift towards a more sustainable construction industry requires cooperation among stakeholders, innovative technologies, and supportive legislation.

7.7.3 Total Households

Household greenhouse gas emissions fluctuate with time, showing a discernible rise from 2017 to 2019, a decline in 2020, and a rise once more in 2021. Beyond the construction site, sustainable construction considers the environmental impact of buildings at every stage of their lives. Essential elements of sustainable construction include boosting sustainable construction designs, encouraging energy-efficient behaviours in homes, and improving the energy performance of residential buildings. Table 7.4 shows the contributions of the household industry to sustainable construction.

To sum up, residential greenhouse gas emissions are essential to the larger idea of sustainable construction. The construction sector can foster a more sustainable built environment through the promotion of sustainable construction designs, the encouragement of energy-efficient activities, and the improvement of energy performance in residential structures. The integrated approach considers the continuous effects of buildings throughout their existence in addition to the development phase.

Table 7.4 Aspects of household industry to the sustainable construction

Aspect	Explanation
Variability in household emissions	The information shows a fluctuating pattern in residential greenhouse gas emissions throughout time. Interestingly, there was a surge in 2021 after a decline in 2020 and an increase in 2019 and 2020. Gaining knowledge of these variances helps to understand the dynamics of environmental effects at the home level
Extended scope of sustainable construction	The concept of sustainable construction goes beyond the construction site to include a building's whole existence, including its operational phase. Acknowledging the ecological influence of households is consistent with the more comprehensive viewpoint of sustainable construction, stressing the significance of sustainable methods even after construction is completed
Encouraging energy-efficient practices	To build sustainably, households must be encouraged to adopt energy-efficient behaviours. This entails encouraging the use of energy-saving equipment, integrating smart home technology, and putting energy-saving measures into action. Providing homeowners with information about energy-efficient options helps to reduce the carbon footprint that comes with residential buildings
Sustainable building designs	One of the core components of sustainable construction is the integration of sustainable building designs into residential structures. This includes using environmentally friendly materials, planning areas that are insulated and light- and ventilation-efficient, and orienting buildings to maximise natural light and ventilation. During a building's operation, sustainable building designs help to lower energy use
Enhancing energy performance	Improving residential buildings' energy efficiency is one of the main aims of sustainable building. This entails putting policies in place like better insulation, energy-efficient HVAC systems, and the production of electricity from renewable resources. The goal of these programmes is to reduce household energy use and related greenhouse gas emissions
Lifecycle considerations	Sustainable construction places a strong emphasis on a lifecycle approach, considering the environmental effects of building, deconstruction, and reuse. Residential building lifecycle issues encompass not only the building materials and techniques but also the energy consumption throughout operation, upkeep procedures and final end-of-life concerns
Behavioural changes and education	A key part of sustainable construction is encouraging homeowners to adopt new behaviours. Residents can be empowered to embrace sustainable habits by being informed about the environmental impact of their daily choices through education and awareness initiatives. This could involve adopting practices for waste minimization, water conservation, and energy conservation
Renewable energy integration	Including renewable energy sources in residential buildings advances the goals of sustainable building. To generate power on-site, this can entail installing solar panels, wind turbines, or other renewable energy technology. It is possible for homeowners to actively take part in the production of clean energy

(continued)

Table 7.4 (continued)

Aspect	Explanation
Government incentives and regulations	Encouraging sustainable practices among households is greatly aided by government rules and incentives. Environmentally conscious decisions are encouraged in homes by policies like tax credits for energy-efficient improvements, building rules that emphasise sustainability, and incentives for the use of renewable energy
Smart home technologies	Adoption of smart home devices improves efficiency and energy management. Energy-monitoring systems, home automation, and smart thermostats all help to maximise energy use, cut down on waste, and create more sustainable living spaces
Community engagement	To promote a shared commitment to living sustainably, sustainable construction also involves community engagement. Part of the shared duty for sustainability is the work of neighbourhood programmes, community initiatives, and resident cooperation

7.7.4 *Water Supply; Sewerage, Waste Management, and Remediation Activities*

There is a general increased tendency in the emissions from waste management, remediation, sewage, and water supply. Waste reduction and water management are essential components of sustainable construction management practices. Reducing the environmental impact of water supply and waste activities can be achieved by practices including rainwater collecting, water recycling, and sustainable waste management. See Table 7.5 for the explanation.

Table 7.5 Measure to reduce the environmental impact of water supply and waste activities

Measure	Explanation
General upward trend in emissions	The information shows that greenhouse gas emissions from waste management, sewage treatment, water supply, and remediation operations are generally on the rise. To create sustainable construction strategies that consider the environmental impact of these vital services, it is imperative to understand the forces that are driving this trend
Critical role of water and waste management in sustainable construction	Two essential elements of sustainable construction are waste management and water use. In addition to having a positive environmental impact, effective and sustainable practices in these sectors support resource conservation and climate change resistance
Water management strategies	Sustainable construction places a strong emphasis on ethical methods to water management. This entails setting up rainwater harvesting systems, which gather and store rainwater for a range of uses, reducing reliance on traditional water sources. Projects that recycle water also help to preserve a more sustainable water supply by lowering water waste

(continued)

Table 7.5 (continued)

Measure	Explanation
Rainwater harvesting	Rainwater harvesting is the technique of collecting and storing rainwater from rooftops and other surfaces for use at a later time. Because they help preserve water resources, ease the burden on municipal water supplies, and minimise the environmental impact of water extraction, rainwater harvesting systems are a useful tool in sustainable construction
Water recycling	Reusing wastewater for non-potable uses, such as irrigation, industrial activities, or even toilet flushing, is known as water recycling. The entire water footprint and the environmental impact of wastewater disposal are reduced when water recycling systems are implemented in construction projects
Sustainable waste management	Conscious waste management is a key part of sustainable construction. This entails implementing policies that reduce waste production, encourage recycling, and guarantee the correct disposal of waste from construction and demolition projects. Emissions from waste breakdowns in landfills are decreased by using sustainable waste management practices
Circular economy principles	A circular economy is a fundamental part of sustainable building practices. Adopting the concepts of the circular economy entails planning infrastructure and buildings to maximise material reuse, recycling, and repurposing. Circular economy techniques aim to reduce the environmental impact of resource extraction and waste generation in the context of water supply and waste activities
Green infrastructure	Sustainable water management is aided by green infrastructure initiatives, such as the addition of green roofs and permeable surfaces to building projects. These components enhance the overall environmental performance of buildings and infrastructure, reduce stormwater runoff, and increase the quality of the water
Community awareness and participation	Initiatives for sustainable waste and water management depend heavily on community involvement and awareness. Public awareness campaigns, community engagement initiatives, and educational programmes can encourage people to adopt water-saving habits and contribute to waste reduction initiatives
Technological innovations in remediation	Technological advancements that reduce their negative effects on the environment might be helpful for remediation projects, such as the cleaning of contaminated sites. Using environmentally friendly, energy-efficient, and successful remediation solutions is a part of sustainable construction practices
Policy support and regulations	Sustainable techniques in waste management, water supply, sewage, and remediation are greatly influenced by government laws and regulations. The overall sustainability of these services is enhanced by regulatory frameworks that promote the adoption of environmentally friendly technologies, set up criteria for water conservation, and offer incentives for sustainable activities
Monitoring and assessment	Evaluating the environmental impact of water and waste management systems requires regular monitoring and assessment. Continuous improvement via data-driven decision-making is a key part of sustainable building, guaranteeing that procedures meet environmental goals

Addressing the general upward trend in emissions from waste management, remediation, sewage, and water supply is necessary for sustainable construction. Through the implementation of waste reduction and water management strategies, adopting circular economy concepts, and engaging the community, the construction sector may contribute to the creation of a more resilient and sustainable built environment. Techniques for managing water and waste sustainably increase resource efficiency, lessen environmental harm, and guarantee the long-term financial success of construction projects.

7.8 Challenges for Achieving Net Zero Carbon in the Built Environment

To fulfil the ambitious carbon reduction targets, various technologies and behavioural practices must be widely used and advanced. The adoption of sustainable technologies and practices typically entails substantial upfront expenses in the shift to net zero emissions. International collaboration and coordination are necessary to achieve worldwide construction net zero carbon emissions. Given the interconnectedness of climate change, cooperation is necessary to solve the disparate problem that different regions face. It is a difficult undertaking to promote behavioural changes in both individuals and organisations. Therefore, comprehensive education, rewards, and a societal shift towards placing a higher priority on environmental stewardship are all necessary to change deeply rooted behaviours and habits towards sustainability. Net zero carbon involves the comprehensive and aggressive strategy for reducing the impact of climate change. It entails a complex plan that includes carbon offsetting, renewable energy adoption, emission reduction, technology advancement, and regulatory aid. Considering the pressing need to address climate change, the idea of net zero carbon serves as a guide for the transition to a future that is more robust and sustainable. Reaching net-zero carbon in the built environment is a challenging task that calls for overcoming several obstacles. Table 7.6 explains the primary obstacles to reaching net-zero carbon in the built environment.

A complex strategy including financial incentives, supportive regulations, public awareness campaigns, technology innovation, and stakeholder participation is needed to overcome these obstacles. The built environment's net-zero carbon aim will not advance without ongoing efforts to solve these issues.

Table 7.6 Constraints facing the achievement of net zero carbon in the built environment

Constraint	Explanation
Existing building stock	Most buildings in the built environment are already existing constructions. It can be difficult to retrofit them to meet net-zero carbon criteria because of obsolete systems, architectural constraints, and the cost of changes
Cost implications	Putting into effect sustainable technology and methods, like renewable energy sources and energy-efficient systems, frequently entails hefty upfront expenses. Limited funds could prevent widespread implementation
Technological limitations	A few technologies that are essential to reaching net-zero carbon emissions are still in the early phases of research and development. Advancement may be hampered by the absence of tested solutions or mature technologies
Supply chain challenges	There may be restrictions on the cost and availability of sustainable building materials. One limitation may be the absence of a reliable supply chain for these materials
Regulatory hurdles	Different regions have different building codes and regulations about sustainability and carbon emissions, which makes it difficult to adopt net-zero carbon standards uniformly and creates inconsistent compliance requirements
Land use and urban planning	Expansive urban regions present greater challenges to achieving net-zero carbon targets. The amount of land that can be used for green spaces and renewable energy installations may be limited in densely populated areas
Behavioural and cultural factors	Convincing stakeholders to embrace new sustainable habits and altering long-standing traditions can be difficult. Progress may be hampered by opposition to change in the construction community and within the sector
Lifecycle assessments	Building lifecycle evaluations need a comprehensive strategy and specialised knowledge. One limitation may be a lack of knowledge or awareness about the long-term effects of the materials and methods used in construction
Energy transition and grid decarbonization	Switching to renewable energy sources is often necessary to get net-zero carbon. The viability of net-zero buildings may depend on the rate at which energy networks are becoming less carbon-intensive and the availability of renewable energy sources
Education and awareness	The widespread adoption of sustainable practices may be hampered by architects, builders, and consumers' ignorance about net-zero carbon concepts and technologies
Financing and investment	It can be difficult to secure funding for sustainable building projects, particularly for smaller businesses or initiatives. Because there are no tested financial models or because they perceive dangers, investors may be apprehensive
Data and monitoring challenges	It might be technically difficult to implement efficient data gathering and monitoring systems to evaluate and confirm carbon emissions throughout a building's existence

(continued)

Table 7.6 (continued)

Constraint	Explanation
Policy and political will	Investors and developers may become uneasy if net-zero carbon targets are not supported by clear and consistent policies. An environment of stable policies is essential for long-term planning
Interdisciplinary collaboration	Collaboration between several academic fields, such as government, engineering, architecture, and urban planning, is necessary to meet net-zero carbon targets. Lack of communication across various disciplines can impede advancement

7.9 Policies and Regulations for Net Zero in the Built Environment

Effective rules and regulations are necessary to promote the adoption of net zero carbon in the built environment; without them, it may be challenging to do so, particularly given the fragmentation within the building industry. In essence, the rules and regulations would support the establishment of a framework that directs the parties involved in the built environment and activities, promotes innovation, and guarantees openness and accountability. In the built environment, policies and laws are key drivers of the shift towards Net Zero Carbon. Table 7.7 lists the justifications for the policies and laws needed to promote the net zero transition:

Table 7.7 Reasons for policies drive net zero for sustainable construction

Reason	Explanation
Setting clear standards and targets	To reduce carbon emissions in the built environment, policies set explicit goals and requirements. Assuring a uniform and quantifiable strategy to reaching Net Zero Carbon, these guidelines offer a framework for the industry
Creating a level playing field	Guarantees that all parties involved in the built environment follow the same set of norms, and regulations help to level the playing field. As a result, unfair competition is avoided, and wider adoption of sustainable practices is encouraged
Providing confidence and predictability	Investments and enterprises can benefit from policies' predictability and clarity. Informed decision-making, future planning, and investment in practices and technology that support Net Zero aims are all eased by industry players' awareness of the regulatory environment
Encouraging innovation and research	Regulations foster innovation by setting up ambitious standards and promoting the creation of novel techniques and technology. This promotes research and development in sustainable construction methods and cultivates a culture of continual improvement
Ensuring accountability	Policies create accountability by delineating the roles and obligations of various parties involved, such as local government bodies, builders, and developers. Ensuring that non-compliance is addressed and progress towards Net Zero is tracked is made possible by explicit accountability procedures

(continued)

Table 7.7 (continued)

Reason	Explanation
Incentivizing compliance	Policies may include inducements to encourage companies and individuals to implement Net Zero activities, such as tax credits, grants, or subsidies. Sustainable building is becoming more appealing and economically workable thanks to these financial incentives
Promoting consistency in planning	Consistency in urban development and planning is ensured by policies. They offer guidance to local government bodies on how to integrate renewable energy sources, encourage compact urban planning, and build sustainable communities
Addressing market failures	Policies that internalise carbon emission-related externalities can address market failures. Carbon pricing methods, including carbon taxes and cap-and-trade schemes, serve to accurately stand for the environmental cost of carbon emissions
Facilitating long-term planning	Is encouraged by stable regulatory environments that are provided by long-term policies. This is especially crucial for the construction sector, where projects often have long schedules, as it enables stakeholders to apply Net Zero initiatives from the outset
Creating a culture of compliance	Building a culture of conformity with sustainability goals is eased by regulations. As Net Zero becomes a legal mandate, it permeates industry procedures and promotes a group commitment to environmentally friendly building techniques
Supporting international commitments	Policies are in line with global accords and obligations regarding climate change. To ensure that local efforts support larger initiatives, governments often enact rules that demonstrate their commitment to the global climate goals
Enhancing public awareness	Regulations raise awareness among the public by highlighting the significance of Net Zero in the built environment. Customers, companies, and the public are all affected by this awareness, which promotes accountability and sustainable decision-making
Catalyzing market transformation	By generating demand for sustainable technology and materials, regulations have the potential to serve as catalysts for market transformation. Consequently, this fosters creativity and the growth of a strong market for Net Zero solutions
Adapting to climate change	Provisions for climate change adaptation may be included in regulations. To endure extreme weather events and contribute to the long-term sustainability of the built environment, may entail resilient building design requirements

Stakeholders must understand the necessity of rules and regulations, as they play a pivotal role in steering, motivating, and implementing the requisite modifications to attain Net Zero Carbon in the built environment. To link actions with more general sustainability goals and ensure a resilient and low-carbon future, the policies and regulations offer the basis for a systematic and coordinated effort across the construction sector.

7.10 Suitable Policies and Regulations for Net Zero Carbon in Sustainable Construction

A thorough set of rules and guidelines is also necessary to achieve Net Zero Carbon in the built environment for sustainable construction. Carefully crafted and executed, these laws and rules have the potential to foster the broad implementation of Net Zero Carbon practices in the built environment, supporting international efforts to slow down global warming. A few current policies and regulations are shown in Table 7.8.

Table 7.8 Policies and regulations for Net Zero Carbon in sustainable construction

Policies and regulations	Explanation
Building energy codes and standards	Establish and promote building energy standards and guidelines that impose stringent restrictions on energy efficiency and promote the use of renewable energy sources
Net zero energy building mandates	Pass laws requiring new construction to either produce as much renewable energy as they consume or reach Net Zero Energy status
Renewable energy integration requirements	Establish rules requiring the inclusion of renewable energy technologies in newly constructed buildings, such as wind turbines or solar panels
Carbon pricing mechanisms	Use carbon pricing mechanisms like carbon taxes or cap-and-trade programmes to internalise the cost of carbon emissions and promote the reduction of carbon footprints
Green building certification programs	Promote, or require, involvement in green building certification schemes that prioritise Net Zero Carbon aims and environmentally friendly building techniques
Mandatory building performance disclosure	Encouraging transparency and increases in energy efficiency, mandate building owners to reveal their properties' energy performance
Energy-efficient retrofit incentives	Building owners that undertake energy-efficient retrofits should be given grants, tax credits, or other financial incentives, with the goal being Net Zero Carbon
Electrification mandates	Establish laws or other incentives to promote building electrification, especially to move away from fossil fuels for heating and air conditioning systems
Circular economy and embodied carbon regulations	To reduce waste, including laws that address the embodied carbon in building materials and advance the ideas of the circular economy

(continued)

Table 7.8 (continued)

Policies and regulations	Explanation
Smart infrastructure requirements	Require the integration of smart building technology and infrastructure to allow for the optimisation, management, and real-time monitoring of energy use
Community planning for sustainability	To decrease the need for energy-intensive commuting, policies promoting sustainable urban planning should be developed. These should include mixed-use zoning, walkability, and access to public transportation
Government procurement policies	Set up procedures for government projects' procurement that give priority to Net Zero Carbon buildings, setting an exemplary model and increasing demand for environmentally friendly building practices
Research and development funding	Provide funds for the advancement of sustainable building technology to promote creativity and the creation of fresh approaches
Energy performance contracting support	To achieve energy efficiency and carbon reduction, building owners should be encouraged to work with energy service firms through the provision of incentives and support for energy performance contracts
Education and training requirements	Establish requirements for education and training for professionals working in the construction sector to guarantee they have the knowledge needed to build sustainable structures
Public–private partnerships	Encourage public–private partnerships that serve as a means of easing cooperation between non-profits, businesses, and government agencies in pursuit of Net Zero Carbon goals
Land use and zoning regulations	Include zoning and land use laws that promote environmentally friendly building techniques, such as those that mandate green areas, low-impact development, and energy-efficient building layouts
Resilience standards	Establish resilience requirements to guarantee that buildings are made to withstand the effects of climate change, thereby promoting sustainability over the long run
Monitoring, reporting, and verification systems	To improve accountability and transparency, set up monitoring, reporting, and verification mechanisms to track the development of projects and buildings towards Net Zero Carbon targets
Financial incentives for green financing	Promote green finance by offering rewards and aid to lenders who grant favourable conditions for environmentally friendly building projects

7.11 Metrics to Measure Net Zero Carbon Within the Built Environment in Sustainable Construction

Metrics are needed to decide whether the goal of net zero carbon has been reached or not. It is crucial to know if the measures are successful, to be aware of the degree of success. Several measures and indicators can be applied to assess and measure the effectiveness of Net Zero Carbon-oriented sustainable construction projects. To calculate the total carbon footprint of a building or development project, several factors must be considered when measuring net zero carbon within the built environment. Customising the measurements, however, can be required depending on the unique aims, circumstances, and features of any sustainable construction project. For instance, it might be needed to decide whether it's for new construction, renovation, refurbishment, or conversion. An all-encompassing perspective that considers operational and embodied carbon in addition to other sustainability aspects offers a thorough grasp of how a building affects the environment and society. Table 7.9 provides some of the relevant metrics.

Table 7.9 Metrics to measure net zero carbon within the built environment

Metrics	Explanation
Embodied carbon	Calculates the overall amount of carbon emissions resulting from the manufacture, shipping, and installation of building materials. This involves deciding the project's material's carbon impact
Operational carbon	Calculate the carbon emissions that come from the building's daily operations, such as the energy used for appliances, lighting, heating, and cooling
Renewable energy generation	Determine the amount of energy generated on-site from sustainable resources, such as solar or wind power, in order to either equal or exceed the building's operating energy consumption
Energy efficiency	Decide the energy consumption intensity, which is the total amount of energy used per structure per square foot. Higher energy efficiency is indicated by lower values of Energy Use Intensity
Water efficiency	Assess the building's water usage efficiency, considering both indoor and outdoor water usage, and put water-saving measures and technologies into place
Waste generation and management	Keep an eye on the quantity of waste produced throughout construction and operation, as well as the rates at which waste is recycled and diverted from landfills
Building life cycle assessment	A comprehensive life cycle assessment should be carried out to identify the effects of building materials on the environment, from extraction and production to recycling or disposal

(continued)

Table 7.9 (continued)

Metrics	Explanation
Transportation emissions	Evaluate the transportation-related carbon emissions, encompassing the commute to and from the building, and advocate for environmentally friendly modes of transportation
Carbon offsetting	To make up for carbon emissions that cannot be completely eradicated, think about using carbon offset credits to fund carbon removal or reduction initiatives in other places
Resilience measures	Evaluate the building's resistance to the effects of climate change, such as extreme weather events, and put resilience-boosting measures into place
Indoor environmental quality	Assess and preserve good indoor air quality with non-toxic materials, efficient ventilation systems, and proper humidity management
Biodiversity and green spaces	Incorporate green areas and evaluate the effects on regional biodiversity to foster ecological harmony in the constructed surroundings
Adaptability and flexibility	To minimise the necessity for major renovations or demolitions, consider how flexible and adaptive the building design is to future changes
Social impact	Evaluate the project's social impact by considering things like the generation of jobs, community involvement, and enhancements to the standard of living in the area
Occupant satisfaction	Get input from building tenants to find out how comfortable they are inside and how their experience was overall
Innovation and technology adoption	Monitor the uptake of innovative building techniques and technology that support Net Zero and sustainability aims
Financial metrics	Analyse sustainable building projects' financial success by calculating their long-term return on investment
Carbon reporting and certification	For transparent reporting and verification of carbon reduction activities, use accredited frameworks and certifications like PAS 2080 or the Carbon Trust Standard

7.12 Metrics to Monitor Net Zero Carbon Within the Built Environment in Sustainable Construction

To evaluate the sustainability performance of the construction projects, tracking the different metrics is necessary for Net Zero Carbon monitoring in the built environment. Table 7.10 lists the parameters to monitor the achievements of Net Zero Carbon in sustainable construction.

By keeping an eye on these indicators, stakeholders can gain a thorough grasp of a building's sustainability performance, which facilitates decision-making, progress towards Net Zero Carbon targets, and ongoing improvement of sustainability practices in the built environment.

Table 7.10 Metrics to monitor net zero carbon within the built environment

Metrics	Explanation
Embodied carbon metrics	Calculate all the carbon emissions that are related to the manufacture, delivery, and assembly of building materials. This includes deciding how much of the project's materials have a carbon footprint
Operational carbon metrics	Decide the building's carbon intensity by calculating the energy used for appliances, lighting, heating, and cooling. The unit of measurement for this metric is often kilogrammes of CO2 equivalent per square metre (kgCO2e/m2) per year
Renewable energy generation metrics	Track the percentage of overall energy consumption that comes from renewable sources and keep an eye on how much renewable energy is produced locally, such as solar or wind power
Energy efficiency metrics	Decide the building area's energy consumption intensity, which is the total energy used per square foot or square metre. Higher energy efficiency is shown by lower values of Energy Use Intensity
Water efficiency metrics	Analyse the building's water use efficiency, considering both indoor and outdoor water usage. Monitor water consumption and put water-saving strategies and technologies into action
Waste generation and management metrics	Keep an eye on the quantity of waste produced throughout construction and operation, as well as the rates at which waste is recycled and diverted from landfills
Transportation emissions metrics	Calculate the transportation-related carbon emissions, such as the emissions from getting to and from the building. Encourage sustainable modes of transportation to cut down on carbon emissions
Building life cycle assessment	To examine the environmental impact of building materials from extraction and manufacture to disposal or recycling, conduct and update a life cycle assessment
Carbon offset metrics	To make up for carbon emissions that cannot be completely eradicated, think about using carbon offset credits to fund carbon removal or reduction initiatives in other places
Resilience metrics	Evaluate the building's ability to withstand the effects of climate change, including harsh weather. Create and monitor a resilience index to assess the building's ability to bear down on and bounce back from shocks
Indoor environmental quality metrics	High indoor air quality can be monitored and maintained with the help of efficient ventilation systems, non-toxic materials, and proper humidity regulation. To guarantee a healthy indoor environment, monitor IAQ parameters
Biodiversity and green spaces metrics	Consider the effects on regional biodiversity while integrating green spaces. Formulate and monitor a biodiversity index to assess the ecological equilibrium in the constructed surroundings
Adaptability and flexibility metrics	Consider how flexible and adaptive the building's design is to allow for future modifications. Create a score for adaptability and monitor it to see how well the building changes over time

(continued)

Table 7.10 (continued)

Metrics	Explanation
Social impact metrics	Evaluate the project's social impact by considering things like the generation of jobs, community involvement, and enhancements to the standard of living in the area. Create and monitor social impact metrics
Occupant satisfaction metrics	Get input from building tenants to find out how comfortable they are inside and how their experience was overall. Create satisfaction metrics and monitor them
Innovation and technology adoption metrics	Monitor the uptake of innovative building techniques and technology that support Net Zero and sustainability objectives. Create and monitor an innovation index to gauge advancement
Financial metrics	Analyse sustainable building projects' financial success by calculating their long-term return on investment. Keep an eye on the financial indicators that show how cost-effective sustainability initiatives are
Carbon reporting and certification metrics	For transparent reporting and verification of carbon reduction activities, use accredited frameworks and certifications like PAS 2080 or the Carbon Trust Standard. Verify that carbon reporting is correct

7.13 Measure to Achieve Energy Efficiency in Sustainable Construction

As was previously explained, producing electricity involves significant carbon emissions, which influence how energy-efficient construction projects and operations are. In the context of sustainable construction, achieving energy efficiency in construction needs a comprehensive strategy that incorporates design, technology, materials, and operational procedures. It is possible to achieve energy efficiency in construction at every stage of the construction process. The possibility to increase energy efficiency from material manufacturing to the operation of built facilities lies in the goal of decarbonising the built environment. The list of strategies that the construction sector, clients, and users can utilise to achieve energy efficiency are contained in Table 7.11.

Stakeholders may greatly aid in the advancement of sustainable construction aims and the achievement of building energy efficiency by combining these tactics and customising them to the unique circumstances of a construction project.

Table 7.11 Strategies to achieve energy efficiency in buildings

Strategy	Explanation
Passive design principles	A building's orientation should be optimised to receive as much natural light and heat as possible. Consider elements like site topography, sun angles, and prevailing winds. To provide a well-insulated building envelope, use high-quality insulation materials to reduce heat transmission through the walls, roof, and floors
High-performance windows and glazing	To minimise heat gain or loss, install energy-efficient windows with thermal breaks and low-emissivity coatings Consider windows with double or triple glazing for better soundproofing and insulation
Energy-efficient lighting	Use LED lighting, which consumes less energy, lasts longer, and generates better-quality illumination To maximise natural light and reduce the need for artificial lighting during the day, implement daylighting strategies
Energy-efficient HVAC systems	Install air conditioning, ventilation, and heating systems that use less energy. Use programmable thermostat controls and zoning to optimise HVAC usage according to building occupancy and individual areas
Renewable energy integration	To generate sustainable and clean power, use on-site renewable energy resources like solar or wind turbines Consider your options for connecting to the grid and enrolling in net metering initiatives
Smart building technologies	Install building energy management systems to track and manage energy use in real-time. Utilise automation and smart sensors to change other systems, such as lighting and temperature, in response to outside factors and occupancy
Efficient water management	Reduce water use by using water-saving fixtures and technology, which will help you save energy that is otherwise used to heat water When using non-potable water, consider rainwater collection and graywater systems
Energy-efficient devices	Select energy-efficient devices and appliances with the Energy Star logo or other comparable certifications Promote energy-efficient practices among building inhabitants with awareness campaigns and educational initiatives
Thermal mass and passive solar design	To help control indoor temperature, include thermal mass components in the building's design. These will store and release heat Incorporate passive solar design concepts, such as placing windows to maximise solar gain and using materials with a high thermal mass

(continued)

Table 7.11 (continued)

Strategy	Explanation
Green Roof and cool roof technologies	Think about vegetated green roofs, which improve energy efficiency by reducing heat absorption, acting as insulation, and providing shade Use cool roof solutions to lessen the building's cooling burden by reflecting more sunshine and absorbing less heat
Lifecycle assessments and sustainable materials	To decide how building materials will affect the environment, perform lifetime evaluations Make using locally produced and sustainable materials a top priority to lessen the impact on the environment and embodied energy
Occupant engagement and education	Encourage building inhabitants to adopt energy-saving behaviours by launching initiatives for awareness and education Promote energy conservation by reminding people to switch off lights and electronics when not in use
Government incentives and certification programmes	Examine grants, tax breaks, and other government incentives available for energy-efficient construction projects. Go after certification schemes to show your dedication to sustainability
Regular maintenance and upgrades	Establish routine maintenance plans to guarantee energy-efficient systems run at their best To take advantage of innovative technology and increase overall building efficiency, schedule recurring updates

7.14 Sustainable Construction as a Means to Achieve Net Zero Carbon in the Built Environment

In the built environment, reaching net-zero carbon requires sustainable construction. Its distinctive significance stems from the all-encompassing strategy it employs to lessen the negative effects on the environment, cut carbon emissions, and build more durable and effective structures. The significance of sustainable construction in reaching the net-zero carbon aims in the built environment is elucidated in Table 7.12.

To achieve net-zero carbon in the built environment, sustainable construction is particularly crucial because it integrates renewable energy sources, addresses carbon emissions at various phases of a project's lifecycle, encourages resource efficiency, and follows extensive green building standards and certifications. It embodies an all-encompassing method of building design and construction that makes a substantial contribution to the more general aim of developing a low-carbon-built environment. Sustainable construction lays the groundwork for a constructed environment that is resilient and carbon neutral by addressing many areas of building design, materials, energy consumption, and environmental effects.

Table 7.12 How sustainable construction can contribute to the achievement of net-zero carbon

Technique	Explanation
Energy efficiency	High-quality insulation and passive design concepts are key components of sustainable building, which lowers the energy needed for heating and cooling. Since energy consumption is a significant contributor to carbon emissions, reaching net-zero carbon depends on this
Renewable energy integration	Sustainable construction practices promote the integration of on-site renewable energy sources. This immediately reduces carbon emissions through the local generation of renewable energy
Low-carbon materials	Use of recycled and low-carbon materials is prioritised in sustainable construction. The emissions connected to the manufacture and delivery of building materials can be decreased by selecting materials with a smaller carbon footprint
Life-cycle assessment	Life-cycle evaluations are a component of sustainable construction, considering a building's environmental impact from the time of construction to its operation and eventual removal. A comprehensive understanding of a building's overall carbon footprint is ensured by this all-encompassing method
Water conservation	Water-efficient fixtures, water-conserving landscaping techniques, and rainwater collection are examples of characteristics found in sustainable buildings. This lessens the environmental effect of water treatment and usage, which tangentially lowers carbon emissions
Waste management	The primary aim of sustainable construction methods is to reduce construction waste through effective project management, material recycling, and reuse. This lowers the embodied carbon linked to the creation of materials as well as the amount of waste dumped in landfills
Green building certifications	Performance benchmarks for sustainable building are set up by certifications. Buildings that follow these standards are guaranteed to meet energy efficiency and environmental impact reduction requirements
Biodiversity and green spaces	Green areas and biodiversity features are often incorporated into sustainable construction, which increases the built environment's overall ecological value, improves air quality, and provides added carbon sequestration
Adaptability and resilience	Climate-responsive design features are included in sustainable construction to improve a building's flexibility and resistance to changing climatic circumstances. This guarantees lifespan and lessens the need for ongoing repairs or construction, which lowers related carbon emissions
Community well-being	The well-being of building inhabitants and the surrounding community is taken into consideration during sustainable development. Sustainable buildings increase tenant satisfaction and productivity by creating cosy and healthful interior spaces

(continued)

Table 7.12 (continued)

Technique	Explanation
Regulatory compliance	To achieve net-zero carbon targets, regulations and policies are constantly changing, and sustainable construction methods are no exception. By meeting or surpassing these criteria, sustainable buildings are positioned to support more general environmental goals
Public awareness and education	Sustainable building initiatives often work as teaching tools, increasing public knowledge of how structures affect the environment. This understanding can spread to sustainable living methods, supporting a larger cultural movement towards more ecologically friendly ways of living
Carbon sequestration	Sustainable construction considers the use of landscaping, plants, and green areas. By absorbing carbon dioxide during photosynthesis, plants contribute to the sequestration of carbon and help balance off emissions from other sources
Behavioural impact	Features that support sustainable living practices, such as bike storage, electric vehicle charging stations, and community areas that foster social interaction are often incorporated into sustainable construction projects, which lessens the need for carbon-intensive activities
Government incentives and policies	Government subsidies and regulations that promote the use of ecologically friendly techniques often support sustainable construction. These laws, which support sustainable building methods, may include tax breaks, grants, and ordinances

7.15 Summary

The construction sector faces significant challenges in its efforts to decarbonize due to its extensive reliance on steel, cement, and fossil fuels. These materials are integral to the construction process, yet their production and use contribute substantially to greenhouse gas emissions. The industry's reliance on traditional energy sources, such as coal and fossil fuels, for generating heat and energy compounds the problem, as these sources are not only prevalent but also embedded in the production processes of construction materials, including those marketed as 'green' or sustainable. Despite advancements in developing more sustainable materials and energy-efficient technologies, a comprehensive solution to reduce carbon emissions across the entire lifecycle of construction projects remains elusive. The challenge is further exacerbated by the current limitations in energy technology, where no existing technique allows for the generation of energy without concurrently producing carbon dioxide. Addressing these issues requires a multi-faceted approach that includes innovations in material science, energy systems, and construction practices, alongside a concerted effort to shift towards low-carbon and renewable energy sources.

Bibliography

1. Almahmoud E, Doloi HK (2015) Assessment of social sustainability in construction projects using social network analysis. Facilities 33(3/4):152–176
2. Atanda JO (2019) Developing a social sustainability assessment framework. Sustain Cities Soc 44:237–252
3. Baldwin C, King R (2018) Social sustainability, climate resilience and community-based urban development: what about the people? Routledge
4. Bamgbade JA, Nawi MNM, Kamaruddeen AM, Adeleke AQ, Salimon MG (2022) Building sustainability in the construction industry through firm capabilities, technology and business innovativeness: empirical evidence from Malaysia. Int J Constr Manag 22(3):473–488
5. Brinkmann R (2016) Introduction to sustainability. Wiley
6. Dujon V, Dillard J, Brennan EM (2013) Social sustainability. Taylor & Francis
7. Edum-Fotwe FT, Price AD (2009) A social ontology for appraising sustainability of construction projects and developments. Int J Project Manage 27(4):313–322
8. Fatourehchi D, Zarghami E (2020) Social sustainability assessment framework for managing sustainable construction in residential buildings. J Build Eng 32:101761
9. Goel A, Ganesh LS, Kaur A (2020) Social sustainability considerations in construction project feasibility study: a stakeholder salience perspective. Eng Constr Archit Manag 27(7):1429–1459
10. Hossain MU, Poon CS, Dong YH, Lo IM, Cheng JC (2018) Development of social sustainability assessment method and a comparative case study on assessing recycled construction materials. Int J Life Cycle Assess 23:1654–1674
11. ISO 45001:2018 Occupational health and safety management systems
12. Kiani Mavi R, Gengatharen D, Kiani Mavi N, Hughes R, Campbell A, Yates R (2021) Sustainability in construction projects: a systematic literature review. Sustainability 13(4):1932
13. Lami IM, Mecca B (2020) Assessing social sustainability for achieving sustainable architecture. Sustainability 13(1):142
14. Mahdavinejad M, Sadraie A, Sadraie G (2014) Social sustainability of high-rise buildings. Am J Civil Eng Archit 2(1):34–41
15. Manzi T, Lucas K, Jones TL, Allen J (eds) (2010) Social sustainability in urban areas: Communities, connectivity and the urban fabric. Routledge
16. McKinsey Global Institute (January 2022) The net-zero transition: What it would cost, what it could bring. www.mckinsey.com/mgi.
17. Montalbán-Domingo L, García-Segura T, Amalia Sanz M, Pellicer E (2019) Social sustainability in delivery and procurement of public construction contracts. J Manag Eng 35(2):04018065
18. Montalbán-Domingo L, García-Segura T, Sanz MA, Pellicer E (2018) Social sustainability criteria in public-work procurement: an international perspective. J Clean Prod 198:1355–1371
19. Ng E (ed) (2009) Designing high-density cities: for social and environmental sustainability. Routledge
20. Ortiz O, Castells F, Sonnemann G (2009) Sustainability in the construction industry: a review of recent developments based on LCA. Constr Build Mater 23(1):28–39
21. Shirazi MR, Keivani R (eds) (2019) Urban social sustainability: theory, policy and practice. Routledge
22. Sierra LA, Pellicer E, Yepes V (2017) Method for estimating the social sustainability of infrastructure projects. Environ Impact Assess Rev 65:41–53
23. Sierra LA, Yepes V, García-Segura T, Pellicer E (2018) Bayesian network method for decision-making about the social sustainability of infrastructure projects. J Clean Prod 176:521–534
24. Smith D, Beeck S, Lommerse M, Metcalfe P (2014) An introduction to social sustainability and interior architecture. Perspectives on social sustainability and interior architecture: life from the inside, pp 1–11

25. Sodangi M (2018) Social sustainability efficacy of construction projects in the pre-construction phase. In: Proceedings of the institution of civil engineers-engineering sustainability, vol 172, no 2. Thomas Telford Ltd, pp 57–67
26. Toole TM, Carpenter G (2013) Prevention through design as a path toward social sustainability. J Archit Eng 19(3):168–173
27. Valdes-Vasquez R, Klotz LE (2013) Social sustainability considerations during planning and design: framework of processes for construction projects. J Constr Eng Manag 139(1):80–89
28. Vallance S, Perkins HC, Dixon JE (2011) What is social sustainability? A clarification of concepts. Geoforum 42(3):342–348
29. Wang H, Zhang X, Lu W (2018) Improving social sustainability in construction: conceptual framework based on social network analysis. J Manag Eng 34(6):05018012
30. Yates JK, Castro-Lacouture D (2018) Sustainability in engineering design and construction. CRC Press
31. Van der Leeuw S (2020) Social sustainability, past and future: undoing unintended consequences for the Earth's survival. Cambridge University Press
32. Yıldız S, Kıvrak S, Gültekin AB, Arslan G (2020) Built environment design-social sustainability relation in urban renewal. Sustain Cities Soc 60:102173
33. Zuo J, Jin XH, Flynn L (2012) Social sustainability in construction—an explorative study. Int J Constr Manag 12(2):51–63

Chapter 8
Sustainable Construction Management

Abstract This chapter examined the principles and practices of sustainable construction management, offering a comprehensive overview of how sustainability is integrated across all phases of construction projects. Positioned at the convergence of environmental responsibility, social equity, and economic viability, sustainable construction management redefines traditional project delivery by embedding sustainability into planning, design, execution, and post-construction evaluation. The chapter introduces key strategies such as the use of sustainable materials, energy-efficient technologies, and waste minimisation practices. It also outlines performance monitoring techniques like life cycle assessments, which are essential for tracking environmental impacts and improving project outcomes. Emphasising a holistic approach, the chapter discusses how construction stakeholders can adopt innovative and ethical methods to ensure that each project contributes positively to both society and the environment. By exploring techniques for evaluating and enhancing sustainability performance, this chapter equips professionals with the knowledge to benchmark projects against sustainability goals. Ultimately, the chapter promotes a paradigm shift in construction management by moving from short-term gains to long-term resilience, which advocates for a future in which construction serves as a driver for sustainable development and social advancement. This approach is applicable across the architecture, engineering, and construction (AEC) sectors, aiming to foster infrastructure that is not only structurally sound but also environmentally and socially responsible.

Keywords Sustainable performance · Project benchmark

8.1 Learning Outcomes

This chapter aims to provide an in-depth understanding of sustainable construction management by exploring its core concepts, guiding principles, and real-world applications. It introduces a comprehensive framework of techniques, strategies, and best practices that are instrumental in integrating sustainability throughout the lifecycle of construction projects. These include methodologies for environmentally

A. Olanrewaju, *Sustainable Construction Management: Research and Practice Companion*, Management in the Built Environment,
https://doi.org/10.1007/978-981-96-4910-5_8

responsible planning, resource-efficient design, ethical procurement, and socially inclusive implementation. The chapter also delves into the evolving roles and responsibilities of construction stakeholders in driving sustainability agendas within their respective domains. In addition, the chapter examines the role of sustainable design as a foundational element of sustainable construction management, highlighting the importance of selecting sustainable materials, minimising environmental impact, and incorporating passive and active energy-saving technologies. It further explores how innovative practices and digital tools can be used to monitor and evaluate the performance of construction projects against sustainability targets.

After studying this chapter, you will be able to:

- Explain sustainable construction management practices.
- Explain the techniques to incorporate sustainability in construction projects.
- Explain techniques for monitoring performance in sustainable construction.
- Explain techniques for evaluating performance in sustainable construction.

8.2 Introduction

The consequences of climate change, which include declining property values, lower labour productivity, health problems from prolonged heat exposure or a lack of clean water and air, and displacement, may make construction projects riskier. Outside construction sites are more vulnerable to weather-related dangers because of their greater susceptibility. Redefining the conventional construction sector landscape, sustainable construction project management sits at the intersection of innovation and responsibilities of the construction players. As the globe struggles with social demands and environmental problem, traditional ideas of project management are changing to reflect a more all-encompassing and sustainable perspective. This paradigm change represents a fundamental rethinking of the way we plan, carry out, and oversee construction projects rather than just a fashion. Sustainable construction project management is fundamentally about striking a balance between social equality, environmental stewardship, and economic success. To simplify, sustainable construction management is the collapsing of (a) economic sustainability enabled construction management practices, (b) social sustainability enabled construction management practices and (c) environmental sustainability-enabled construction management practices. Sustainable construction management considers the wide-ranging effects of construction activities on the planet and its inhabitants, going beyond the constrained boundaries of conventional project management. Fundamentally, this method acknowledges that construction projects are essential parts of a wider ecosystem, and their effects go beyond the temporary physical structures they erect. Sustainable construction project management includes a range of tactics aimed at creating structures that endure structural integrity and positive contributions to the environment and communities. These tactics range from the careful selection of eco-ethic materials to the strategic deployment of innovative technologies.

Through an examination of its foundational ideas, navigation of the many obstacles it poses, and an exploration of the numerous chances it affords for constructive change, this introduction aims to delve into the complex layers of sustainable construction project management. As this investigation is examined, it becomes clear that the field of sustainable construction project management is dynamic and ever-evolving, extending beyond the activities on the construction sites. This concept opposes the current quo by calling for a reexamination of accepted conventions and igniting a vision in which every construction project serves as a spark for resilient building techniques, sustainable practices, and the advancement of society. To create a future where every project stands as a tribute to both engineering proficiency and a dedication to environmental credentials, this trip asks us to rethink the role of construction in the larger context of sustainability.

8.3 Sustainable Design

A step in designing physical built environment that align with sustainability standards is designing sustainable construction projects. In contrast to traditional design methods, which frequently prioritise practicality and aesthetics, sustainable design prioritises reducing environmental impact, maximising resource efficiency, and promoting social well-being. Construction project often start as thoughts, which would later transform to design. Table 8.1 presents the sustainable design principle.

Designing a sustainable construction project is a dynamic, non-linear and iterative process that goes beyond conventional design methods. It represents a dedication to construction structures that serve communities and the environment in addition to satisfying functional and aesthetic criteria effectively. Design plays a crucial part in creating a physical environment that aligns with sustainability principles as the construction sector adopts these principles.

8.4 Sustainable Construction Processes and Techniques

Environmental stewardship, resource efficiency, and societal well-being are prioritise in sustainable construction processes and techniques, which offer a revolutionary approach to construction that goes beyond conventional practices. An in-depth investigation of the fundamental ideas, creative approaches, and state-of-the-art methods that define sustainable construction will be conducted to move the sector in the direction of a more environmentally conscious and socially conscious future. Sustainable construction methods and practices represent a paradigm-shifting approach to construction that places a premium on resource economy, environmental ethics, and social welfare. Every aspect of construction, from the first phases of site preparation to the last touches of interior finishes, is carefully considered and optimised to promote long-term sustainability. This thorough investigation explores the fundamental ideas,

Table 8.1 Principle of sustainable design

Principle	Technique
Holistic integration of sustainability principles	Fundamentally, designing a construction project with sustainability requirements entail incorporating sustainability concepts into the building's planning and conceptualization. To guide design decisions, a thorough evaluation of environmental, economic, and social concerns must be conducted. Architects and designers work together from the outset to develop buildings that not only fulfil functional requirements but also have a positive impact on the surrounding community and ecosystem
Material selection and life cycle assessment	A fundamental aspect of sustainable design is the precise choice of materials. These days, architects examine the nuances of material life cycles and assess how they will affect sustainability throughout the process, from extraction to disposal. Selecting materials with a small ecological imprint, high recycling potential, or renewable sources is the aim. Tools for life cycle assessments enable measurement of the sustainability impact of materials, enabling designers to make decisions in line with sustainability objectives
Energy efficiency and passive design strategies	Energy efficiency is given priority in sustainable construction project design by using creative design techniques. To reduce the need for mechanical systems, passive design techniques are used, such as maximising daylighting, integrating natural ventilation, and optimising building orientation. In order to generate clean energy on-site, architects should incorporate renewable energy sources like wind turbines or solar panels
Water conservation measures	Water conservation techniques must be incorporated into designs due to worries about water scarcity. This entails putting in place rainwater harvesting equipment, water recycling technology, and water-efficient landscaping. Water consumption is reduced during a building's life cycle by using sustainable design, which considers both the building's construction and operating stages
Biodiverse and resilient landscaping	Beyond the building's physical footprint, sustainable development considers the surrounding environment. To create ecosystems that support biodiversity, designers incorporate biodiverse landscaping techniques that support the indigenous flora and fauna. This method improves the project's attractiveness while also boosting the environment's general resistance
Innovative technologies and smart design	The integration of innovative technologies is essential to the design of sustainable construction projects. Augmented reality, virtual reality, and building information modelling are used to improve teamwork, mimic environmental conditions, and optimise design choices. These tools enable designers to make data-driven decisions and see how their decisions will affect the world

(continued)

Table 8.1 (continued)

Principle	Technique
Community engagement and inclusivity	Engaging with local communities to understand their needs, attitudes, and concerns is a crucial component of sustainable design. When projects are designed with inclusivity requirements, the cultural context is resonated, and the social fabric is strengthened. Throughout the design process, community feedback is encouraged and sought after, which promotes a sense of shared responsibility and ownership
Compliance with green building certifications	The design of sustainable construction projects frequently complies with established green building certifications, such GBCN, BREEAM, GBI, and LEED. By striving to fulfil the requirements of these certifications, designers demonstrate their dedication to the highest performance and environmental standards

tactics, and innovative methods that define sustainable construction processes. The explanation for the principle of sustainable construction management is listed in Table 8.2.

As presented, sustainable construction methods and procedures introduce a new era of construction practices. The construction sector can significantly contribute to the creation of a built environment that is more sustainable by adopting sustainable ideas and practices. To achieve sustainability in the built environment, construction players must be steadfast in their pursuit of investigating, modifying, and using

Table 8.2 Processes and techniques of sustainable construction

Principle	Technique
Site selection and management	Carrying out in-depth site surveys while considering environmental sensitivity and ecological effects Putting plans for habitat restoration and biodiversity preservation into practice Utilising innovative GIS technologies to assess sites precisely Incorporating clever land-use planning to reduce disturbance to the environment
Sustainable materials selection	Giving materials with environmental certifications and low embodied energy priority Supply networks can be made transparent and traceable by utilising blockchain technology Investigating innovative materials, such as composites made of recycled wood and engineered wood products Continuing research and development to find new sustainable materials

(continued)

Table 8.2 (continued)

Principle	Technique
Energy-efficient design and construction	Utilising Building Information Modelling to provide comprehensive energy modelling Putting new passive design techniques into practice for the best possible thermal performance Investigating the application of artificial intelligence to energy management in real-time Adopting innovative HVAC and lighting control systems that use less energy
Waste reduction and recycling	Imposing strict recycling goals and waste reduction regulations Experimenting with construction waste exchange schemes to promote material recovery Partnering with neighbourhood groups to recycle building debris for community initiatives
Water conservation	Putting in place weather-based controls for intelligent irrigation systems Using localised reuse of decentralised water treatment systems Using biomimicry ideas in landscaping to reduce water usage
Green building certifications	Keeping up with the latest certifications and standards for green buildings Pursuing many certifications approaches to validate sustainability in its entirety Regenerative design ideas are being integrated for projects that aspire to higher than traditional certifications Working together with certifying organisations to shape the development of industry standards
Construction waste management	Utilising the Internet of Things devices to implement innovative waste tracking solutions Developing cradle-to-cradle building methods through the piloting of circular economy projects Working together with waste-to-energy plants to find sustainable disposal options Utilising data analytics to continuously improve processes to reduce waste
Green roofs and urban agriculture	Investigating vertical farming techniques for effective food production in cities Utilising drone technology for green roof upkeep and surveillance Collaborating with the agencies in charge of urban planning to incorporate green spaces Putting in place intelligent irrigation systems for urban farming

(continued)

Table 8.2 (continued)

Principle	Technique
Life cycle assessment	Investigating blockchain-based applications for sharing transparent life cycle data Including innovative LCA software for analysis and optimisation in real-time Working along with academics to do further research on life cycle assessment techniques Cradle-to-cradle project piloting to reduce sustainability effects across a project
Community engagement and social equity	Create engaging community involvement experiences with virtual reality Implementing participatory design approaches for inclusive decision-making Working together to promote cultural preservation with regional craftspeople and artisans Putting a focus on inclusion and diversity in the construction sector
Smart technologies and building automation	Investigating how blockchain technology may be used to handle safe data in smart buildings Ise augmented reality to monitor construction sites in real time Launching smart city projects to create a networked and effective urban infrastructure Using predictive analytics to anticipate building system maintenance
Post-occupancy evaluation and continuous improvement	Putting in place tools for continuous observation in building performance evaluation Gamifying a tenant-led collaborative method to solicit feedback instantly Utilising machine learning to investigate predictive analytics in building performance Establishing cooperative networks to share best practices and lessons discovered

innovative methods in preparation for the day when construction projects will act as indicators of sustainability needs.

8.5 Project Monitoring and Evaluation

A project is a temporary endeavour undertaken to create a unique product, service, or result. Unlike ongoing operations, projects have a defined beginning and end, with specific objectives, deliverables and milestone resources, and constraints in place to achieve a set goal within a particular timeframe. Projects vary widely in scope and complexity, from constructing buildings to developing software, building ships, marriages, organizing events, or implementing new business processes. Project objectives are the specific goals a project aims to achieve, such as completing a sustainable housing project within 18 months or reducing construction waste by 30%. Project targets, on the other hand, are the measurable milestones set to achieve these objectives, such as using at least 60% recycled materials or ensuring 90% of site workers receive sustainability training. Clearly defining objectives and targets is essential for effective project monitoring, as they provide a concrete basis for tracking progress, evaluating performance, and making informed decisions to keep the project aligned with its intended outcomes. Project monitoring and evaluation are the required methodical procedures for tracking, evaluating, and comprehending the effectiveness, efficiency, and results of the projects. The following is a summary of project monitoring and evaluation specific to construction projects:

Project Monitoring in Construction:

i. Progress Tracking

Maintaining a close look on and updating the construction project schedule to guarantee that activities and deliverables are finished on time are crucial. Throughout their existence, sustainable construction projects seek to balance social, economic, and environmental factors. Monitoring these deliverables is crucial to guaranteeing that sustainability targets are fulfilled and proving responsibility to interested parties. The monitoring of the status of sustainable construction projects may start by outlining the project's expectations, indicators, and the quantification of sustainability goals. These aims ought to coincide with certifications for sustainable construction, industry norms, and the objectives of the project's stakeholders. Common indicators include reducing waste, conserving water, increasing energy efficiency, and considering social equality. Determine and set up key performance indicators that correspond with the specified sustainability goals. KPIs are quantifiable measurements used to assess development. KPIs may be used to track performance criteria like waste diversion rates, water usage efficiency, energy consumption reduction percentages, and the use of sustainable products. Incorporate BIM into the undertaking. BIM as a computerised depiction of a construction's structural and operational features, it makes real-time cooperation possible, makes collision detection easier, and helps with the analysis and visualisation of sustainability elements.

Utilise technologies and software specific to sustainability to track and evaluate project outcomes. Monitoring resource usage, carbon emissions, and other environmental effects can be aided by these techniques. For example, life cycle assessment technologies can assess the environmental impact of construction materials and processes. Maintain constant vigilance on energy efficiency during the building phase. This includes monitoring energy usage, fine-tuning HVAC systems, and making sure renewable energy sources are integrated. Energy management programmes and smart metres can offer real-time data to support well-informed decision-making. Establish a system to monitor the procurement, application, and disposal of construction supplies. This entails confirming the sustainability certifications of the materials, encouraging the use of locally or recycled products, and keeping an eye on the waste produced during building. Track water use and conservation efforts. Use water-saving fixtures, install effective irrigation systems, and integrate rainwater collection technologies. The project's compliance with water conservation objectives is guaranteed by tracking water usage against predetermined benchmarks. Incorporate social impact evaluations into the system for monitoring progress. Keep an eye on employees' well-being, make sure fair labour practices are followed, and evaluate how the project will affect the neighbourhood. One can use feedback mechanisms, interviews, and surveys to assess social sustainability.

Keep thorough records of all the sustainability techniques and associated results as well as the lessons learned. Share project updates with stakeholders by regularly using reports, slideshows, or specialised sustainability dashboards. Open communication informs stakeholders and fosters trust. Through post-occupancy assessments, continue the monitoring procedure into the post-construction phase. Evaluate the projects's performance in real-world scenarios and get input from the users of the projects. This action will advance the ongoing development of sustainable practices and features. Perform routine sustainability audits to ensure adherence to defined guidelines and certifications. Getting certified is a common goal of sustainable construction projects. Continuous monitoring and documentation are needed to keep certificates current. Incorporate lessons learnt from continuous monitoring to foster a culture of continuous improvement. Determine areas that require innovation and improvement, then incorporate suggestions into upcoming projects. The development of sustainable construction methods is aided by this iterative process.

ii. Quality Control

Putting quality control procedures into practice to make sure construction operations and results adhere to set requirements. Ensuring that the built environment satisfies strict criteria for durability, safety, and environmental performance is critically crucial for sustainable construction projects. Quality control monitoring plan for sustainable construction projects all-encompass strict procedures, innovative technologies, and a dedication to achieving sustainability goals for the projects. The following is a comprehensive manual that will help organisations to keep an eye on the quality control of sustainable building projects, see Table 8.3.

A comprehensive strategy that incorporates sustainability standards into each stage of the building process is needed to monitor the quality control of sustainable construction projects. Project teams may make sure that sustainability objectives are not just reached but surpassed by integrating technology, reliable procedures, and a dedication to excellence. This tactical strategy advances the larger goal in addition to helping specific projects succeed.

iii. Budget Monitoring

Monitoring the use and distribution of financial resources to detect overspending and guarantee budgetary compliance. Successful sustainable construction projects depend on effective budget monitoring, which makes sure that funds are distributed

Table 8.3 Techniques to monitor quality control in sustainable construction

Technique	Explanation
Define sustainable quality standards	Start by outlining quality requirements that are unique to sustainability. This entails setting standards for waste minimization, material sustainability, energy efficiency, and adherence to green building certifications like BREEAM, LEED, GBCN, or GBI. Stakeholder expectations and project objectives should be in line with these criteria
Incorporate building information modeling	To improve the coordination and visualisation of the construction process, we incorporate building information modelling into the project. With the use of BIM, the project may be represented digitally, which makes it easier to identify conflicts, lowers mistake rates, and guarantees that sustainable design features are applied correctly
Establish quality assurance and control plans	Create thorough plans for quality control and quality assurance that particularly address sustainability requirements
Implement advanced materials testing	Make use of innovative materials testing methods to evaluate the performance and sustainability of building materials. Testing for durability, resilience to environmental conditions, and recycled content are a few examples of this. Make sure the products choosen adhere to project criteria and sustainability certifications
Conduct energy modeling and simulation	To evaluate the building design's energy performance, we use energy modelling and simulation technologies. To satisfy sustainability goals, this stage enables the identification of energy inefficiencies and the optimisation of building systems. Update models often to account for evolving sustainability objectives and design modifications
Integrate smart technologies for quality monitoring	Use sensors, Internet of Things devices, and smart technologies to monitor quality in real-time. Data on material performance, environmental factors, and structural integrity can be obtained from smart building materials, allowing for proactive quality-maintenance actions
Implement lean construction practices	Use lean construction techniques to streamline operations and cut waste. Minimising resource use, cutting down on inefficiencies, and optimising value are all benefits of sustainable construction projects. Lean methodologies enhance project quality overall and promote environmental sustainability
Perform regular site inspections	Verify that construction activities comply with sustainability guidelines by conducting routine site inspections. This entails making certain that waste is managed appropriately, confirming that sustainable solutions are installed correctly, and examining the calibre of the job to avoid errors
Monitor indoor environmental quality	Pay close attention to elements like lighting, thermal comfort, and air quality while assessing indoor environmental quality. The health and comfort of inhabitants are given priority in sustainable projects, and indoor environmental quality monitoring makes sure that these criteria are met

(continued)

Table 8.3 (continued)

Technique	Explanation
Establish traceability for sustainable materials	When using sustainable materials in construction, traceability measures are in place. To make sure the materials fulfil sustainability standards, this entails monitoring the materials' place of origin, transit, and installation. Construction technologies like blockchain technology has the potential to improve supply chain transparency
Implement third-party certifications and audits	To get outside confirmation that the project is adhering to sustainable quality requirements, involve third-party organisations in certifications and audits. Such certifications can increase the project's legitimacy
Continuous training and education	To guarantee that project personnel are knowledgeable about and proficient in sustainable construction techniques, make ongoing educational and training investments. Maintaining the team's knowledge of the most recent tools, supplies, and techniques helps them maintain exacting standards of quality
Document and analyze quality metrics	Keep thorough records of all quality control procedures and results. Analyse quality indicators regularly to spot patterns, development opportunities, and effective strategies. Data-driven insights promote a continuous improvement culture and assist well-informed decision-making
Post-construction evaluation	Assess the performance of sustainable features under real-world situations by conducting post-construction studies. Obtain input from tenants, examine energy usage trends, and pinpoint areas that could benefit from improvements or modifications to maximise sustainability

wisely to achieve sustainability objectives. Institute strong budget monitoring procedures into place is essential for upholding sustainability goals, encouraging stakeholder transparency, and preserving financial discipline. Detailed instructions on how to keep an eye on the budget for sustainable construction projects are provided in Table 8.4.

To monitor budgets effectively in sustainable construction projects, proactive and integrated approaches that balance sustainability goals with financial considerations are necessary. Through technology, best practices, and open communication with stakeholders, project teams can manage budgets while achieving social and environmental objectives. Strategic budget monitoring promotes financial stewardship and enhances the effect and long-term success of sustainable construction projects.

iv. Risk Management

A risk is the possibility of an event or condition that may have a negative or positive impact on objectives. It involves uncertainty and can result in potential losses, damages, or gains depending on how it unfolds. In sustainable construction, negative risks (threats) can be mitigated through careful planning, while positive risks (opportunities) should be strategically exploited to maximize project

Table 8.4 Techniques to monitor budget control in sustainable construction

Technique	Explanation
Establish detailed budgetary allocation	Create a thorough budget that is in line with the project's sustainability goals to start the budget monitoring process. Divide the budget into several sections for things like labour, supplies, waste management, energy-efficient devices, and other environmentally friendly projects. Precise allocation helps with focused observation and guards against mistakes
Integrate sustainability costs	Make sure the budget includes a clear integration of sustainability costs. This entails setting aside funds for environmentally friendly products, energy-saving appliances, waste management strategies, and accreditations. Budgetary adjustments later in the project are avoided when sustainability costs are integrated from the start
Implement project management software	Use budget tracking capabilities in project management software. Real-time monitoring of spending, variations from the budget, and resource allocation is made possible by this technology. Tools for automated reporting improve openness and help stakeholders, and project managers make well-informed decisions
Regularly review and update budgets	Review and revise the budget regularly to account for changes to the project's scope, shifting market dynamics, or unanticipated difficulties. Budgets must be adjusted appropriately because sustainable construction projects frequently use materials and technologies that are constantly changing. Regular reviews help to avoid financial surprises and enable early corrections
Use earned value management	Utilise Earned Value Management strategies to evaluate project outcomes about the allocated budget. Scope, schedule, and cost are all integrated by EVM to give a complete picture of project development. This approach makes it possible to detect cost overruns or underruns early on, allowing for prompt remedial action
Monitor resource efficiency	To assess the efficiency of your resources, keep an eye on how much labour, energy, and materials are being used. The optimisation of resources is a key component of sustainable building, and tracking efficiency measures reveals areas in need of development. Making educated decisions to improve sustainability and save costs is made easier with the help of real-time data on resource consumption
Conduct life cycle cost analysis	To assess the overall cost of ownership over a building's lifecycle, integrate life cycle cost analysis into budget monitoring. LCCA considers startup costs, ongoing costs, and maintenance expenditures. This method offers a comprehensive understanding of the financial effects of sustainability choices
Establish contingency plans	Create backup plans in case there are unforeseen issues that could affect the budget. Projects involving sustainable construction may be subject to hazards associated with implementing new materials or technology. It is crucial to have backup plans ready to handle unanticipated events without compromising the sustainability goals of the project

(continued)

Table 8.4 (continued)

Technique	Explanation
Collaborate with suppliers and contractors	Keep lines of communication open with contractors and suppliers to learn about changes in the market, shifts in prices, and new developments in sustainable solutions. Establishing collaborative ties facilitates the negotiation of advantageous terms and can lead to cost savings without sacrificing sustainability objectives
Analyze cost–benefit ratios	Keep lines of communication open with contractors and suppliers to learn about changes in the market, shifts in prices, and new developments in sustainable solutions. Establishing collaborative ties facilitates the negotiation of advantageous terms and can lead to cost savings without sacrificing sustainability objectives
Conduct regular audits	To make sure that budgetary restrictions and sustainability goals are being followed, conduct routine financial audits. An impartial evaluation of financial procedures is provided by audits, which can reveal inconsistencies or areas in need of development. Accountability is improved and stakeholder trust is increased through transparent audits
Engage stakeholders	By giving stakeholders frequent information on financial performance, you may involve them in the budget monitoring process. Open communication encourages cooperation since it informs stakeholders about the financial effects of sustainability decisions and allows them to offer insightful comments
Embrace lean construction principles	Adopt lean building techniques to save waste and boost productivity. The application of lean techniques, which prioritise optimising value over resource consumption, can help sustainable building projects achieve cost savings
Monitor certification costs	Keep monitoring the certification procedure costs if the team plan to pursue green building certifications. Fees for evaluation, paperwork, and compliance are included in this. Maintaining a record of certification expenses guarantees that the financial plan considers sustainability certifications

success. A well-structured risk management plan ensures that sustainable construction projects achieve their economic, environmental, and social deliverables effectively. Recognising, evaluating, and mitigating problem-risks associated with sustainability, unforeseen events, or safety. With a focus on sustainability problem, sustainable construction projects provide potential and challenges. Sustaining the achievement of sustainability objectives in these projects necessitates proactive risk monitoring that is integrated and comprehensive. The guidelines for efficiently monitoring hazards in sustainable construction projects are provided in Table 8.5.

It takes a dynamic, multifaceted approach to risk monitoring in sustainable construction projects that incorporates sustainability concerns into all facets of risk management. Through technology, keeping up with sectordevelopments, and encouraging a proactive culture of risk-taking, project teams can overcome obstacles and grasp chances for advancement and innovation. In addition to ensuring the success

Table 8.5 Risk management techniques in sustainable construction

Technique	Explanation
Identify and assess risks	A comprehensive process of identification and assessment is the initial stage in risk monitoring. Identify any risks unique to sustainable construction by including stakeholders, such as project managers, sustainability specialists, and crucial team members. This includes changes in the regulatory landscape, supply chain interruptions, uncertainty in the environment, and difficulties implementing innovative technologies
Prioritize risks	After problems have been identified, order them according to likelihood and potential impact. Project teams can more efficiently deploy resources when risks are prioritised because not all risks are equally significant. Examine how crucial risks are to project schedules, sustainability goals, and success in general
Establish a risk management plan	Create a thorough risk management plan that includes methods for transferring, accepting, or minimising hazards that have been recognised. Roles and duties, communication procedures, and backup plans must be spelt out in the plan. Stress the need to include sustainability issues in risk management plans
Implement building information modeling	Optimise risk monitoring by leveraging BIM. Through the use of a digital representation of the project, BIM makes it possible to visualise potential problems and conduct cooperative risk assessments. BIM's clash detection features can spot conflicts during the design and construction stages, lowering the possibility of mistakes that could be dangerous
Utilize risk assessment tools	To assess risks and their effects quantitatively, make use of specialised risk assessment tools. Project teams can better comprehend the financial, environmental, and social ramifications of individual risks by using these tools, which can model different scenarios. Update risk assessments often to account for evolving project requirements
Monitor environmental and regulatory compliance	Strict sustainability standards are frequently used for sustainable construction projects. Provide monitoring systems to guarantee continued adherence to sustainability laws and guidelines. This entails keeping an eye on waste management procedures, pollutants, and compliance with green building standards
Assess supply chain vulnerabilities	Examine the project's supply chain's weaknesses, especially concerning where sustainable materials and technology are sourced. Keep an eye on your suppliers' robustness and dependability while considering market volatility, geopolitical risks, and disruptions. To reduce supply chain risks, diversify your suppliers wherever you can
Monitor technological risks	Innovative technologies are commonly adopted in sustainable construction projects. Keep an eye out for technological hazards when implementing new materials, digital tools, and energy-efficient systems. To guarantee the effective integration of sustainable technologies, keep up with the most recent developments and obstacles

(continued)

Table 8.5 (continued)

Technique	Explanation
Regularly review project timelines	Project delays may result in new hazards and affect sustainability goals. Examine project schedules and objectives regularly to spot any delays or bottlenecks. Put policies in place to maintain consistency with sustainability objectives and respond quickly to problems
Incorporate social risks	Think about the social hazards connected to labour practices, community connections, and stakeholder participation. Continue to monitor the social issues like stakeholder comments, worker happiness, and community impact assessments. Take proactive measures to mitigate social risks to preserve goodwill and respect social sustainability norms
Implement continuous monitoring systems	Install ongoing surveillance technologies to identify risks in real time. Automated sensors, data analytics, and monitoring solutions that offer real-time input on safety parameters, environmental conditions, and other crucial indicators may fall under this category. Ongoing monitoring would prompt be reacting quickly to new dangers in the projects
Integrate climate change adaptation strategies	Projects involving sustainable construction are susceptible to the effects of climate change. Therefore, incorporate measures for adapting to climate change into risk monitoring, considering variables including harsh weather, increasing sea levels, and shifting temperature patterns. Goals for sustainability should be in line with adaptation strategies
Foster a culture of risk awareness	Encourage the project team to have a risk-aware culture. Promote honest dialogue, the disclosure of hazards, and a proactive attitude to risk control. The team's capacity to recognise and manage hazards as a group can be improved with regular training and awareness initiatives
Establish a contingency fund	Establish a contingency fund to handle unforeseen problem-risks in the projects. This financial cushion permits rapid reaction and adjustment to unforeseen difficulties while maintaining sustainability goals

of sustainable construction projects, strategic risk monitoring advances robust and conscientious construction practices.

v. Health and Safety Monitoring

Ensuring that construction sites follow safety guidelines and requirements to avoid mishaps, injuries and fatalities. In any construction project, ensuring the health and safety of stakeholders and workers is of utmost importance. Sustainable construction initiatives bring with them additional considerations on this discourse. A thorough and proactive approach is needed to monitor health and safety in sustainable construction projects to maintain sustainability objectives and provide a safe working environment. The manual for efficiently overseeing health and safety in sustainable construction projects comprises. The methods for keeping an eye on health and safety in sustainable construction are listed in Table 8.6.

A dynamic, integrated strategy that considers risks specific to sustainable practices, as well as those associated with conventional construction, is necessary for

Table 8.6 Techniques to monitor health and safety in sustainable construction

Technique	Explanation
Comprehensive risk assessment	Start by carrying out an extensive risk assessment tailored to sustainable construction's health and safety requirements. Determine any risks associated with the application of innovative technologies, eco-friendly materials, and distinctive building techniques. Both conventional building hazards and risks specific to sustainable methods should be considered in this assessment
Integrate health and safety into sustainable design	When making decisions on sustainable design, such as selecting building materials and layouts, the health and welfare of employees and users should come first. Make sure that the design reduces problems and that sustainable features do not jeopardise safety standards
Establish robust health and safety policies	Create succinct and unambiguous health and safety policies that support the objectives of sustainable construction. The hazards connected to sustainable practices-such as managing environmentally friendly products, utilising renewable energy sources, and putting green construction technology into practice-should be covered by these policies
Regular training and education	Make frequent investments in education and training programmes to guarantee that all employees are knowledgeable about the health and safety procedures involved in sustainable building. This entails educating staff members about the dangers connected to sustainable practices and offering direction on how to use innovative technologies appropriately
Implement building information modeling	Use BIM to improve the tracking of health and safety. With the use of BIM, a digital depiction of the construction project may be created, facilitating virtual safety inspections, clash detection, and the early detection of dangers
Use wearable technologies	Present wearable technology, such as smart vests and helmets with sensors that can track an employee's health and safety in real-time to them. These gadgets can monitor variables such as body temperature, heart rate, and exposure to dangerous compounds. Patterns and possible hazards can be found by analysing the gathered data
Regular safety audits	As construction progresses, carry out routine safety inspections. Safety audits should analyse how well the sustainable practices reduce hazards in addition to determining whether health and safety policies are being followed. It is imperative to maintain a safe working environment through continuous improvement, informed by audit findings

(continued)

Table 8.6 (continued)

Technique	Explanation
Monitor ergonomics	Take ergonomics into account while designing work areas and procedures. Innovative construction methods and materials are frequently used in sustainable construction, which may call for the adoption of new or modified work procedures. Make sure ergonomics are considered to avoid musculoskeletal problems and improve employee wellbeing
Address mental health	Acknowledge and deal with mental health problems that arise among construction workers. Upskilling requirements and strict deadlines are two additional stressors that sustainable construction projects may bring. Encourage employees to talk about their mental health in an open and supportive workplace by raising awareness of the issue and offering tools for support
Emergency preparedness and response	Create and update emergency action and readiness plan regularly. Projects involving sustainable construction should take special materials or technology into consideration in case of catastrophes. Make certain that employees receive emergency procedures training and hold frequent drills to strengthen readiness
Environmental monitoring	Include environmental monitoring in your policies for health and safety. Certain environmentally compliant materials may be used in sustainable construction projects; however, improper handling of these materials may result in health concerns. To protect the health and safety of employees, keep an eye on the air quality, noise levels, and other environmental aspects
Collaborate with contractors and suppliers	Keep lines of communication open with suppliers and contractors to guarantee that health and safety regulations are followed through the supply chain. Establish a shared commitment to health and safety objectives, cooperate on risk assessments, and exchange best practices
Encourage reporting and feedback	Promote a reporting and feedback-giving culture among employees. Employees should have the confidence to voice any safety concerns or hazards without fear of retaliation. To foster an open and responsible reporting culture, set up procedures for anonymous reporting
Post-occupancy evaluation (for buildings)	Continue to monitor health and safety in buildings after they are occupied, even after they have finished construction. Examine whether the sustainable measures put in place are still improving the health and well-being of the tenants. Users' feedback on the building can help shape future initiatives
Benchmarking and certification	Make use of health and safety certification programmes and industry benchmarking. OHSAS 18001 and ISO 45001 certifications, for example, offer globally accepted guidelines for managing occupational health and safety. Getting certified and keeping it shows that you follow best practices

monitoring health and safety in sustainable construction projects. By utilising technology for in-the-moment monitoring and incorporating health and safety concerns into the project's architecture, regulations, and procedures.

vi. Material and components management

To avoid delays and inefficiencies, monitor the ordering, delivery, and use of construction supplies and equipment. The efficient handling of materials and components is essential to sustainable construction initiatives. Achieving sustainability goals is facilitated by efficient management, which also guarantees timely resource availability. Sustainable construction projects necessitate a systematic approach to material and component management that incorporates environmental considerations, resource efficiency, and innovative technology. Table 8.7 contains examples of material and component management techniques in sustainable construction.

A comprehensive and proactive approach is required for the effective monitoring of material and component management in sustainable construction projects.

Table 8.7 Material and component management techniques in sustainable construction

Technique	Explanation
Life cycle assessment	Consider life cycle assessment while choosing materials. LCA assesses how materials will affect the environment at each stage of their life cycle, from extraction to disposal. Using a comprehensive approach aids in selecting materials that support sustainability goals. Update life cycle assessments frequently to take new materials and industry standards into account
Sustainable procurement practices	Obtain products from vendors who follow ecologically friendly policies to implement sustainable buying methods. Provide precise guidelines for sustainable procurement, considering things like energy efficiency, recyclable material, and conformity to accredited eco-labels. Evaluate vendors' sustainability performance regularly and consider forming alliances with certified providers
Implement building information modeling	To manage materials and components effectively. Use Building Information Modelling for real-time tracking of materials, quantities. This makes it easier to create a digital picture of the project. It improves supply chain transparency, lowers errors, and streamlines communication
Lean construction principles	Use the concepts of Lean Construction to reduce waste and maximise material use. Just-in-time delivery and effective on-site handling are two examples of lean techniques that help cut down on excess inventory and increase resource efficiency. Track and evaluate material flow to pinpoint areas in need of improvement
Waste reduction strategies	Encourage recycling, repurposing, and reuse of construction waste to implement waste reduction measures. Track recycling rates, keep an eye on waste generation throughout the project, and work with waste management partners. Set waste reduction targets and evaluate your efforts regularly

(continued)

Table 8.7 (continued)

Technique	Explanation
Inventory management systems	Install innovative inventory management systems to keep an eye on the availability and quantity of materials. Real-time stock-level data is provided by these systems, allowing for proactive buying and averting shortages or surplus inventory. On-time material delivery is ensured via integration with project time limits
Digital tracking technologies	Use barcoding RFID tags or other digital tracking technologies to monitor the flow and use of materials on-site. Digital tracking improves traceability, lowers the possibility of theft, and offers useful information for better material logistics planning
Circular economy principles	Adopt the ideas of the circular economy by planning for reuse and disassembly. During the phases of demolition or renovation, monitor the possibility of salvaging materials. Determine whether it is feasible to recycle materials for use in future building projects or to repurpose them within the project
Collaboration with contractors	To align material and component management processes, and establish cooperative partnerships with contractors. All parties in the supply chain are guaranteed to follow sustainability objectives when there is regular contact and collaboration with contractors. Planning together can increase productivity and lower the possibility of mistakes
Quality assurance and control	Put strong quality control and assurance procedures in place for incoming materials. Monitor the sustainability of the materials to make sure they live up to the required certifications and requirements. Conduct frequent testing and inspections to aid in the prevention of flaws and guarantee that the materials used in the project contribute to its overall durability and sustainability
Track embodied carbon	Monitor and record the materials' embodied carbon. Consider the carbon footprint resulting from material extraction, manufacture, transportation, and installation. Making educated judgements to reduce the project's total impact on sustainability is made easier with the tracking of embodied carbon
Energy-efficient materials	Make it a priority to use materials that are energy-efficient and improve the project's overall energy performance. Track developments in energy-efficient technologies and material science to continuously raise the project's sustainability rating
Real-time data	Use real-time data analytics to track trends in the use of materials and components. Data analysis facilitates proactive decision-making, pinpoints opportunities for development, and optimises material management plans by project needs
Certification and compliance	Verify that the project's materials adhere to all applicable sustainability certifications and standards. Update certification requirements regularly considering industry developments
Continuous improvement practices	Review material and component management procedures regularly to foster a culture of continuous improvement. To find lessons learnt and apply changes in the next initiatives, conduct post-project reviews. To improve tactics, solicit input from vendors and project teams

Through the integration of environmental factors, adoption of technology innovations, and promotion of cooperation throughout the supply chain, project teams may maximise resource efficiency, reduce waste, and facilitate the overall accomplishment of sustainable construction endeavours. The constant search for novel and sustainable material management techniques guarantees that projects not only fulfil but also surpass social and environmental standards.

vii. Plant and Equipment Management

To execute sustainable construction projects with the least possible impact on the environment, plant and equipment are essential. A sophisticated strategy that maximises performance and is in line with sustainability goals is necessary for monitoring plant and equipment management. A thorough manual on monitoring plant and equipment management in sustainable construction projects may be found in Table 8.8.

A comprehensive approach that considers efficiency, environmental impact, and technological improvements is necessary for effective plant and equipment management monitoring in sustainable construction projects. Construction projects can minimise their environmental impact while optimising equipment performance through the integration of sustainable practices into fleet management, utilisation of innovative technologies, and a dedication to continual development. This methodical and comprehensive approach not only supports sustainability objectives but also enhances the general prosperity and durability of sustainable construction projects.

viii. Stakeholder Communication

A comprehensive approach that considers efficiency, environmental impact, and technological improvements is needed for effective plant and equipment management monitoring in sustainable construction projects. Sustainable construction projects can minimise their environmental impact while optimising equipment performance through the integration of sustainable practices into fleet management, utilisation of innovative technologies, and a dedication to continual development. This methodical and comprehensive approach will not only support sustainability objectives but also enhances the general prosperity and durability of sustainable construction projects. See Table 8.9 regarding how to communicate sustainable construction practices with stakeholders.

In sustainable construction projects, tracking stakeholder communication is a dynamic, continuing activity. Through the implementation of culturally sensitive and transparent communication strategies, the utilisation of technology, and proactive stakeholder feedback gathering, project teams may foster greater collaboration, establish credibility, and guarantee the accomplishment of sustainability objectives. In addition to helping projects succeed, a thoughtful and attentive approach to communicating with stakeholders may also enhance the overall benefits of sustainable building projects for the environment and local people.

Table 8.8 Techniques to monitor plant and equipment in sustainable construction

Technique	Explanation
Sustainable fleet selection	Make sure that the construction fleet is in line with sustainability goals before starting the monitoring procedure. Give top priority to equipment that complies with the most recent environmental regulations, has reduced emissions, and is more fuel-efficient. Update the fleet often to take advantage of new developments in environmentally friendly technologies
Emission monitoring	To detect and analyse emissions data, install real-time emission monitoring systems on construction equipment. These data are helpful in pinpointing areas that require improvement, making the best use of the equipment, and guaranteeing that emission reduction targets are met. Emissions monitoring supports larger environmental and sustainability objectives
Utilization tracking	To maximise productivity and minimise downtime, keep an eye on how plants and equipment are being used. Employ tracking and telemetry technologies to collect information on equipment utilisation trends. Examine the data to find ways to appropriately scale the fleet and reduce operating expenses and the impact on the environment
Maintenance planning for efficiency	Formulates an anticipatory maintenance plan with an energy-saving focus. To minimise the chance of malfunctions and maximise efficiency, keep an eye on the equipment maintenance schedules to guarantee prompt servicing. Equipment that has been well-maintained runs more smoothly, supporting sustainability goals
Integration of IoT and sensors	Utilise digital technologies to collect data on equipment performance in real-time. Sensors can keep an eye on things like energy efficiency, engine health, and fuel use. Predictive maintenance is made possible by the integration of such technologies, which decreases downtime and improves sustainability
Renewable energy integration	Examine ways to include renewable energy sources in the functioning of construction machinery. This could entail lowering dependency on traditional energy sources by powering equipment with solar panels or hybrid systems. Observe how well renewable energy solutions are working in real-world situations
Noise reduction measures	When operating machinery and plants, keep an eye on and put noise reduction measures into practice on sites. Reducing the negative effects on the environment is frequently given top priority in sustainable construction projects. Put in place barriers to block out noise, use quieter equipment, and plan noisy operations for when people are less sensitive
Lifecycle analysis for equipment	To evaluate the overall environmental impact of construction equipment, conduct lifecycle studies. Consider aspects like production, delivery, consumption, and disposal. Keeping an eye on the equipment's lifecycle facilitates decision-making on replacement, disposal, and equipment selection

(continued)

Table 8.8 (continued)

Technique	Explanation
Sustainable fuel options	Investigate and keep an eye on the use of sustainable fuel alternatives for construction machinery. Electricity, biodiesel, and other alternative fuels might be examples of this. Analyse the viability, economics, and environmental advantages of converting the construction fleet to sustainable fuel sources
Operator training programs	Provide equipment operators with training that emphasises sustainable practices. Make sure operators are knowledgeable on how to reduce emissions, operate equipment in an energy-efficient manner, and use it sustainably. Operators are kept up to date on the most recent developments through regular training updates
Smart charging and energy storage	Use energy storage devices and clever charging strategies if you are using electric machinery. Track how well these technologies work to optimise charging schedules and reduce energy consumption. In addition to reducing energy waste, smart charging increases the project's overall sustainability
Recycling and repurposing equipment	Find ways to recycle or repurpose outdated machinery. Evaluate the viability of reusing machinery for various building phases or projects and track the success of equipment component recycling programmes. This strategy fits with the ideas of the circular economy
Carbon offsetting programmes	To offset the emissions brought on by the operation of plants and equipment, think about taking part in carbon offsetting programmes. To guarantee accountability and openness in reaching carbon neutrality targets, track and document the success of offsetting projects
Compliance with environmental standards	Conduct routine audits of plant and equipment operations to verify adherence to regional and global environmental regulations. Keep an eye on compliance with laws about emissions, noise levels, and environmentally friendly procedures. It is imperative to swiftly handle non-compliance issues to prevent adverse environmental effects
Continuous improvement culture	Encourage the management of plants and equipment to be a culture of continual improvement. To find areas that could be improved, ask operators and maintenance teams for their opinions. Review and improve sustainable practices regularly, taking advantage of feedback and observational insights

ix. Environmental Compliance

Observing and guaranteeing adherence to environmental laws. One of the most crucial components of sustainable construction projects is ensuring environmental compliance. Throughout a project, compliance monitoring entails using a methodology to track adherence to sustainability objectives, environmental standards, and best practices. The steps listed in Table 8.10 would guide organisations on strategies for environmental compliance in sustainable construction projects.

Table 8.9 Techniques of communication with stakeholders in sustainable construction

Technique	Explanation
Stakeholder identification and analysis	Determine and evaluate all pertinent parties involved in the sustainable construction project first. This includes members of the project team, contractors, local communities, regulatory agencies, and environmental organisations. Recognise their positions, expectations, and worries to adjust communication tactics
Stakeholder engagement plan	Create a thorough plan for engaging stakeholders that specifies communication tactics, distribution methods, and frequency. Clearly state the main points, goals, and expected results of all communications with the various stakeholder groups. Feedback systems must be included in the strategy to provide a two-way communication channel
Utilize communication technologies	Utilise innovative communication tools to improve stakeholder involvement. Establish communication applications, social media sites, and project websites to disseminate updates, milestones, and pertinent data. Keep an eye on how well these channels are reaching various stakeholder groups
Regular progress reports	Create and send stakeholders progress reports regularly. The main accomplishments, sustainability benchmarks, and any project plan modifications ought to be emphasised in these reports. Project teams can adjust their communication techniques in response to stakeholder feedback by keeping an eye on how progress updates are received
Transparency and openness	Stress honesty and transparency in your communication. Make certain that all relevant parties have access to current and accurate information regarding the project's social contributions, environmental effects, and sustainability activities. To foster confidence among stakeholders, keep an eye on how transparency is perceived
Feedback mechanisms	Create feedback channels and keep an eye on them so that interested parties can voice any questions, comments, or concerns. Online polls, neighbourhood gatherings, or special hotlines for communication are a few examples of this. Analyse the comments you receive regularly to find areas that need work and deal with the latest problems
Cultural sensitivity	Recognise and honour the stakeholders' diverse cultural backgrounds. Consider linguistic preferences, communication methods, and cultural quirks while designing culturally sensitive communication strategies. Track how satisfied stakeholders are with the project's cultural sensitivity strategy
Community meetings and workshops	Organise frequent community gatherings and seminars to interact with nearby communities. Provide updates on the status of the project, respond to queries, and solicit feedback. To determine the success of community engagement, keep an eye on community sentiment, attendance, and participation at these events

(continued)

Table 8.9 (continued)

Technique	Explanation
Collaborative decision-making	Invite stakeholders to participate in crucial project choices to promote cooperative decision-making processes. This could involve decisions made on design, methods for sustainability, or programmes for community development. Track the level of stakeholder participation and evaluate how cooperative decision-making affects project results
Crisis communication plan	Decide for crisis communication to deal with unforeseen issues or disagreements. Keep an eye on how well crisis communication tactics are controlling expectations from stakeholders, preserving confidence, and reducing any harm to the project's reputation for sustainability
Sustainability workshops and training	Plan training sessions and workshops on sustainability for all interested parties, such as contractors and the project team. These seminars can improve knowledge of sustainability objectives, optimal procedures, and the significance of individual efforts. After these activities, keep an eye on how sustainable practices are being adopted
Regular project website updates	As the primary source of project information, keep an updated project website. Update information frequently, highlighting community participation projects, environmental impact analyses, and sustainability accomplishments. Track user activity and website traffic to evaluate the online platform's usability and accessibility
Compliance with regulatory communication requirements	Make sure that the regulations of communication about sustainable construction projects are followed. Keep an eye on compliance with reporting requirements, environmental impact evaluations, and public disclosure duties. Ensure that information is provided to stakeholders promptly and accurately, adhering to regulatory requirements
Stakeholder surveys	Survey stakeholders regularly to learn more about their perceptions, awareness, and satisfaction with sustainability initiatives. Examine survey data to find patterns, areas that need work, and chances to improve stakeholder engagement tactics
Continuous improvement feedback loop	Create a feedback loop for ongoing improvement in stakeholder communication. Review and evaluate communication tactics regularly, applying best practices and lessons discovered to next initiatives. Keep an eye on how stakeholder relationships are changing and modify your communication strategies, as necessary

A proactive, multifaceted strategies are necessary for the effective monitoring of environmental compliance in sustainable construction projects. Construction activities can be made to comply with environmental requirements and favourably impact sustainability goals by project teams through the integration of sophisticated technologies, stakeholder involvement, and continual strategy improvement. The monitoring procedure improves the project's overall environmental performance and has

Table 8.10 Techniques to monitor environmental compliance in sustainable construction

Technique	Explanation
Regulatory framework understanding	Get a thorough grasp of the laws and regulations controlling environmental compliance at the project site beforehand. Determine whether laws, rules, and guidelines apply to building operations, environmental preservation, and sustainable development. Keep this knowledge base up to speed on regulatory changes by regularly updating it
Environmental impact assessment (EIA)	To detect potential environmental problems, conduct a thorough EIA at the beginning of the project. As the project moves forward, keep an eye on and update the EIA to make sure that mitigation strategies are applied successfully and that unanticipated effects are quickly handled
Compliance tracking tools	Use software and digital tools created especially for tracking environmental compliance. By centralising data, automating reporting, and offering real-time insights into compliance status, these technologies help simplify the monitoring process. To improve monitoring skills, regularly assess these instruments' efficacy and efficiency
Site inspections and audits	To ensure adherence to environmental laws and project-specific sustainability objectives, conduct routine site inspections and environmental audits. Keep an eye on the results of these inspections to spot patterns, areas that could use improvement, and possible remedial measures. Make certain that certified environmental professionals perform the checks
Environmental management plans	Create and carry out Environmental Management Plans (EMPs) that specify the steps needed to comply with environmental regulations. Keep an eye on the execution of EMPs, monitoring the status from time to time and modifying plans in response to evolving environmental issues or project conditions
Stakeholder involvement in compliance monitoring	Engage all relevant parties in the monitoring process, such as neighbourhood associations and environmental organisations. Promote citizen science projects and community-led audits as ways to involve the public in compliance monitoring. To resolve problems before they arise, keep an eye on stakeholder comments and concerns
Real-time environmental monitoring systems	Install real-time environmental monitoring devices to keep an eye on crucial metrics like noise levels, water quality, and air quality. These systems give quick access to information about possible environmental effects, enabling prompt intervention and modification of construction-related activities. Examine the precision and dependability of the monitoring apparatus regularly
Training and capacity building	To increase project teams', contractors', and pertinent stakeholders' capacity for environmental compliance, invest in training programmes. Assessments, comments, and performance reviews are used to track the success of training programmes. Programmes for training should be updated frequently to reflect new environmental requirements

(continued)

Table 8.10 (continued)

Technique	Explanation
Waste management tracking	To monitor the production, handling, and disposal of construction waste, put in place a reliable waste management tracking system. Establish goals for recycling and reusing waste, then evaluate your progress regularly. Keep an eye on contractors' compliance with waste management plans and, if necessary, offer advice for improvement
Green building certifications	Monitor the adherence to certification standards if the teams are going for green building certifications. This could involve monitoring indoor air quality, water conservation, energy efficiency, and the use of sustainable materials. Update the reports and paperwork needed for certification regularly
Emergency response drills	To make sure that project teams are equipped to handle environmental crises like spills, leaks, or accidents, conduct frequent emergency response drills. Using drills, assess the efficacy of emergency response processes and modify them based on lessons discovered
Erosion and sediment control	Follow environmental standards when implementing erosion and sediment control methods. To stop soil erosion and sedimentation in water bodies, erosion control structures should be routinely inspected and monitored. Control methods should be modified in response to building activity and weather
Biodiversity conservation monitoring	Use biodiversity protection strategies and keep an eye on their success if the project influences natural environments. Monitor biodiversity, evaluate efforts to restore habitat, and keep an eye out for protected species. Based on monitoring results, modify conservation strategies
Carbon footprint tracking	During the project's lifecycle, keep an eye on and record the carbon footprint. Make routine evaluations of emissions, energy use, and the adoption of environmentally friendly transportation choices. Implement tactics to lower the project's carbon footprint and advance general sustainability objectives
Continuous improvement and reporting	For environmental compliance, create a cycle of continual improvement. Review monitoring data regularly, carry out post-project assessments, and communicate lessons discovered. Provide thorough reports on environmental compliance for stakeholders, the public, and regulatory agencies. Make use of these reports to highlight accomplishments, discuss difficulties, and layout plans

a long-lasting beneficial effect on the surrounding ecology in addition to adhering to regulatory requirements.

Project Evaluation in Construction

Construction project evaluation is the methodical and thorough process of evaluating a construction project at distinct phases to ascertain its viability, advancement, and overall success. Examining the project from financial, technical, environmental, and social-will help you make well-informed decisions and make sure the project accomplishes its goals. There are three stages to a project evaluation: pre-construction, during construction, and post-construction. Project evaluation seeks to guarantee that the construction project adheres to quality standards, satisfies legal requirements, and provides value to stakeholders in line with its original goals. Throughout the project, monitoring, assessment, and adaptation are all part of the dynamic and continual process of evaluating a construction project. The evaluation components for construction projects differed.

Construction Quality Evaluation

Evaluating the finished construction project to make sure the quality fulfils the required criteria. Beyond conventional metrics of utility and longevity, quality evaluation in sustainable construction projects is a complex process. Sustainable construction prioritises social responsibility, economic feasibility, and environmental responsibility. For such projects, assessing quality necessitates a comprehensive strategy that considers several factors, from long-term performance to material selection. A manual for assessing quality in environmentally friendly building projects is contained in Table 8.11.

Table 8.11 Techniques for evaluating quality in sustainable construction

Technique	Explanation
Sustainable materials selection	Evaluation criteria: when choosing materials, consider their effect on the environment, their capacity to use resources efficiently, and their potential for recycling Standards and certifications: verify that the materials adhere to accepted sustainability certifications and standards Do a life cycle assessment to find out how materials affect the environment at each stage of their life, from extraction and manufacture to disposal and recycling
Energy efficiency and performance	Assess the project's design for energy efficiency, considering renewable energy sources and passive design techniques Performance metrics: throughout the project, track and evaluate energy performance, considering elements such as insulation, HVAC systems, and lighting efficiency Obtain and assess green building certifications, such as GBI, BREEAM, LEED, or GBCN, which place a focus on environmental performance and energy efficiency
Construction processes and techniques	Sustainable construction practices: assess the building procedures to make sure that sustainable practices—such as minimising waste, preventing pollution, and using water responsibly—are followed Innovative technologies: evaluate the application of innovative building technologies that maximise productivity while reducing their negative effects on the environment Training and certification: ascertain that workers in the construction industry have received training in sustainable building methods and have obtained the necessary certificates

(continued)

Table 8.11 (continued)

Technique	Explanation
Water conservation and management	Water-efficient fixtures: to reduce water use, install and assess water-efficient fixtures, irrigation systems, and landscaping techniques Evaluate the viability and efficiency of rainwater harvesting devices for usage other than potable water Protection of water quality: assess building methods to avoid soil erosion, sedimentation, and contamination of water sources
Indoor environmental quality	Ventilation and air quality: evaluate indoor air quality and use low-emission materials and suitable ventilation systems to enhance occupant health and comfort Natural lighting: use natural lighting techniques to lessen your dependency on artificial lighting and make your home healthier Thermal comfort: to maximise occupant well-being, evaluate thermal comfort solutions such as insulation, energy-efficient windows, and shading devices
Resilience and adaptability	Climate resilience: consider how resilient the project is to the effects of climate change, such as harsh weather and sea level rise Adaptability: create structures that can be retrofitted or adjusted to accommodate changes in the future or advances in technology Risk assessment: to identify dangers associated with climate change and to design mitigation and adaptation plans, conduct risk assessments
Post-construction evaluation	Feedback from occupants: find out what they think about the energy efficiency, indoor environment, and general level of satisfaction with the sustainable elements Monitoring and maintenance: create post-construction monitoring programmes to evaluate sustainable systems' continuing performance and carry out routine maintenance to guarantee their sustained effectiveness
Social and economic considerations	Impact on the community: assess how the project will affect the local community in terms of employment growth, social inclusion, and overall quality of life enhancements Economic viability: evaluate the long-term savings, return on investment, and initial costs when determining if sustainable activities are economically feasible
Certification and compliance	Green building certifications: aims for and assess certifications to make sure they meet accepted sustainability guidelines Regulatory compliance: keep an eye on adherence to national and international laws about green building and sustainable building techniques
Continuous improvement	Establish systems for gathering feedback from all relevant parties, including local communities, construction crews, and tenants, to pinpoint areas that require improvement Benchmarking: to keep up with developments in sustainable building, regularly compare the project's performance to industry norms and best practices

Assessing the quality of sustainable construction projects necessitates a thorough and integrated method that takes the social, economic, and environmental aspects into account. Through the integration of sustainable concepts across all project phases, ranging from material selection to post-construction monitoring, stakeholders may guarantee that the construction project fulfils superior quality standards and also makes a good impact on the environment and occupant well-being. A dedication to changing sustainability practices and a constant state of improvement are essential components of an effective quality evaluation process in sustainable construction.

Performance Evaluation

Assessing the quality of sustainable construction projects requires a thorough and integrated method that considers the social, economic, and environmental aspects. Through the integration of sustainable concepts across all project phases, ranging from material selection to post-construction monitoring, stakeholders may guarantee that the project fulfils superior quality standards and makes a good impact on the environment and occupant well-being. A dedication to changing sustainability practices and a constant state of improvement are essential components of an effective quality evaluation process in sustainable construction. Table 8.12 contains the techniques for evaluating the performance of sustainable construction.

An integrated strategy that considers the environmental, social, and economic aspects is necessary to assess the effectiveness of sustainable construction projects.

Table 8.12 Techniques for evaluating project performance in sustainable construction

Technique	Explanation
Sustainability goal alignment	Establishing aims: let us start by outlining the construction project's sustainability aims and objectives. These could include following green building guidelines, social justice, resource conservation, and energy efficiency Conformity to international standards: make sure that the project complies with internationally recognised sustainability certifications and standards, such as BREEAM, GBI, GBCN, LEED, etc.
Key performance indicators	Creating KPIs: identify the sustainability-related key performance indicators that are precise, quantifiable, and attainable. These could include waste production, water use, energy use, and social impact metrics Benchmarking: to compare the project's performance to predetermined norms, and develop benchmarks based on industry standards and best practices
Energy and resource efficiency	Energy performance: analyse data on energy consumption to determine how energy-efficient the project is. Analyse performance gaps between design expectations and actual performance to find areas for improvement Resource utilisation: evaluate how well resources and materials were used during the building process. Think about reducing waste, using recycled materials, and using sustainable procurement practices

(continued)

Table 8.12 (continued)

Technique	Explanation
Life cycle assessment	Environmental impact analysis: to examine how the construction project would affect the environment from the extraction of raw materials to their disposal, do a life cycle evaluation. Think about the amount of water used, the embodied carbon, and other environmental aspects Implement mitigation efforts to lessen identified environmental impacts and keep an eye on how well these are working overtime
Social impact assessment	Community interaction: using feedback systems and stakeholder interaction, assess how the initiative is affecting the local community. Evaluate the creation of jobs, social inclusion, and the general well-being of the community Feedback from tenants can be used to determine how satisfied they are with the built environment, indoor air quality, and general living or working conditions
Economic viability	Financial performance: examine the sustainable building project's financial performance by contrasting actual and planned expenses. To ascertain long-term financial benefits, consider life cycle cost analyses ROI: To determine the ROI, assess the financial gains made possible by the project's sustainable characteristics and practices
Regulatory compliance	Monitoring compliance: ensure that local, national, and international laws about sustainable building methods are followed regularly. Keep up with any modifications to regulatory requirements so that you may modify project tactics appropriately
Green building certifications	Examine the project's conformance with the standards established by certifying bodies if you are interested in obtaining green building certifications. Make sure you have all the supporting documentation and proof needed to apply for certification Continuous improvement: by addressing areas that have been highlighted for improvement, use certification processes as a chance for ongoing development
Post-occupancy evaluation	Behaviour of occupants and well-being: post-occupancy assessments should be carried out to determine how tenants engage with the sustainable elements and how that affects their well-being. Consider elements like user behaviour, comfort, and happiness Operational performance: assess the continuous performance of sustainable systems, such as water management, lighting, and HVAC, to determine what needs to be optimised or maintained
Resilience and adaptability	Assessment of climate resilience: determine how resilient the project is to the effects of climate change, considering variables like harsh weather and rising temperatures Adaptability: evaluate how well-suited the built facility is to deal with upcoming modifications, innovative technology, and changing sustainability requirements

Project teams should make certain that sustainability goals in projects are accomplished and that the project positively impacts the environment and the communities it serves by using key performance indicators, life cycle assessments, and continual stakeholder involvement. Adaptability, a dedication to learning from every project, and constant development are factors that make sustainable building programmes successful overall.

Safety Performance Assessment

Evaluate the safety performance of project by evaluating safety occurrences, records, and compliance with safety procedures. When it comes to sustainable construction, where the health and safety of both workers and the environment are of utmost importance, the importance of ensuring safety in construction projects is crucial. In sustainable construction projects, safety performance assessment entails a thorough analysis of procedures, guidelines, and results to ensure a safe working environment while reducing the project's environmental impact. The methods listed in Table 8.13 are used to assess safety performance in sustainable construction projects:

Table 8.13 Techniques for evaluating safety in sustainable construction

Technique	Explanation
Safety culture and leadership	Leadership commitment: examine how well a strong safety culture has been developed and communicated to gauge how committed project leadership is to safety. To create a culture where employees feel encouraged to voice concerns without fear of retaliation, leadership should actively prioritise safety Safety training: evaluate the efficacy of the management and employee safety training initiatives. Make sure the training includes specifics about sustainable construction techniques in addition to general safety procedures
Safety planning and risk assessment	Safety plans: examine and evaluate the project's safety plans to make sure they meet industry standards and are thorough. Incorporate sustainable practice considerations, such as managing eco-friendly materials and reducing pollutants during construction Risk evaluations: make routine risk assessments that consider the hazards connected to sustainable methods as well as the risks inherent in conventional construction. Determine any risk-problems associated with sustainable materials, renewable energy installations, and green technologies
Safety performance metrics	Event reporting and analysis: examine safety event reporting and analysis to find trends, underlying issues, and development opportunities. Establish a structure for in-depth incident investigations and make sure that the lessons learnt are used to stop future occurrences Leading and lagging indicators: both leading and lagging indicators can be used to assess safety performance. Leading indicators focus on preventive measures (e.g., training hours, safety meetings), while trailing indicators include historical data (e.g., injury rates). Strive for balance in order to provide a proactive approach to safety

(continued)

Table 8.13 (continued)

Technique	Explanation
Emergency preparedness and response	Evaluate the efficacy of emergency drills and reaction strategies in different scenarios. Take sustainability factors into account when responding to environmental emergencies or situations involving green technologies Communication protocols: during emergencies, make certain that everyone involved is aware of and follows the established communication protocols. Establish a transparent communication and command structure to improve the effectiveness of your responses
Worker involvement and empowerment	Worker participation: assess the degree of worker involvement in decision-making concerning safety. Employees should be given the opportunity to actively participate in safety programmes, offer feedback, and recommend changes Encourage safety-conscious behaviour by putting in place a system for acknowledging and rewarding it. Encourage a positive safety culture by praising individuals and groups for their dedication to safety
Sustainable material handling and construction practices	Safe handling of sustainable materials: evaluate the procedures for installing and handling sustainable materials safely. Consider the special qualities of sustainable products and make sure employees have received the necessary training to handle them properly Construction methods: assess construction methods, particularly those about sustainable technologies like solar or green roofs. Make sure that safety procedures are customised to these sustainable construction components
Health and well-being programmes	Evaluation of occupational health programmes: determine whether there are any occupational health programmes in place and how successful they are in addressing both physical and emotional well-being. Comprehensive health and safety protocols should be given top priority in sustainable building Fatigue management: assess methods for addressing worker fatigue, as it is essential to sustain concentration and avert mishaps. Put policies in place including sufficient rest intervals, shift work arrangements and efforts to raise awareness of the value of sleep
Technology integration for safety	Use of safety technology: assess how well safety performance is monitored and enhanced by integrating safety technology, such as wearables or sensor-based systems. Use technology to monitor safety indicators in real time and to take immediate action, when possible, risks arise Digital reporting platforms: make use of digital platforms to report events, near misses, and safety issues. Make sure these platforms allow for anonymous reporting to promote candid discussion on safety-related matters
Continuous improvement	Safety audits: perform routine safety audits to evaluate the efficacy of safety efforts and compliance with safety requirements. To continuously improve safety performance, put audit results into practice by implementing remedial actions Mechanisms of feedback: provide channels via which employees and other interested parties can comment on safety procedures. Ask for feedback on how safety procedures might be made better and utilise this information to motivate ongoing development
Regulatory compliance and certification	Regulatory compliance: consistently examine and verify adherence to safety requirements, encompassing those particular to sustainable building practices. Keep abreast of revisions to safety regulations and modify procedures, as necessary Certifications for Safety: Seek safety certifications while prioritising environmentally friendly building practices

To create a safe and sustainable conscious workplace, safety performance assessment is essential in sustainable construction g projects. Through the implementation of an all-encompassing strategy that considers worker involvement, leadership commitment, risk assessment, and continuous improvement, project teams may guarantee that safety is given top priority during the entire duration of the project. Adopting actions based on sustainable construction methods improves the overall safety culture and advances the project's and the participant's well-being.

Cost–Benefit Analysis

Evaluating cost-effectiveness of the construction projects by weighing the returns on investment against the resources used. The elements and factors listed in Table 8.14 when performing a cost–benefit analysis for sustainable construction projects.

For sustainable construction projects, a thorough cost–benefit analysis necessitates a close evaluation of the social and environmental factors in addition to the financial ones in decision making. By carefully balancing the costs and advantages of sustainability features, stakeholders may both make informed decisions that support economic goals and contribute to a more sustainable built environment. Achieving a sustainable construction project that is both economically feasible and successful requires striking a balance between initial costs and long-term rewards.

Table 8.14 Components of cost–benefit analysis in sustainable construction

Component	Explanation
Initial costs	Investing in sustainable features: determine the up-front expenses related to incorporating sustainable components into the building endeavour. This could include costs for sustainable materials, energy-efficient technology, and green building certifications Costs of construction and implementation: determine how much it will cost to use sustainable construction techniques during the construction stage. This includes labour, supplies, and any other costs necessary to comply with environmental regulations
Operational and maintenance costs	Energy and water efficiency: evaluate the possible operating cost reductions from the implementation of water- and energy-saving technologies. Examine the financial advantages of lower utility costs and upkeep requirements over the long run Life cycle assessment: perform a life cycle cost analysis to determine the overall cost of ownership for the project
Environmental and social benefits	Emissions reduction: calculate how much less energy, water, and carbon emissions will assist the environment. Consider the cost of carbon or the impact on the environment and translate these reductions into possible financial savings or societal gains Better indoor air quality: consider the possible health advantages of better indoor air quality, which can result in higher productivity and lower medical expenses

(continued)

Table 8.14 (continued)

Component	Explanation
Long-term resilience	Evaluate the financial benefits of constructing climate change resilience. Ponder on the possible financial savings from preventing flooding, harsh weather, or other risks related to climate change Flexibility and adaptability: consider the financial advantages of planning for future changes and flexibility, as this may lower the costs of maintenance, refurbishment, renovations or retrofits
Social and community impact	Community engagement: consider how supportive and involved the community can be sustainably. Property values and community satisfaction may rise in response to projects that improve the well-being of the surrounding communities Job creation: assess the financial advantages of job creation and the local employment prospects brought about by sustainable construction methods

Client Satisfaction

Gathering thoughts and experiences about how satisfied customers and end users are with the finished project is crucial. In the world of sustainable construction, client happiness has even more importance as a key indicator of project success. In sustainable construction, client satisfaction goes beyond standard project metrics and includes evaluating how well the project complies with the customer's social and environmental values, fulfilling sustainability elements that were promised, and meeting or surpassing expectations. The methods for assessing client satisfaction in sustainable construction projects are as follows, see Table 8.15.

Table 8.15 Techniques to evaluate client satisfaction in sustainable construction

Technique	Explanation
Understanding client expectations	First consultation: during the first consultation, ascertain the client's expectations to start the evaluation process. Determine their preferences, particular sustainability goals, and any special needs about green building techniques Open and honest communication establish effective channels of communication to go over project goals, deadlines, and any obstacles. Make sure the client is aware of the sustainable elements included in the project
Pre-construction engagement	Design development: include the client in the process so they may offer suggestions on energy-efficient systems, sustainable features, and material selections. Get input on design ideas and make plans appropriately Initiatives for education: give your clients educational materials that explain the advantages and ramifications of sustainable building. Make sure they understand the possible financial savings and long-term benefits of using green building techniques

(continued)

Table 8.15 (continued)

Technique	Explanation
Sustainable features implementation	Enforce stringent quality assurance protocols at the construction stage to guarantee the accurate installation of sustainable elements. To keep clients' trust, keep the clients informed about progress regularly and respond to any problem right away Site visits and updates: to keep clients updated on the status of construction, schedule site visits or send out frequent updates. This encourages participation and gives clients a chance to see how sustainable features are integrated directly
Post-construction evaluation	If appropriate, instruct owners on how to use and maintain sustainable features. Make sure your customers are prepared to reap the full advantages of water-saving devices, energy-efficient systems, and other environmentally friendly building elements Conduct a final walkthrough with the customer to demonstrate the finished sustainable features. Answer any queries or worries you may have and get opinions on the entire building process
Performance monitoring	Monitoring of energy and resource use: put in place mechanisms to monitor how well sustainable features are working after construction. To show the continued benefits of resource-efficient solutions, this involves monitoring water and energy use Provide a way for tenants to offer their opinions on how well-functioning and efficient sustainable features are. Consider focus groups, polls, and internet channels to get insightful information
Long-term resilience	Evaluation of climate resilience: determine the extent to which the project satisfies the client's expectations in terms of climate resilience. Assess how well using sustainable construction methods has improved the building's ability to withstand climate-related problems Adaptive techniques: talk about and put into practice adaptive techniques that improve the project's long-term resilience and support the client's wish for a building that is both sustainable and ready for the future
Economic and social impact	Return on investment: determine the return on investment to appraise the sustainable characteristics' economic impact. Emphasise any financial gains or cost savings from less maintenance, energy efficiency, or other sources Community impact: evaluate how the construction project will affect the neighbourhood socially. Get input on the project's perceived benefits whether it involves creating jobs or engaging the community
Post-occupancy survey and interviews	To obtain in-depth client feedback, conduct interviews or post-occupancy survey sessions. Find out how satisfied they are with the project overall, how well the sustainable aspects work, and if there is anything that may be improved Provide clients with the opportunity to share their experiences, difficulties, and particular areas of the project that met or surpassed their expectations using open-ended questions

(continued)

Table 8.15 (continued)

Technique	Explanation
Continuous improvement	Knowledge acquired workshops: plan discussions or workshops evaluate the project's lessons learned. Invite clients to share their perspectives on what went well and where future sustainable construction projects might be improved Integration of feedback: include customer feedback in the cycle of continual improvement for the project. Make better use of the knowledge acquired to improve project delivery in general, communication tactics, and sustainable construction methods
Certification and recognition	Consider awarding clients that enthusiastically engage in and endorse sustainable construction projects. Recognise their dedication to eco-friendly construction methods and emphasise how they help create a built environment that is more sustainable Accomplishment with certifications: if appropriate, commemorate the attainment of sustainability honours or green building certifications. Highlight the project's benefits for the client and the larger community

To assess client satisfaction in sustainable construction projects, a comprehensive method that extends beyond the physical structure's completion is required. It includes continuous feedback mechanisms, performance monitoring of sustainable elements, and customer involvement throughout the construction phase. Through emphasizing open communication, meeting client expectations, and consistently searching for ways to improve project performance.

Post-occupancy Evaluation for Building Projects

A crucial stage in the life cycle of sustainable construction projects is post-occupancy evaluation, which offers insightful information on how well the sustainable features function in actual use and how they affect occupants. This assessment extends beyond conventional measurements, exploring environmental, social, and economic aspects to guarantee that objectives related to sustainability are fulfilled. Methods for carrying out a post-occupancy assessment on sustainable construction projects are contained in Table 8.16.

An all-encompassing strategy is necessary for the complex process of evaluating post-occupancy performance in sustainable construction projects. Project teams can obtain a thorough understanding of the sustainability features' performance in real-world situations by considering environmental, social, and economic elements. For sustainable construction projects to be successful over the long run, they must be continuously improved, flexible, and transparent in their interactions with stakeholders and occupants.

Table 8.16 Techniques for conducting POE in buildings

Technique	Explanation
Assessing environmental performance	Energy efficiency: to evaluate the real energy usage of the adopted sustainable features, use building performance data. The efficiency of energy-efficient devices can be assessed by comparing the actual energy usage to the initial forecasts Water conservation: compare patterns of water consumption to assess the effectiveness of water-saving measures. Evaluate the effectiveness of irrigation systems, rainwater harvesting, and water-efficient fixtures to see how they affect total water use
Indoor environmental quality evaluation	Use air quality monitoring to evaluate the quality of the interior environment. Examine the efficiency of air filtering, ventilation systems, and other methods for preserving hygienic indoor air quality while considering particulate matter and VOC levels Thermal comfort: find out what the clients think about their level of thermal comfort. Evaluate the efficiency of passive design techniques in preserving pleasant interior temperatures, as well as the performance of heating, ventilation, and air conditioning systems
Occupant satisfaction and well-being	Surveys and Interviews: To find out how satisfied clients are with the general building atmosphere, conduct surveys and interviews with them. Examine elements like lighting, sound, and room arrangement, and consisder how they affect the productivity and well-being of occupants Post-Occupancy Workshops: To encourage practical conversations regarding clients' experiences, set up workshops or focus groups. Request input on features, comfort levels, and facilities of the property
Functional performance of sustainable features	Green technologies assessment: check the efficiency and usefulness of sustainable elements, like rainwater collection systems, green roofs, and solar panels. Evaluate if these features are serving the intended functions and advancing the overall project objectives Smart building technologies: evaluate the functionality of integrated smart building technologies. Analyse how well occupancy sensors, automation systems, and other technologies work to maximise energy efficiency and improve client comfort
Economic impact and cost–benefit analysis	Operational cost analysis: examine how much the sustainable features will cost to operate. The long-term economic impact of these elements should be evaluated by considering the costs of replacement, repair, and ongoing maintenance Cost savings recognition: determine and measure any financial savings attained by conserving water, energy, and other sustainable measures. In a thorough cost–benefit analysis, emphasise the monetary advantages of these aspects

(continued)

Table 8.16 (continued)

Technique	Explanation
Community and social impact	Assess the effects of the construction project on the local community through community engagement. Evaluate the extent to which the sustainable construction approaches have resulted in beneficial social consequences, employment development, and community engagement Surveys of public perception: to find out how the public feels about the sustainable construction project, conduct surveys within the community. Examine whether the project has improved the well-being and quality of life in the community
Regulatory compliance and certifications	Verify compliance: ascertain whether the sustainable construction project is still adhering to legal requirements. Ensure that all environmental rules, building codes, and any certifications earned during the construction phase are followed Certification maintenance: evaluate how these certifications are maintained if the project resulted in green building certifications. This entails continuing to follow the certification guidelines and the project's dedication to environmentally friendly procedures
Adaptability and resilience	Flexibility assessment: determine how well the project can accommodate evolving requirements. Examine whether the design can be expanded or modified to meet changing occupancy requirements or sustainability goals Evaluate the construction's ability to withstand the effects of climate change. Evaluate the extent to which the sustainable elements enhance the resilience of the building, considering extreme weather events, temperature changes, and other climate-related issues
Continuous improvement	Architects, engineers, quantity surveyors, construction managers, contractors, and sustainability experts should be included in the post-occupancy assessment sessions that you arrange with the project team. Identify areas for improvement in the next sustainable construction projects and talk about the lessons learnt Feedback combination: the project's continuous improvement plan should incorporate input from building managers, residents, and other stakeholders. Refine future project strategies, address any concerns, and improve sustainable procedures with the help of this knowledge
Communication and transparency	Establish platforms for continuous contact to allow tenants to offer comments. Make use of online communities, suggestion boxes, and digital platforms to promote open dialogue about their experiences Public reporting: consider issuing a report detailing the main conclusions and lessons discovered during the occupation. For the sake of upholding accountability and openness, distribute this report to the tenants, the neighbourhood, and other relevant parties

8.6 Integration of Sustainability into Construction Projects

Sustainability in construction projects has moved from being a desired feature to a critical success factor. Sustainability needs to be included in a project's strategic definition from the opening to the end for it to be effectively integrated. This thorough guide examines the essential ideas, procedures, and factors to consider when incorporating sustainability into the strategic planning of construction projects. Table 8.17 lists the methods used to include sustainability in building projects.

Table 8.17 Techniques for integrating sustainability into construction projects

Practice	Technique
Defining sustainable objectives	Start with a clear vision: clearly define the project's sustainability goals. This entails creating broad sustainable goals that are in line with organisational, regional, and worldwide sustainability objectives Include stakeholder input: engage crucial stakeholders early on to collect a range of viewpoints and make sure that sustainability goals align with the expectations and values of all parties concerned
Sustainable project charter	Create a sustainable project charter: write a project charter that clearly states the objectives, targets, and obligations related to sustainability. During the project, this charter acts as a foundational document and guide for decision-making Include the fundamentals: to establish a well-rounded and comprehensive strategy, incorporate fundamental sustainability concepts, such as the triple bottom line (economic, social, and environmental factors), into the project charter
Lifecycle thinking and assessment	A lifecycle view that considers the project's effects on the environment, society, and economy from inception to decommissioning should be adopted The environmental impact of construction materials, technology, and procedures can be assessed by integrating lifecycle assessments. This helps in making judgements about the production of waste, the usage of resources, and energy
Risk and opportunity analysis	Determine and evaluate sustainability-related risks that could affect the project. These can include modifications to regulations, the accessibility of resources, and reputational hazards related to social and environmental activities Identify ways to improve sustainability performance, such as by putting new green technologies to use, adopting circular economy principles, or making use of renewable energy sources
Regulatory compliance and certification	Recognise the regulations' needs: keep up with regional, national, and international construction rules. To reduce the risk of legal issues and compliance, make sure that the project's design and implementation comply with these requirements Go for certifications: if you are seeking to verify and achieve sustainable construction practices, you might think about going for well-known sustainability certifications like GBCN, LEED, or BREEAM

(continued)

Table 8.17 (continued)

Practice	Technique
Integrated design processes	Adopt integrated design practices: encourage cooperation from the outset between contractors, engineers, architects, and other stakeholders. Sustainable design principles are incorporated into the project's DNA through integrated design processes, which impact choices on materials, energy efficiency, and spatial layout Use BIM: use BIM technology to support a data-driven and cooperative design process. BIM would allow project member collaborate and share information in real-time, which promotes more sustainable decision-making
Material selection and procurement	Set criteria for material selection based on social, environmental, and life cycle assessment factors to make sustainable materials a priority. Give preference to materials with low embodied carbon, recyclable content, and ethical sourcing methods Interact with sustainable vendors: join forces with vendors who are dedicated to eco-friendly operations. Provide open lines of communication to suppliers to make sure they follow moral and environmental guidelines
Energy-efficient design and renewable energy	Design for energy efficiency: include energy-saving technologies and passive design techniques in the project. To cut down on energy use, consider elements like insulation, natural lighting, and building orientation Include renewable energy sources: examine methods for integrating renewable energy sources, such solar or wind. To find the project's most efficient and sustainable energy options, do feasibility studies
Water conservation strategies	Install water-efficient systems: install water-saving devices and frameworks, such as rainwater collection, smart irrigation, and low-flow plumbing fixtures. Give top priority to landscape designs that use as little water as possible Examine water recycling: to reduce the impact on nearby water supplies, investigate the viability of water recycling technologies. For example, putting in place greywater systems can help with sustainable water management
Social considerations and community engagement	Social impact assessment: to determine the advantages and disadvantages for the neighbourhood, conduct a social impact assessment. Think about things like inclusiveness, the production of jobs, and community well-being Programmes for community engagement: create initiatives for community engagement that include the local populace in the project. Ask for feedback, respond to worries, and explain the good things the project hopes to do for the community
Resilience and adaptability	Evaluate the project's susceptibility to extreme weather events and climate change as part of your climate resilience planning. To guarantee the project's long-term viability, incorporate resilience measures into the design and construction process Design for adaptability: include features in your design that can be changed in the future. This guarantees that the project will continue to adapt to new technical developments and changing sustainability criteria

(continued)

Table 8.17 (continued)

Practice	Technique
Monitoring and performance measurement	Create measurable KPIs to track sustainability performance throughout the project: establish key performance indicators. KPIs might include social effect indicators, waste reduction goals, and energy efficiency metrics Put real-time monitoring systems in place: incorporate real-time monitoring tools to monitor and evaluate the progress of your projects. Because these systems offer immediate feedback, proactive decision-making and ongoing development are made possible
Stakeholder communication and reporting	Transparent communication: keep open lines of communication with all relevant parties about the objectives, developments, and results of sustainability. Provide regular updates to stakeholders via a range of platforms, such as meetings, reports, and online resources Sustainability reporting: draft thorough reports that track the project's results in terms of the environment, society, and economy. These reports provide an open and honest assessment of the project's sustainability contributions

Construction projects must take a proactive and diversified strategy to incorporate sustainability into their strategic concept. Construction teams may design structures and infrastructure that not only satisfy present requirements but also favourably impact the social, economic, and environmental well-being of present and future generations by incorporating sustainability principles into every phase of project development, from goal setting to post-project evaluation. Sustainable construction projects can be role models for responsible and progressive development via strategic planning, teamwork, and a dedication to ongoing change.

8.7 Integrating Sustainability into the Preparation and Briefing of Construction Projects

In construction projects, sustainability is not an optional feature; rather, it is a basic component that ought to be woven into the planning and briefing process of the project. The first phases of a construction project establish the framework for the remainder of its lifecycle, from identifying project objectives to establishing sustainability targets. Table 8.18 contains a thorough reference that examines the essential ideas, tactics, and factors to consider when incorporating sustainability into construction project planning and briefing.

It takes a thorough and proactive strategy to include sustainability in the planning and briefing of construction projects. Construction teams may design a successful project plan that gives priority to social, economic, and environmental factors by incorporating sustainability concepts into the project's basis. Strategic planning,

Table 8.18 Techniques incorporating sustainability during the preparation and briefing stage

Practice	Technique
Defining sustainability objectives	Include sustainability from the outset: the project's inception marks the start of the road towards sustainability. Make sure that from the start, sustainability is ingrained in the project's DNA Describe unambiguous sustainability goals: set measurable, unambiguous sustainability goals that are in line with the overarching organisational, regional, and global sustainability objectives. Decision-making will be guided by these goals for the duration of the project
Stakeholder engagement	Determine the principal players: acknowledge the wide spectrum of parties engaged in the project, including local communities, architects, and financiers. Early on, interact with stakeholders to learn about their goals, worries, and expectations concerning sustainability Take stakeholder feedback into account: incorporate stakeholder input into the project's sustainability plan in a proactive manner. This guarantees that the project is in line with the objectives and values of people who will be affected by it, both directly and indirectly
Sustainability briefing	Create a comprehensive sustainability brief: draft a paper outlining the project's sustainability commitment in a sustainability briefing style. The main objectives of sustainability, performance metrics, and tactics should be included in this paper Alignment with organisational values: make sure the sustainability briefing is in line with the larger commitments and values of the organisation. To develop a coherent and integrated strategy for sustainability, this alignment is essential
Legal and regulatory compliance	Recognise regulatory frameworks: make sure the project is designed and carried out by these regulations to minimise risks related to compliance and the law Expect future modifications to regulations: plan for flexibility and be prepared for any changes that may occur to sustainability-related regulations. Create the project with the intention of meeting and surpassing future sustainability criteria
Certifications and standards	Examine sustainable certifications: investigate and choose sustainability standards and certifications pertinent to the project's objectives. A formal framework is offered by certifications like GBCN, GBI, LEED, or BREEAM for achieving and confirming sustainable building practices Embrace certification conditions: include in the project's briefing documents the prerequisites for the chosen certifications. This guarantees that sustainability goals are verified and accepted by established criteria, in addition to being met

(continued)

Table 8.18 (continued)

Practice	Technique
Integrated design practices	Promote collaboration within the project team, comprising architects, engineers, contractors, and sustainability specialists, through the use of a collaborative design methodology. Using integrated design methodologies influences decisions on project design, materials, and energy efficiency, promoting a holistic approach to sustainability Use BIM into practice: make use of BIM technologies to support data-driven decision-making and teamwork. Sustainability problems are effortlessly integrated into the design process using BIM's ability to facilitate real-time information sharing and coordination
Material selection and procurement	Set criteria for material selection based on life cycle assessments, ethical concerns, and environmental effects to prioritise sustainable materials. Give priority to materials with low embodied carbon, recyclable content, and ethical sourcing methods Involve ethical vendors: join forces with vendors who are dedicated to eco-friendly operations. Keep lines of communication open and honest to make sure vendors follow moral and environmental guidelines
Energy-efficient design	Passive design techniques: to maximise energy efficiency, incorporate passive design techniques into the project. To cut down on energy use, consider elements like insulation, natural lighting, and building orientation Examine alternatives for renewable energy: integrate renewable energy sources, including wind turbines or solar panels. To find the project's most efficient and sustainable energy options, carry out feasibility studies
Water conservation strategies	Install water-efficient systems: install water-saving devices and frameworks, such as rainwater collection, smart irrigation, and low-flow plumbing fixtures. Give top priority to landscape designs that use as little water as possible Think about recycling water: determine whether water recycling options are feasible in order to reduce reliance on external water sources. For instance, putting in place greywater systems can support sustainable water management
Social considerations and community engagement	Perform social impact assessments: evaluate how the project may affect local communities socially. When designing and executing the project, consider aspects like inclusion, the creation of jobs, and the health of the community Participate in local communities: create initiatives for community involvement that include the locals in the project. Ask for feedback, respond to worries, and explain the good things the project hopes to do for the community

(continued)

Table 8.18 (continued)

Practice	Technique
Resilience and adaptability	Evaluate the project's susceptibility to extreme weather events and climate change as part of your climate resilience planning. To guarantee the project's long-term viability, incorporate resilience measures into the design and construction process Adaptable design elements that enable future adjustments should be incorporated. By doing this, the project is guaranteed to remain adaptable to changing environmental standards and technological improvements
Risk and opportunity analysis	Determine sustainability risks: identify and evaluate sustainability-related risks, such as reputational, resource availability, and regulatory changes. Create plans to reduce these risks and take advantage of possibilities connected to sustainability Include prospects for innovation: determine where innovation in sustainable construction can be applied. Accept novel approaches and technology that can improve sustainability performance and help the project succeed
Communication and reporting	Provide open lines of communication to all parties involved so they are aware of the objectives, developments, and results of sustainability. Provide stakeholders with frequent updates using various channels, such as meetings, reports, and online resources The environmental, social, and economic performance of the project should be documented in detail in the sustainability reports that should be prepared. The project's contributions to sustainability are transparently reported in these reports

teamwork, and dedication to ongoing development can make sustainable construction projects role models for conscientious and progressive development.

8.8 Integration of Sustainability into the Concept Design Phase of Construction Projects

Sustainability is a design philosophy that needs to be integrated into the core concept of a construction project proactively. It should not be an afterthought. Forming the project's identity and functioning starts with the concept design phase. The guidelines for implementing sustainability in building at the concept design phase are listed in Table 8.19.

The basis of a sustainable construction project is laid during the concept design phase. Construction teams may provide designs that not only satisfy present demands but also have a good impact on the environment, society, and economy by adopting sustainability from the project's start. Concept Design becomes a potent catalyst for responsible and revolutionary construction projects through creative design

Table 8.19 Techniques incorporating sustainability during the concept stage

Practice	Technique
Foundation of sustainable principles	Establish the sustainability vision: creating a distinct sustainability vision is a key opportunity during the concept design process. Describe the broad guidelines that will direct the project's decision-making process Adopt the triple bottom line incorporate the triple bottom line strategy, taking the environment, society, and economy into account. This all-encompassing viewpoint guarantees sustainability is included in every facet of the project
Stakeholder collaboration	Encourage a cooperative design process with input from many stakeholders by implementing an inclusive design methodology. Include end users, environmental experts, engineers, and architects to provide a range of viewpoints Community engagement: incorporate the neighbourhood during the concept design stage. To ensure that the project is in line with and benefits the community, it is crucial to comprehend their requirements, worries, expectations, and goals
Sustainable design charrettes	Arrange design charrettes: arrange for intensive brainstorming sessions with key stakeholders by holding sustainable design charrettes. Establish performance objectives, investigate creative concepts, and produce a cohesive sustainability vision Investigate alternative designs: promote the investigation of alternative designs that emphasize energy conservation, resource efficiency, and a smaller environmental effect
Site selection and planning	Select a place that will have the least negative effects on the environment and the most possibilities for sustainability. Think about things like accessibility to public transportation, biodiversity, and the health of the current ecosystems Planning for smart land use: to maximise the project's footprint, employ intelligent land use planning. Give green areas top priority, encourage biodiversity, and reduce interference with natural environments
Energy-efficient design	Adopt passive design techniques that maximise the energy sources found in nature. To improve energy efficiency, consider elements like building orientation, window placement, and shading Integration of renewable energy: examine the integration of renewable energy sources and go on to eexamine whether producing renewable energy on-site is feasible to power the project sustainably
Water conservation strategies	Efficient water use: give water-efficient design priority by including fixtures and systems that use the least amount of water possible. To save water, we use rainwater collection, intelligent irrigation, and low-flow fixtures Stormwater management: to reduce pollution from runoff and encourage groundwater recharge, design with stormwater management. Include green infrastructure options and porous surfaces

(continued)

Table 8.19 (continued)

Practice	Technique
Material selection and lifecycle analysis	Sustainable Material Selection: Make material selections considering lifecycle performance and environmental impact. Give top priority to materials with minimal embodied carbon, a high percentage of recycled content, and certifications proving sustainable sourcing The life cycle assessment: to determine how selected materials will affect the environment, do life cycle studies. Think about the full lifespan, including extraction, production, consumption, and disposal at the end of the useful life
Biomimicry and nature-inspired design	Biomimicry principles: utilise natural sources of inspiration and incorporate biomimicry concepts into your concept design. To produce designs that are more robust, sustainable, and effective, emulate natural systems and processes Accept nature-inspired aesthetics, which will improve the project's aesthetic appeal while also advancing its ecological and sustainable functionality
Green infrastructure and landscaping	Include green infrastructure: include components of green infrastructure, like living walls, permeable surfaces, and green roofs. These characteristics support overall sustainability, improve biodiversity, and lessen the consequences of heat islands Native plants: when choosing landscaping, choose native plants to help local ecosystems and lessen the need for excessive watering and upkeep
Flexibility and adaptability	Design with flexibility: make sure your design is adaptable to future modifications and developments in sustainable technology. Make sure the project is flexible enough to adjust to changing consumer demands and environmental standards Planning for resilience: evaluate the project's resistance to the effects of climate change. Include design elements that improve the project's resilience to environmental stresses and help it recover from them
Daylighting and natural ventilation	Create with flexibility: ensure that your design can be modified to accommodate new advancements in sustainable technologies in the future. Ascertain that the project can adapt to evolving customer needs and environmental regulations Making resilience plans: assess the project's ability to withstand the consequences of climate change. Incorporate design features that strengthen the project's ability to withstand and recover from environmental pressures
Innovative technologies and systems	Investigate and incorporate innovative sustainable technology into the concept design by exploring sustainable technologies. This could include sophisticated water management systems, energy-efficient HVAC systems, and smart building systems Performance testing: during the concept design stage, evaluate the functionality of sustainable technologies using tests and simulations. Make sure these technologies complement the project's sustainability objectives

(continued)

Table 8.19 (continued)

Practice	Technique
Cost–benefit analysis for sustainability	Cost–benefit analysis: evaluate the sustainability measures included in the concept design by conducting a thorough cost–benefit analysis. Highlight the return on investment and long-term financial advantages of sustainable features Emphasise the value proposition. Stakeholders should be made aware of the value proposition of sustainability, with a focus on the financial, project resilience, and environmental benefits
Documentation and communication	Comprehensive sustainability documentation: make sure that the concept design includes comprehensive documentation of all sustainability elements and tactics. Throughout the building and operation phases, project teams and stakeholders can refer to this material Communicate design intent: make sure that everyone who needs to know knows about the concept design and understands its sustainable intent. Project teams, contractors, investors, and end users are all included in this

approaches, creative stakeholder collaboration, and a dedication to sustainable principles.

8.9 Integrate Sustainability into the Spatial Coordination Phase of Construction Projects

In the construction project lifecycle, spatial coordination is a crucial stage where design concepts are translated into concrete structures. Sustainability must be smoothly incorporated into spatial coordination to optimise the benefits to the environment, society, and economy. The following are the guidelines, tactics, and what to assess about when integrating sustainability into construction projects' spatial coordinating phase as listed in Table 8.20.

When sustainable design ideas are translated into the tangible construction projects of today, spatial coordination is a crucial stage. Construction teams may design environments that not only satisfy the demands of the present but also improve the health of people and the environment by implementing a thorough and integrated approach to sustainability during the spatial coordination process. Spatial coordination is a revolutionary approach that enables construction projects to be in line with sustainability principles by carefully considering energy efficiency, material selection, waste reduction, and community engagement.

Table 8.20 Practices for incorporating sustainability during the preparation and briefing stage

Practice	Technique
Holistic approach to spatial coordination	Establish specific and measurable sustainability objectives: before any spatial coordination, clearly define your sustainability goals. This may include waste minimization, resource conservation, energy efficiency, and general environmental stewardship Working together to make decisions: encourage the participation of architects, engineers, project managers, and sustainability specialists in cooperative decision-making
Green building certifications	Choose the right certifications: determine whether green building certifications are pertinent to the project's sustainability objectives. A formal framework for long-term spatial cooperation is provided by certifications Comply with integrate certification standards: verify that the spatial coordination satisfies the prerequisites for the selected certifications. This entails fulfilling additional requirements specified in the certification standards, considering eco-friendly certified products, and optimising spatial layouts for energy efficiency
Energy-efficient spatial layouts	Optimise building orientation: to maximise exposure to natural light and passive solar gain, optimise building orientation by utilising spatial coordination. This improves energy efficiency by lowering the need for artificial heating and lighting Make use of open and flexible areas: give priority to open and flexible areas while designing spatial layouts. In addition to improving occupant comfort, this makes it possible to implement lighting, heating, and cooling systems that are more effective
Resource-efficient material selection	Embodied Carbon Considerations: When coordinating space, consider the embodied carbon of the building components. To reduce the project's environmental impact and to promote resource efficiency, choose materials with lower embodied carbon Local and recycled materials: make using locally sourced and repurposed resources a priority. These would boost the regional economy in addition to lowering emissions associated with transportation
Waste reduction strategies	Design for deconstruction: to enable future material reuse or recycling, incorporate deconstruction ideas into spatial coordination. This design strategy lessens waste associated with building while advancing the ideas of the circular economy Material efficiency planning: reduce overordering and maximise material consumption to plan for material efficiency during spatial coordination. This helps to cut down on waste during the building process
Natural ventilation and daylighting	Improving the indoor environment: apply spatial coordination techniques that improve the indoor environment. To build healthier and more comfortable indoor environments, prioritize natural ventilation and daylighting Cut down optimise spatial arrangements to reduce the requirement for mechanical HVAC systems. In addition to lowering energy usage, this improves the project's overall sustainability

(continued)

Table 8.20 (continued)

Practice	Technique
Smart technology integration	Use BIM to improve stakeholder participation and information exchange during spatial coordination. BIM makes data-driven decision-making easier, which makes it possible to analyse sustainability concerns more precisely Smart building systems: examine how to incorporate smart building systems while coordinating space. This covers automation, sensors, and other technology that maximise occupancy comfort, energy efficiency, and overall building performance
Water-efficient spatial planning	Effective plumbing layouts: to reduce water usage, maximise plumbing layouts during spatial coordination. To improve water efficiency, think about where fixtures go, use water-saving devices, and add greywater systems Integrate rainwater harvesting: investigate ways to gather and store rainwater by creating spatial arrangements that make this possible. You can use this collected water for irrigation or other non-potable uses
Urban greening and biodiversity	Effective plumbing design: make the most of plumbing design during spatial coordination to minimise water consumption. Consider where fixtures will go, employ water-saving tools, and install greywater systems to increase water efficiency Incorporate rainwater collection: investigate methods for collecting and storing rainfall by designing spaces that facilitate this. Numerous non-potable uses, such irrigation, are possible for this collected water
Accessibility and inclusivity	Apply universal design principles to encourage inclusivity and accessibility in the planning and arrangement of spaces. Make sure areas and spaces are built considering the requirements with people with disabilities to promote inclusivity and social equality Participation of the community in spatial decisions: participate in spatial decision-making processes with the community. Get community feedback on walkability, public areas, and general spatial layouts to design spaces that accommodate a wide range of needs
Life cycle assessment	Integrate Life Cycle Thinking: Throughout the project's lifecycle, consider the environmental effects of design choices by using life cycle thinking to inform spatial coordination. This all-encompassing method considers the processes of material and system extraction, manufacturing, usage, and disposal Continuous LCA analysis: as the project moves further, we do continuous life cycle assessment analysis. This makes it possible to continuously optimise spatial decisions depending on changing sustainability

(continued)

Table 8.20 (continued)

Practice	Technique
Adaptability and future-proofing criteria	Design for adaptability: consider how easily spatial layouts may be modified to meet evolving user requirements, technological advancements, and sustainability requirements. By using this future-proofing strategy, the project will be sustained and relevant for a long time Evaluate how adaptable spatial configurations are to possible environmental difficulties to practise resilient spatial planning. Create environments that are resilient to the effects of climate change and enhance the built environment's overall resilience
Public transport integration	Encourage sustainable mobility: combine sustainable mobility measures with spatial planning. Make sure that spaces are planned with accessibility to bicycle infrastructure, public transportation, and to be provide pedestrian-friendly pathways Decreased parking footprint: make the most of spatial arrangements to lessen the requirement for substantial parking infrastructure. Give preference to other modes of mobility and encourage a decrease in personal vehicle use
Community collaboration and education	Collaborative design workshops: arrange for community people, local government representatives, and sustainability specialists to participate in collaborative design workshops. This cooperative strategy makes sure that spatial coordination is in line with the goals of sustainability and community values. Implement educational programmes to enlighten stakeholders about sustainable design concepts during the spatial coordinating process. This raises awareness of sustainable practices and cultivates a sense of accountability for them

8.10 Integrating Sustainability into the Technical Design Phase of Construction Projects

In construction projects, the technical design phase serves as a link between theoretical concepts and real-world applications. Including sustainability at this critical stage guarantees that technical solutions are in line with economic efficiency, social welfare, and environmental stewardship. The guidelines, tactics, and factors to consider when incorporating sustainability into building projects' technical design phase are listed in Table 8.21.

Construction projects' main engineering component, the Technical Design phase, can become a catalyst for responsible and revolutionary outcomes by incorporating sustainability into its core. The Technical Design stage makes sure that the built asset not only satisfies functional requirements but also makes a beneficial impact on society and the environment by carefully considering eco-friendly materials, water conservation, energy-efficient technologies, and community involvement. By embracing this comprehensive methodology, engineering teams assume a crucial function in moulding a sustainable trajectory for the constructed surroundings.

Table 8.21 Techniques incorporating sustainability during the technical stage

Practice	Technique
Clear sustainability objectives	Set project-specific objectives: specify sustainability goals unique to your project so that they can inform technical design choices. These objectives might include material optimisation, water conservation, carbon footprint reduction, and energy efficiency Quantifiable metrics: provide quantitative indicators to gauge how well sustainability goals are being met. These measurements serve as a foundation for assessing how technical design decisions will affect society, the environment, and the economy
Lifecycle thinking	Think about the entire lifecycle: when deciding on the technological design, use a lifecycle perspective. Analyse the effects that systems and materials have on the environment at every stage of their existence, from extraction and manufacture to use and disposal Optimise maintenance and operation: to increase the built asset's longevity and efficiency, the design should be made as simple to maintain as possible. Throughout its whole existence, this strategy promotes sustainability
Energy-efficient systems and technologies	High-efficiency HVAC systems: when designing technically, use HVAC systems with high efficiency. Both operating expenses and energy usage can be decreased by these solutions Examine how renewable energy sources, such solar panels, wind turbines, or geothermal systems, might be included into technology design. The shift to sustainable energy lowers carbon emissions
Smart building management systems	Build automation by incorporating intelligent building management systems that maximise energy efficiency. These technologies improve the overall performance of buildings by enabling real-time monitoring, predictive maintenance, and adaptive control Lighting controls and occupancy sensors: To reduce wasteful energy use, include lighting controls and occupancy sensors. Energy efficiency is increased by smart lighting systems that change their settings according to occupancy and the availability of natural light
Water-efficient design	Low-flow fixtures: to cut down on water usage, low-flow fixtures should be specified in the technical design. This includes water-efficient faucets, toilets, and irrigation systems that nonetheless perform well Rainwater harvesting systems: the integration of rainwater harvesting systems into technical designs is recommended for the collection and storage of rainwater for non-potable uses. This environmentally friendly method lessens dependency on conventional water sources while conserving water resources

(continued)

Table 8.21 (continued)

Practice	Technique
Material selection and eco-friendly practices	Materials with eco-friendly certifications and little environmental impact should be specified under the sustainable material specifications. To reduce the construction project's carbon footprint, give priority to materials that are recycled, reclaimed, or rapidly renewable Techniques for reducing waste: create technical solutions with the least amount of waste in construction. Reducing waste and conserving resources is facilitated by prefabrication, modular construction, and effective material utilisation
Optimized structural design	Material efficiency in structural design: material efficiency should be maximised in structural designs. Reduce embodied carbon by using sophisticated structural analysis techniques to make sure the selected materials can match strength requirements with the least amount of usage Creative techniques in construction: investigate innovative building techniques such as advanced prefabrication, modular construction, and 3D printing. These techniques frequently lead to shorter construction schedules and less wasteful use of materials
Natural ventilation and passive design	During technical design, take advantage of passive design techniques and natural ventilation. To maximise natural ventilation and reduce dependency on mechanical ventilation, optimise building orientation, window location, and shading Thermal mass utilisation: to control indoor temperature, include thermal mass concerns in technical designs. This may reduce the demand for forced air conditioning and heating, saving energy
Biodiverse landscaping and green infrastructure	Native plant selection: to encourage biodiversity and sustain regional ecosystems, include native plant species in landscape designs. Because native plants are more acclimated to the climate where they grow, they need less water and upkeep Enhance green infrastructure by incorporating green walls and roofs into technological designs. These characteristics support stormwater management, thermal insulation, and environmental sustainability in general
Innovative stormwater management	Permeable surfaces: to encourage effective stormwater management, design permeable surfaces. Rainwater can seep into the earth through bioswales, green spaces, and permeable pavements, minimising runoff and decreasing water contamination Innovative stormwater detention and retention technologies should be incorporated to regulate the amount and quality of runoff. These systems support environmentally friendly methods of managing water resources

(continued)

Table 8.21 (continued)

Practice	Technique
Resilient and adaptive design	Environment-responsive design: create architectural designs that adapt to the environment in your area. Resilient design elements that improve the project's adaptability over time should be incorporated, considering the possible effects of climate change Flexible space planning: to account for evolving requirements and potential changes, incorporate flexible space planning into technological designs. This flexibility guarantees that the created asset continues to be useful and pertinent in changing environments
Community engagement in design decisions	Public input sessions: to get feedback from the community, organise public input sessions during the technical design phase. Involve stakeholders, such as nearby companies and citizens, to make sure the project fits in with the needs and values of the neighbourhood Educational outreach: to educate the public about the sustainable elements incorporated into the technical design, hold educational outreach events. The public gains awareness and a sense of ownership as a result
Life cycle assessment	Continuous LCA analysis: throughout the technical design process, carry out continuous life cycle assessment analysis. Evaluate design choices' effects on the environment regularly, making modifications and enhancements possible in response to changing sustainability standards Minimise environmental impact: make wise decisions that put sustainability first for the duration of the asset's existence to reduce the overall environmental impact of the developed asset
Digitalization for efficient design	Building information modelling (BIM): use BIM technology to improve data-driven decision-making and teamwork. During the technical design phase, accurate visualisation, simulation, and analysis are made possible by the development of a digital twin made possible by BIM

8.11 Integrating Sustainability into the Manufacturing and Construction Phase of Construction Projects

The transformation of design concepts into tangible buildings occurs during the manufacturing and construction stages. Using sustainable manufacturing techniques, reducing the impact on the environment while building, and guaranteeing the longevity of the built environment are all crucial components of incorporating sustainability into this crucial stage. The methods used in construction projects to include sustainability in manufacturing and construction often comprise the practice listed in Table 8.22.

For buildings to be constructed that serve as both useful structures and beneficial contributions to the environment and society, sustainability must be incorporated into the manufacturing and construction phases. Achieving a sustainable constructed

Table 8.22 Techniques incorporating sustainability during the manufacturing and construction stage

Practice	Technique
Sustainable manufacturing practices	Sustainable material selection: using resources sourced responsibly and sustainably is given top emphasis during the manufacturing process. Select materials that meet project-specific sustainability objectives, have minimal embodied carbon and have eco-certifications Materials that can be recycled and recyclable: stress the need to use recyclable and recycled materials in production operations. As a result, there is less of a need for virgin resources, and waste in the construction supply chain is reduced
Energy-efficient manufacturing facilities	To power manufacturing plants, incorporate renewable energy sources. Making the switch to sustainable energy lowers the carbon emissions brought on by manufacturing operations Energy-efficient machinery: make investments in manufacturing technology and equipment that uses less energy. Modern machinery can lower total energy usage and increase production efficiency
Waste reduction and recycling	Lean manufacturing concepts: use these concepts to streamline operations and reduce waste production. Simplify production processes to cut down on resource usage and environmental damage Programmes for material recycling: create programmes for the recycling of materials in manufacturing plants. Keep production-related waste out of landfills by gathering and recycling it
Prefabrication and modular construction	Prefabrication techniques: use prefabricated techniques to produce building materials off-site. This lowers the amount of time spent building on-site, saves material waste, and improves construction efficiency overall Examine the methods for assembling pre-manufactured modules on-site using modular building. This method shortens project timelines, decreases waste in building, and enables precision in production
Green certification for products	Product eco-labels: search for eco-labels and green certifications on manufactured building supplies. Life cycle evaluation (LCE): To determine the environmental impact of manufactured goods, do life cycle assessments. Improve production procedures due to LCA results to improve sustainability
Water conservation in manufacturing	Water-efficient practices: make use of water-efficient procedures in factories. Implement closed-loop water systems, repurpose process water, and make investments in technology that lower water usage in manufacturing Stormwater management: to stop manufacturing activities from contaminating water, create efficient stormwater management plans. Put into practice strategies like permeable surfaces and sedimentation basins

(continued)

Table 8.22 (continued)

Practice	Technique
Supply chain sustainability	Engage suppliers: work together with vendors who are dedicated to environmentally friendly methods. Assess suppliers according to responsible manufacturing standards, ethical sourcing practices, and environmental considerations Prioritise local sourcing to lessen the transportation sector's carbon footprint. Purchase supplies and parts from nearby vendors to help local businesses and communities
Energy-efficient construction practices	On-site renewable energy: when building, incorporate on-site renewable energy options. Renewable technologies such as wind turbines or solar panels may be used to power building operations Energy-efficient equipment: make use of machinery and equipment for building that uses less energy. Use technology that minimises emissions, maximises fuel efficiency, and improves overall operational performance
Construction waste management	Waste segregation and recycling: on building sites, put into practice efficient waste segregation and recycling procedures. Allocate specific spaces for distinct waste categories and give precedence to recycling over disposing of waste in landfills Circular economy principles: adopt the circular economy by using repurposed or recycled materials in your building projects. Identify ways to recycle items that have been rescued from building or demolition projects
Low-impact construction techniques	Reduce site disturbance: employ building methods that reduce disturbance to the site and effects on the ecosystem. When a building is underway, take care to conserve natural features, avoid needless excavation, and protect existing plants Green construction matting: to preserve the earth and lessen soil erosion, choose a material that is favourable to the environment. When it makes sense, select reusable or biodegradable matting
Water-efficient construction practices	Install intelligent watering systems for the landscape while the building is underway. Water waste is decreased by these systems, which optimise water usage depending on plant requirements and meteorological circumstances Temporary sediment management: to stop soil erosion and safeguard neighbouring water bodies, use temporary sediment management techniques. Put in place sediment basins, silt barriers, and other erosion control measures
Community engagement in construction	Maintain open communication with nearby communities during the development period to foster stakeholder engagement. Assist with issues, include the community in decision-making, and provide information on construction activity Community benefits: find ways that construction projects can have a beneficial influence on the community. This could involve initiatives to create jobs, increase skill sets, or upgrade local infrastructure for the good of the community

(continued)

Table 8.22 (continued)

Practice	Technique
Biodiversity protection	Restoration of the ecosystem: to preserve and improve biodiversity, construction sites should use ecological restoration techniques. Restore ecosystems impacted by construction activities, replant native vegetation, and establish habitats for wildlife Environmental monitoring: to evaluate the effects of construction on the local biodiversity, carry out routine environmental monitoring. To reduce adverse consequences, modify construction methods by monitoring data
Continuous improvement and innovation	Feedback loops: to continuously evaluate and enhance sustainability performance, set up feedback loops. Gather feedback from stakeholders, construction teams, and environmental monitoring to determine areas that need improvement Technological innovation in construction: accept innovative building technology that improves sustainability. This might include environmentally friendly and efficient robotics, 3D printing, or innovative building materials

environment can be achieved by stakeholders through the implementation of eco-friendly manufacturing techniques, waste minimization, responsible sourcing, and a focus on community and environmental well-being during construction. The basis for a robust, resource-efficient, and socially conscious constructed environment is laid by this all-encompassing approach to manufacturing and construction.

8.12 Integration of Sustainability into the Handover Phase of Construction Projects

The handover phase signifies the change from development to occupancy, and including sustainability at this crucial juncture guarantees that the built environment will keep improving the health of its residents and the community at large. The following are some tactics and factors to consider when integrating sustainability into building projects during the handover stage are provided in Table 8.23.

Table 8.23 Techniques incorporating sustainability during the handover stage

Practice	Technique
Comprehensive documentation	As-built paperwork: provide thorough as-built paperwork that contains in-depth details on the environmentally friendly elements incorporated into the building process. This documentation guarantees the effective operation of sustainable systems and acts as a reference for future maintenance and repairs
Occupant training and awareness	Training programmes: provide educational materials on sustainable features, energy-efficient systems, and water-saving techniques to building inhabitants. Tenants who feel empowered are more likely to take an active role in sustaining sustainability programmes
Smart building systems integration	Sustained surveillance: incorporate intelligent building systems to conduct ongoing assessments of water and energy use, indoor air quality, and general building functionality. Proactive modifications to maximise effectiveness and sustainability are made possible by real-time data
Waste management protocols	Clear procedures should be provided for the sorting and disposal of waste within the building. To reduce the impact on the environment, install recycling facilities and teach residents proper waste management techniques
Regular maintenance programs	Scheduled maintenance: to guarantee the continuous effectiveness of sustainable systems, such as HVAC, lighting, and water-saving technology, establish regular maintenance schedules. Preventive maintenance keeps systems from degrading and extends the life of equipment
Renewable energy utilization	Teach residents about renewable energy: inform them about the building's use of solar or wind power integrated into the structure. Increase awareness of these elements' positive environmental and energy-saving impacts
Water conservation measures	Promote water-saving behaviours: urge tenants to embrace water-saving practices, like watering plants sparingly and utilising low-flow faucets. Offer instructional resources that emphasise the significance of conserving water
Biodiversity and green spaces	Maintenance of green spaces: carry out the plans for upkeep of the building's surrounding biodiversity elements and green spaces. This includes landscaping techniques that promote the native flora and wildlife and aid in the preservation of biodiversity
Indoor air quality management	Establish rules for maintaining the quality of the air within buildings. These should include appropriate ventilation techniques and routine HVAC system inspections. Inform residents on the significance of indoor air quality for their health and welfare
Community engagement programs	Community events: plan programmes for building inhabitants to become involved in neighbourhood sustainability activities. Encourage involvement in environmental activities and programmes to cultivate a sense of community responsibility

(continued)

Table 8.23 (continued)

Practice	Technique
Energy-efficient practices	Encourage residents to adopt energy-saving habits by teaching them to switch off lights and electronics when not in use. Encourage an energy-conscious culture by running awareness campaigns and offering rewards for eco-friendly behaviour
Continuous feedback mechanisms	Feedback platforms: provide ongoing channels of communication for tenants to voice concerns on sustainable features. Create online forums or other avenues for contact to promote candid conversations between building management and tenants
Renewable energy credits and offsets	Inform on offsetting measures: educate building tenants on the building's renewable energy credits and offsets. Tell them about these actions' beneficial effects on the environment and how they help achieve sustainability objectives
Post-occupancy evaluation	Get input from occupants: after a building has been occupied, conduct assessments to get input on its sustainable features. Utilise this data to pinpoint problem areas and adjust in response to tenant feedback
Social equity initiatives	Community support programmes: launch community-beneficial social equality projects. This could be assisting regional companies, generating employment, or funding community development initiatives that uphold social responsibility and sustainability
Accessibility and inclusivity	Features of an accessible design: Make sure the structure is inclusive and accessible to all users. To support a diverse workforce and foster inclusivity, accessibility features should be evaluated and updated regularly
Environmental education programs	Conduct educational seminars aimed at improving the residents' comprehension of sustainability through environmental education. Topics including responsible consumption, energy conservation, and waste reduction may be covered in these workshops
Certifications and recognition	Certifications to be displayed: throughout the building, make a point of showcasing sustainability certifications and awards. This inspires pride in the residents and conveys the building's dedication to sustainability
Community collaborations	Work together with local organisations: to support community initiatives that are in line with sustainability objectives, work together with nonprofits and local organisations. Positive transformation in the larger community can be sparked by building management
Adaptive design for future changes	Flexibility in design: make sure the structure's layout permits modifications that may be made to adapt to changing sustainability requirements. To improve overall sustainability, design elements should allow for future updates and alterations

The handover phase marks the foundation of a new chapter in the built environment's existence rather than its conclusion. Through the inclusion of sustainability in this stage, involved parties can guarantee that the consequences of conscientious construction go well beyond its conclusion. A sustainable handover lays the foundation for a durable, adaptable, and socially conscious building that benefits its residents as well as the neighbourhood through efficient operations, constant maintenance, and community involvement.

8.13 Integration of Sustainability into the Use and Operation Phase of Construction Projects

The use and operation phase are a crucial stage in the construction project lifecycle where sustainability is realised. Optimising water and energy efficiency, preserving green building elements, and encouraging an environmentally conscious culture among tenants are all necessary for the successful integration of sustainability during this stage. When integrating sustainability into the usage and operation phase of construction projects, there are several tactics and factors to consider. Table 8.24 lists the practice of incorporating sustainability during the operational pages of buildings.

The Use and Operation Phase, where the principles of responsible building practices are implemented daily, is the beating heart of sustainable construction. Stakeholders may make sure that the building not only achieves but surpasses sustainability targets over its lifetime by continuously prioritising energy efficiency, water conservation, waste reduction, and occupant participation. This stage establishes a legacy of responsible building operation and community impact, demonstrating the dedication to environmental care.

Table 8.24 Techniques incorporating sustainability during the use and operational stage

Practice	Technique
Energy-efficient operation	Smart building technology: use smart building technology to monitor and manage energy use in real-time. Incorporate automated lighting, HVAC, and other utility systems to maximise efficiency according to occupancy and surrounding conditions The purpose of occupant awareness programmes is to inform building occupants about energy-efficient measures. Promote sensible actions like shutting off lights, utilising energy-saving appliances, and cutting back on needless energy use
Water conservation Measures	Smart water management systems: to maximise water use inside the building, implement smart water management systems. Make use of automation and sensors to keep an eye on water usage, find leaks, and promote effective irrigation techniques Engage occupants: inform tenants about water-saving techniques, such as using water-efficient fixtures and using water wisely. Stress the value of conserving water in day-to-day activities

(continued)

Table 8.24 (continued)

Practice	Technique
Indoor environmental quality maintenance	Frequent HVAC maintenance: to guarantee the best possible indoor air quality (IAQ) and energy efficiency, schedule regular maintenance for HVAC systems. A healthy indoor atmosphere can be maintained by changing filters, checking ventilation systems, and taking quick action on problems Programmes for occupant comfort: seek input from residents on how comfortable it is indoors. Establish initiatives that enable residents to voice their preferences for air quality and temperature, promoting a cosy and adaptable interior atmosphere
Sustainable landscaping practices	Native plant maintenance: to conserve water, emphasis on native and drought-resistant species. Use landscaping techniques that enhance biodiversity and create an aesthetically pleasing, environmentally friendly outdoor space Eco-friendly pest management: reduce the amount of hazardous chemicals used by implementing eco-friendly pest management techniques. Put into practice integrated pest management techniques that give natural and non-toxic solutions priority
Waste reduction and recycling initiatives	Comprehensive recycling programmes: keep up and improve the building's recycling initiatives. Assist residents in properly separating their waste, provide recycling containers with labels visible, and evaluate the success of recycling programmes regularly Cut down on single-use items: urge tenants to utilise fewer single-use products. Put into practice strategies like offering reusable substitutes, encouraging a zero-waste attitude, and reducing the amount of non-recyclable waste produced
Renewable energy integration	Solar power optimisation: if appropriate, maximise the utilisation of solar energy by doing routine upkeep and observation. To optimise the production of renewable energy, make sure your solar panels are clear and operating well Instruction of occupants in renewable energy: inform tenants about the building's renewable energy sources and the advantages of using clean energy for the environment. Encourage residents to feel proud of and accountable for their participation in sustainable activities
Adaptive reuse and flexibility	Adaptability in space design: consider variations in occupancy and purpose while designing spaces. Because of its adaptability, there is less need for extensive remodelling or brand-new projects, which promotes sustainability through adaptive reuse Renovation for efficiency: throughout the building's existence, give energy-efficient improvements and renovations top priority. As sustainability criteria change, new systems and technology meet those standards
Monitoring and reporting	Reporting on water and energy use: put in place procedures for routinely reporting water and energy use. Transparent reports that highlight the facility's performance and the effects of sustainable actions should be given to building occupants Feedback mechanisms for occupants: provide a way for tenants to voice their opinions about building performance and sustainability initiatives. Act on insightful suggestions to continuously raise tenant happiness and operational effectiveness

(continued)

Table 8.24 (continued)

Practice	Technique
Community engagement programmes	Local community initiatives: reach out to the neighbourhood to encourage sustainability outside the walls of the building. Take part in or start environmental seminars, activities, or programmes that improve the community's general well-being Partnership with local organisations: to maximise the building's beneficial effects, partner with regional environmental organisations. Encourage programmes that advance community development and sustainability objectives
Continuous education and training	Programmes for occupant training: continually instruct and train tenants in sustainable behaviours. Provide tenants with information on energy saving, waste reduction, and water efficiency so they can make decisions that will benefit the environment Frequent sustainability workshops: provide building management and maintenance employees with regular sustainability workshops. Give them the abilities and information required to maintain and improve sustainable systems
Lifecycle assessment and continuous improvement	Periodic lifecycle evaluations should be carried out to analyse the environmental impact of building operations. Utilise the data to pinpoint problem areas and put plans into action to raise sustainability levels all around Innovation acceptance: accept new, innovative methods and technologies as they appear. remain up to date on sustainability developments and contemplate implementing fresh approaches that complement the building's ecological objectives
Accessibility and inclusivity maintenance	Conduct regular accessibility audits: to make sure the building continues to be inclusive of all occupants, conduct regular accessibility audits. As recent problems arise, take quick action to fix them and adjust to increase accessibility Programmes for community inclusion: put in place initiatives that encourage diversity within the building community. Encourage a feeling of understanding and community among residents, regardless of their specific skills or backgrounds

8.14 Summary

This chapter explores the complex ways in which sustainability principles are integrated into various stages of building projects. It starts by highlighting how crucial it is to incorporate sustainability into project planning and acknowledging it as a fundamental component. The integration also includes sustainable design methods, which prioritise the use of environmentally friendly materials, energy-saving devices, and ecologically favourable architectural strategies. The chapter clarifies approaches for reducing environmental effects, embracing green building practices, and putting responsible construction practices into practice as it examines the construction processes and techniques. The topic of discussion includes the choice of environmentally friendly building materials, eco-friendly construction methods, and the importance of recycling and waste reduction during the building process. A major theme is the function of project monitoring and assessment in guaranteeing the sustainability of building projects. The chapter emphasises how crucial it is to continuously monitor the environmental impact of construction activities using techniques like life cycle assessments. The ongoing operating phase is also monitored, highlighting the significance of tracking and optimising waste management, water use, and energy usage. This monitoring goes beyond completion. Given the circumstances; the chapter offers a thorough overview of how sustainable building project management incorporates a holistic methodology. The incorporation of sustainability concepts is described as essential to creating resilient and environmentally responsible built environments, from the commencing stage of project planning through design, construction, and continuous evaluation. The methods, procedures, ideas, and approaches covered in this chapter apply not only to construction projects but to the architectural, engineering, and construction industry. By utilising these methods, engineering projects can be planned to have a positive impact on the economy, society, and environment, resulting in the creation of resilient, sustainable, and socially responsible infrastructure.

Bibliography

1. Ahn YH, Pearce AR, Wang Y, Wang G (2013) Drivers and barriers of sustainable design and construction: the perception of green building experience. Int J Sustain Build Technol Urban Dev 4(1):35–45
2. Baldwin A, Bordoli D (2014) Handbook for construction planning and scheduling. Wiley
3. Bal M, Bryde D, Fearon D, Ochieng E (2013) Stakeholder engagement: achieving sustainability in the construction sector. Sustainability 5(2):695–710
4. Dasović B, Galić M, Klanšek U (2020) A survey on integration of optimization and project management tools for sustainable construction scheduling. Sustainability 12(8):3405
5. Ebekozien A, Aigbavboa CO, Ramotshela M (2024) A qualitative approach to investigate stakeholders' engagement in construction projects. Benchmarking Int J 31(3):866–883
6. McLennan JF (2004) The philosophy of sustainable design: the future of architecture. Ecotone Publishing
7. Ofori G (1998) Sustainable construction: principles and a framework for attainment-comment. Constr Manag Econ 16(2):141–145
8. Ogunbiyi O, Goulding JS, Oladapo A (2014) An empirical study of the impact of lean construction techniques on sustainable construction in the UK. Constr Innov 14(1):88–107
9. Plank R (2008) The principles of sustainable construction. IES J Part A: Civ Struct Eng 1(4):301–307
10. RIBA Plan of Work 2020 Overview glossary and set in bold type
11. Williams DE (2007) Sustainable design: ecology, architecture, and planning. Wiley
12. Yu J, Leung MY (2018) Structural stakeholder model in public engagement for construction development projects. J Constr Eng Manag 144(6):04018046

Chapter 9
The Critical Success Factors for Sustainable Construction Management

Abstract High energy costs, extreme weather, shortages of construction materials, client dissatisfaction, low productivity, urbanisation, and waste are confronting the construction sector. The combination of these is leading to a demand for a paradigm shift in the management of construction projects. Sustainable construction management is introduced to curtail the inadequacies and inefficiencies in conventional construction management practices. However, less is known about the scope and nuances of sustainable construction management. Thus, decision criteria for sustainable construction management are still largely based on anecdotal evidence. Therefore, this research aims to evaluate the critical success factors for sustainable construction management. This research is guided by pragmatism as a research philosophy because it requires multiple research methods. The questionnaires involved closed and open-ended questions and were administered to more than 600 construction companies based on the snowballing technique. The surveys were conducted online. The questionnaire comprises 48 critical success factors for sustainable construction management practices. The data revealed that the main critical success factors for sustainable construction management practice are to 'prevent pollution associated with construction projects', 'provide adequate information to the construction team', 'use a waste management plan', 'provide adequate information to the design team', 'use materials from sustainable sources', 'use renewable materials', and 'ensure proper health and safety management on sites. After studying this chapter, you would be to:

- Explain critical success factors for sustainable construction management
- Apply the sustainable critical success factors for construction projects
- Evaluate the attainment of the application of the critical success factors.

Keywords Business case · Sustainability targets · Pedagogy · Whole life appraisal · Greenwashing

A. Olanrewaju, *Sustainable Construction Management: Research and Practice Companion*, Management in the Built Environment,
https://doi.org/10.1007/978-981-96-4910-5_9

9.1 Learning Outcomes

The previous chapters have provided us with information on the principles, strategies, techniques, and procedures of sustainable construction management. This chapter presents findings from an international survey on the critical success factors for sustainable construction management. Sets out to provide explanations on sustainable construction economics. The critical success factors identified, presented, justified, and explained in previous chapters. The critical success factors for sustainable construction management are the set of predictions that if implemented in construction projects would result in sustainable outcomes. For the construction sector to contribute to the SDGs, there is a need to rethink construction management. Sustainable construction management goes beyond site management and the incorporation of sustainable features into the design of the projects. It is challenging to provide an account of the nature of sustainable construction management because many exciting questions and paradoxes come to attention when discussing sustainable construction management. Primarily the aim is to provide practical insight into the theoretical basis in the previous chapters. The data were collected based on a survey questionnaire administered over 4 months. The initial findings from the survey entitled 'Foundations of Sustainable Construction Management Research' were presented at the International Conference on Sustainable Technology and Development, 31 October–3 November 2021. Although this chapter represents the empirical evaluation of the main strategies, practices, processes, procedures, techniques, etc., in the previous relating to sustainable construction management, the chapter is designed as a stand-alone like other chapters that come before and after it.

9.2 Introduction

This research integrates and draws upon two distinct theoretical frameworks: construction management and sustainability. The literature on construction management examines the nature and problems of the construction management discipline and practice. The literature on sustainability looks at how applicable the theory and idea are to the fields of construction, architecture, engineering, and surveying (SAEC), specifically in the construction management discipline and practice. SAEC handles conceiving, designing, constructing, and maintaining built assets. Construction management could be a procurement strategy, an academic discipline, or a practice. As a discipline, it is like architecture, surveying, engineering, facilities management, and other fields in the AEC sectors. Construction management as a practice is performed as a coordinated effort by the design, construction, and operation teams. Construction management aims to execute a construction project that meets the project's deliverables. However, the inclusion of "sustainable" into "construction management" changes the original landscape of construction management discipline and practice. In this research, high-performance construction, livable buildings, green construction, and sustainable construction are synonymous with sustainable construction. The market for sustainable construction is huge and increasing. The global market for green construction is estimated to increase to US$425.4 billion by 2027 (Research and Markets 2020). This is an increase of 84.75% from 2020 to 2027. A survey involving 13 countries with respondents drawn from design firms, contractors/builders, specialists/consultants, owners/developers, and engineering firms revealed that most of the surveyed respondents expect that 60% of the buildings will be green buildings by 2018 and that global green construction projects will continue to double every three years [6].

While the demand and supply for sustainable construction will continue to rise exponentially, the development, design, and construction of sustainable buildings are faced with many challenges [2, 15]. For instance, sustainable buildings are delayed by more than 15% and cost overruns by more than 20% [16]. The research also revealed that rework in sustainable buildings is about 15%, and the correlation results revealed that rework will increase the cost of green building projects by up to 29% and delay completion time by 27%. Also, compared to conventional buildings, green building construction required an average of 8% more time, and green buildings were 4.8% completed behind schedule [8]. Metz and Cheatham [20] surmised that claims and disputes in green building construction are higher compared to conventional buildings. Cases of delays and cost overruns are more common in sustainable construction [5]. Many members of the design and construction teams have been sued for greenwashing [23]. The profit margins of sustainable construction projects are low. According to Robichaud and Anantatmula [22] and Ismael and Shealy [10], contractors who are involved in sustainable construction face higher [negative] risks compared to conventional construction. However, the problems facing the performance of sustainable construction projects have been attributable to poor construction management practices [8, 11].

Although there is some literature on sustainable project management, it is obvious that the emphasis has been on environmental sustainability and economic sustainability [7, 19], and disregard social sustainability. This is probably because green buildings are easier to understand than sustainable buildings. Deductively, there is a lack of explicit engagement with sustainable construction management theories. Extant research lacks detailed knowledge of the concepts of sustainable construction management. Without this, however, the conceptualization of what it is may not be explicit. This inadequacy is seen in the inclusion of unnecessary factors and the exclusion of critical factors in the earlier literature on sustainable construction management frameworks. Therefore, the first question is: What can we know about sustainable construction management? To answer this question, there is a need to examine the ontological question. The second question is what are the critical success factors for sustainable construction management? Despite the growing popularity of sustainable construction practices, there is still a lack of consensus on the critical success factors for implementing sustainable construction management practices. However, the success or failure of a project depends on the measurement of its critical success factors [3, 24]. So far, there has not been any empirical research conducted that evaluates the essential CSFs from a global viewpoint. Therefore, there is a need to investigate the critical success factors for sustainable construction management from an international perspective. In response, the purpose of this study is to assess the CSF that supports the effective application of sustainable building management techniques in various nations and areas. The findings of this study will provide insights into the best practices and strategies to implement sustainable construction management practices. Also, the findings will help policymakers and industry practitioners make informed decisions for sustainable development.

9.3 Background and Theoretical Considerations in Sustainable Construction

The Intergovernmental Panel on Climate Change (IPCC) Working Group I report, Climate Change 2021, termed "a code red for humanity," found that the earth's climate is changing at unprecedented, irreversible, and unpredictable rates. The IPCC report found that human activity handles 1.1 °C of warming. In other words, in the current state of the world, people, groups, organisations, and countries are effectively consuming more than the planet can sustain in a productive way and are producing and releasing pollution and waste much beyond what the planet can handle. The goal of introducing sustainable development was to lessen the effect of human activity on climate change. The definition of sustainable development that is most frequently used is: development that satisfies current demands without jeopardising the ability of future generations to satisfy their own needs [4]. Practically speaking, sustainable development refers to the process of integrating multiple criteria into the built environment, or SAEC, such as energy efficiency, longevity, waste minimization, social values, a healthy indoor environment, longevity, pollution control, life-cycle cost, user-friendliness, user comfort, and others [13, 14]. The SAEC sector is a major threat or solution to the achievement of Sustainable Development Goals (SDGs). The SAEC sector significantly impacts at least 10 of the 17 SDGs. To illustrate, buildings consumed more than 40% of the world's energy, released 1/3 of CO_2, used about 25% of harvested wood, released about 50% of fluorocarbons, produced 40% of landfill materials, used 45% of energy in operations, and emitted 40% of greenhouse gases. The construction sector uses 15% of the world's usable water. Building interiors have five times more pollutants than the air outdoors [23]. The impact of the sectors on climate changes and global warming will be even greater when additional CO_2 emissions related to buildings, such as those from the manufacturing and transportation of construction and demolition materials, are considered. This is because the sectors have multiplier effects on other sectors. Rising sea levels, altered rainfall patterns, erratic weather, melting glaciers, extended rainy seasons, protracted droughts, floods, landslides, scarcity of water, deforestation, etc. are all effects of climate change that have a big impact on project performance and construction management techniques.

While sustainable construction is undoubtedly beneficial to achieving the SDGs [3, 14, 23], the projects are completed behind schedule [8]. Sustainable construction projects are associated with multiple disputes and claims [20, 16], and the profit margin is low [22]. The associated construction risk is high, and the rate of greenwashing is increasing unabated [15]. Reworks are rampant in sustainable construction as compared to conventional construction projects. Green buildings are expensive, and delayed, and quality is poor. The underlying reason is that the construction management practices that have been used for sustainable construction projects are incompatible with the requirements of sustainable construction projects. As an example, sustainable construction promotes the use of off-site components, industrialised building systems, recycled materials, and local resources. These force construction companies to meet a certain set of standards, competencies, and capabilities. The primary essential components of sustainable construction management that any logical theory should be able to describe are examined in this study.

9.4 Framing the Critical Success Factors for Sustainable Construction Management

Construction performance is measured differently. While some units of measurement involve the finished projects, others involve the processes and activities that took place along the project supply chain. The critical success factors of a project are the key factors needed to make it achieve its designed functions. The term "factor" in the phrase implies a factor, activity, event, process, or action. Despite several studies on critical success factors for sustainable construction, some crucial factors, particularly those relating to social sustainability, have been excluded. Firstly, the earlier critical success factors are not comprehensive enough to accommodate the significant aspects of sustainable construction globally. Examples of social sustainability include child labour, human rights, labour relations, community involvement, diversity, equity, benefits, and compensation. The health and safety of construction workers is directly impacted by two of the 17 Sustainable Development Goals (SDGs). Based on an international survey involving 125 countries, Workwear Guru (2021) found that construction workers are not well paid. Only nine of the 125 countries surveyed (e.g., 7.2%) revealed an accident rate of less than 10%. In other words, in most countries, more than 1 in 10 workers have been injured while working on sites. The accident rate is higher than 20% in about 40% of the countries. The survey also revealed that, in most countries, construction sites are not safe, and many stays in dangerous places to work. Current empirical research on sustainable construction pays little or no attention to social sustainability [16] or social value. For construction management to add social sustainability to the stakeholders (i.e., construction companies), in addition to the living conditions of the workforce, it should cover social capital, wellness, health, safety, justice, community development, resilience, diversity, equity, fair labour practices, empowerment, work-life balance, and community participation as CSF. It is unjustifiable to deliver construction projects that are energy-efficient, cheaper, generate less waste, use fewer materials, or are healthy for occupants if construction workers are exposed to accidents and infectious diseases. Secondly, the existing frameworks are specific to a country.

However, knowledge from an international perspective is needed both in practice and theory. Many sustainable materials and components are sourced internationally, and many international green rating tools are used outside of the country where they were first introduced. For instance, LEED and BREEAM are used in more than 100 countries each. Furthermore, most of the frameworks are still theoretical. In addition, in most developed and advanced developing countries, construction workers are foreign labour. The CSFs involved in sustainable construction management represent the diverse activities that make up sustainable construction management processes and activities, supporting decision-making at each step of the delivery of processes for construction projects.

9.5 Overview of Research Methodology

This research is pragmatistic in nature. We understand the significance of using a realistic strategy that integrates a variety of research techniques to handle the complicated issues facing the construction sector. We seek to produce useful knowledge that may guide and enhance sustainable construction in actual construction contexts by utilising a pragmatic research awareness. While research on the critical success factors for sustainable construction management is receiving attention among academics, there is a need to reconsider the critical success factors, more so at the international level. CSFs analysis would make it possible to identify the most crucial project areas that, in order for the project to succeed, demand a great deal of interest from the stakeholders. To achieve this objective, convenience sampling and snowballing were used to collect primary data. These methods were combined because there was a lack of sufficient information about the respondents. Respondents who were willing to participate, available, and accessible were given the surveys. The contacted responders were requested to assist in forwarding the form to their associates who possess the necessary skills to reply. However, the findings may not be generalizable, but the results can be representative of the population if the response rate is relatively large enough (Sekaran and Bougie 2016). This is consistent with the application of central limit theory. The criteria addressed to respondents were based on a literature review and extensive discussion with those involved in the management of sustainable construction projects in Malaysia, Singapore, and the UAE. Five of those involved in the discussion also participated in the pilot study. The surveys were administered online between July 17 and September 14, 2021, and reopened again between December 13 and January 14, 2022. The survey was administered by (The Chartered Institute of Building, (Malaysia and UK) and (Quantity Surveyors

International) QSi. Both organizations enjoy a vibrant reputation in the construction sector and have membership across the globe.

The criticality of each critical success criteria in the practice of sustainable construction management was inquired of the respondents. A five-point rating system was employed, with 1 signifying the least critical, 2 less critical, 3 somewhat crucial, 4 highly critical, and 5 extremely critical. The factors had a favourable wording. In which case, higher scores indicate a higher criticality, and a low score demonstrates a lower criticality. An average critical index (ACI) was calculated based on a weighted mean (Eq. 9.1) [16]. In terms of interpretation, an ACI score of 1.00–20.00 denotes the least critical, 21.00–40.00 less critical, 41.00–60.00 moderately critical, 61.00–80.00 very critical, and 81.00–1000.00 extremely critical. There is a common difference of 1 between each of the scales. The mode technique was used to analyze the demography of the respondents and determine the distribution of respondents concerning the scales. For data analytics, all collected data was analyzed using IBM SPSS Statistics Data Editor 25. The other statistical tests computed are one-way tests, Cronbach's Alpha reliability, t-tests, Omega, and commonalities, skewness and kurtosis. However, it has been argued that analytical analysis is insufficient for a multidimensional situation of the relationship between multiple constructs. However, analyses of the interaction among factors or instances are unsupervised machine learning problems. Such a problem could be an association or clustering problem. Some of the algorithms that may be applied are principal component analysis, K-mean, C-mean, and factor analysis. For this research, factor analysis was selected as it is consistent with the aim of this research. Factor Fig. 9.1 illustrate the flow chart for the research process.

$$ACI = \left(\sum_{(i=0)}^{5} [[a\_i x\_i]] \right) / \left(5 \sum_{(i=0)}^{5} x\_i \right) \times 100 \tag{9.1}$$

Open-ended data analysis was performed using Bag of Words (BoW). When attempting to extract features from textual data, a bag of words is utilised. Language modelling and document categorization issues are resolved using this strategy.

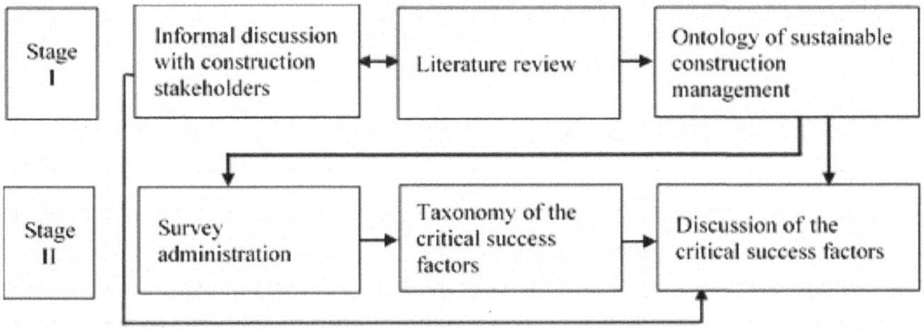

Fig. 9.1 Two-step research process

Orange software was used in the model's development. The method searches the number and regularity of words in a document, no matter where they are located within it. For document feature encoding or extraction, it is helpful. One of the many methods for textual mining natural language is BoW, which transforms texts with varying lengths into vectors with fixed lengths. BOW does not, however, understand a document's word semantics.

9.6 Results

The results of the findings are explained in the following sections:

9.6.1 Demographics of Respondents

Figure 9.2 displays the respondents' country of residency, which demonstrates that they are from countries where the market for sustainable construction projects is strong and growing extremely fast. Many of the respondents were also from 'other' countries, including Canada, Nigeria, Qatar, Germany, Oman, Ghana, and China. Some 56% of the respondents obtained BSc degrees, 27% earned MSc degrees, and 10% earned PhDs, while the rest earned other qualifications. More than 85% have more than 5 years of work experience, and about 40% of the respondents have more than 20 years of work experience. Most of the respondents are Green Building Index certified, and many are BREEAM certified (Table 9.1). The respondents were involved in the conceptualization, design, construction, and operation of sustainable construction projects. Figure 9.3 illustrates that the respondents held strategic positions in their organisations. Table 9.2 shows the types of sustainable projects executed by the respondents. These profiles imply that respondents are competent to provide valid and reliable responses to evaluate the critical success factors for sustainable construction management practice. Figure 9.4 revealed that the respondents are involved in both the downstream and upstream segments of the construction sector. The bag word model is in Fig. 9.5.

9.6.2 Analysing the Taxonomy of the Critical Success Factors

The results of the goodness of fit of the CSFs are place in Table 9.3. Table 9.4 contains the descriptive and explanatory statistics of the critical success factors. About 5% of the respondents alleged that the CSFs were the least critical to sustainable construction management. Similarly, 9% of respondents believed that the factors had less impact on sustainable construction management. However, 26% affirmed that the factors have an average impact on the successful practice of sustainable construction

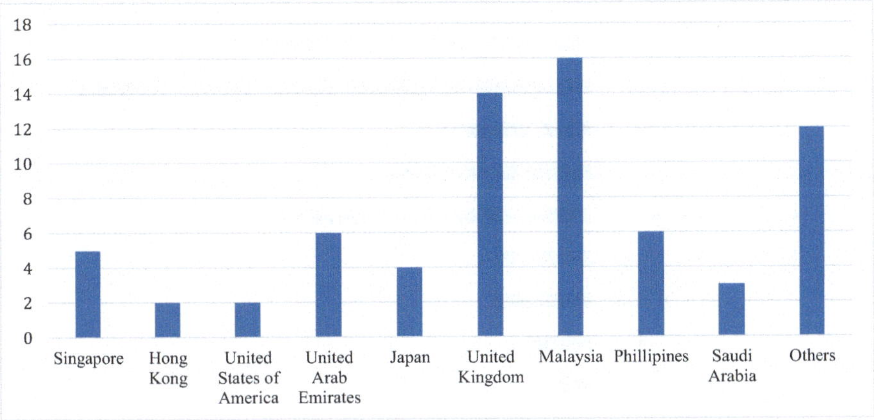

Fig. 9.2 Respondent's location

Table 9.1 You are a facilitator/assessor for?

Rating tool	Size
Green building index (GBI)	23.26
Green marks (Singapore)	4.65
Building research establishment assessment method (BREEAM)	6.98
Leadership in energy and environmental design (LEED)	18.60
Others	46.51

management. Furthermore, 60% of respondents rated all the factors as strongly or extremely crucial for practising and implementing sustainable construction management. The cumulative ACI and standard deviation are 73.23% and 22.46, respectively. Some 57% of the critical success factors are rated above the average ACI. Based on the classification in Sect. 9.5, the factors can be grouped into three main classes: moderately critical, strongly critical, and extremely critical. Six of the CSFs are extremely critical, 42 are strongly critical, and the remaining two are moderately critical. The average difference between the upper and lower is approximately 11% lower than the average standard deviation. The confidence interval value range is narrow, which implies that ACI reflects the population's measurements. Furthermore, the results show that the standard error of the mean is low, and the kurtosis or skewness is low. In general, values of + 1 or − 1 indicate substantial skewness. Similarly, in kurtosis, + 1 stands for high weakness and − 1 implies high flatness. Data are normally distributed if the standard deviation is less than half of the mean score. By inference, the data are representative of the population.

Fig. 9.3 Position in the company

Table 9.2 Type of construction projects

Project	Residential	Commercial	Industrial	Office	Educational	Others
Size	23.776	20.280	13.986	18.182	16.084	7.692

9.7 Discussion of Findings on the Critical Success Factors

Considering space limitations, the discussion will focus specifically on the first 15 CSFs moving forward. Construction contributes 23% of air pollution and 40% of drinking water pollution (UK Green Building Council 2016) and accounts for a significant amount of PM 2.5 and PM 10 emissions in the air. Runoff from construction sites contaminates water bodies and soil. Construction site pollution in cities damages adjoining properties and endangers the health of construction workers. Although waste and pollution seem similar, they are not the same. While pollution can be generated as a by-product, it may not be in all cases. Waste is a by-product. While pollution is harmful, waste may not be, even though it may be unwanted by a specific stakeholder. Contractors and clients may be fined for poor pollution control on sites. Hence, it is not surprising that controlling environmental pollution is a major criterion for sustainable construction management practices. A target of sustainable construction management is to avoid or reduce the generation and spread of pollution during construction. Pollution control measures should be recognised during the

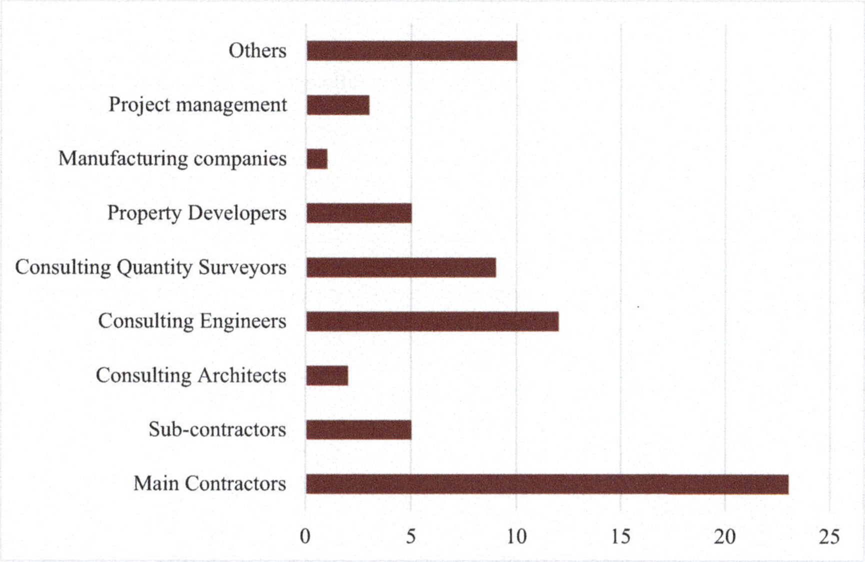

Fig. 9.4 Type of organization

Fig. 9.5 Bag of words model of emerging sustainable construction management practice

Table 9.3 Results of one-sample test for the critical success factors

	t	Sig. (2-tailed)	Mean difference	95% Confidence interval of the difference	
				Lower	Upper
Make use of building information modelling	4.431	0.000	0.500	0.27	0.73
Make use an intelligent building system	5.701	0.000	0.586	0.38	0.79
Make use local materials	7.920	0.000	0.857	0.64	1.07
Use drone technology for site supervision and safety management	− 2.339	0.022	− 0.343	− 0.64	− 0.05
Use renewable materials	9.455	0.000	1.114	0.88	1.35
Make use of construction technology (e.g., 3D scanners)	− 2.228	0.029	− 0.357	− 0.68	− 0.04
Make use of a waste management plan	10.563	0.000	1.243	1.01	1.48
Use industrialised building system	1.826	0.072	0.286	− 0.03	0.60
The design team should have a green/ sustainable assessor	6.596	0.000	0.886	0.62	1.15
Contractors should have green/ sustainable certified staff	5.007	0.000	0.714	0.43	1.00
Green /sustainable consultants should be engaged	5.523	0.000	0.743	0.47	1.01
The design team should be given adequate information	13.229	0.000	1.343	1.14	1.55
Adequate information should be provided to construction team	14.992	0.000	1.414	1.23	1.60

(continued)

Table 9.3 (continued)

	t	Sig. (2-tailed)	Mean difference	95% Confidence interval of the difference	
				Lower	Upper
Presence of a method statement	9.289	0.000	1.043	0.82	1.27
Use value engineering/ management techniques	5.460	0.000	0.757	0.48	1.03
Employ skilled workers	7.792	0.000	0.957	0.71	1.20
Ensure proper health and safety management on sites	7.962	0.000	1.043	0.78	1.30
Engage local operatives	5.445	0.000	0.671	0.43	0.92
Reduced parking footprint	0.984	0.329	0.157	− 0.16	0.48
Conduct an indoor air quality assessment after the project is completed	6.191	0.000	0.786	0.53	1.04
Conduct environmental impact assessments	11.910	0.000	1.286	1.07	1.50
Avoid environmental pollution caused by the projects	14.585	0.000	1.443	1.25	1.64
Make political accountability a priority	2.795	0.007	0.443	0.13	0.76
Use materials / components from sustainable sources	12.472	0.000	1.214	1.02	1.41
Develop project governance	7.348	0.000	0.743	0.54	0.94

(continued)

Table 9.3 (continued)

	t	Sig. (2-tailed)	Mean difference	95% Confidence interval of the difference	
				Lower	Upper
Ensure that the construction company makes a profit	4.035	0.000	0.600	0.30	0.90
Use plants and equipment that use alternative energy source	4.283	0.000	0.600	0.32	0.88
Implement logistic management	6.439	0.000	0.729	0.50	0.95
Offer corporate social responsibility	2.848	0.006	0.386	0.12	0.66
Minimising traffic disruption to the local neighbourhood	8.148	0.000	0.829	0.63	1.03
To resolve disputes, provide the proper claim management structure	3.419	0.001	0.529	0.22	0.84
Use design and build procurement strategy	1.969	0.053	0.314	0.00	0.63
Use construction management procurement strategy	2.125	0.037	0.343	0.02	0.66
Use the traditional design, tender, and build procurement strategy	0.761	0.449	0.129	– 0.21	0.47
Use management contracting procurement strategy	2.039	0.045	0.329	0.01	0.65
Use measurement contract strategy	1.328	0.189	0.214	– 0.11	0.54

(continued)

Table 9.3 (continued)

	t	Sig. (2-tailed)	Mean difference	95% Confidence interval of the difference	
				Lower	Upper
Respect labour rights and engagement	6.200	0.000	0.843	0.57	1.11
Prevent anti-social behaviour among participants	3.796	0.000	0.586	0.28	0.89
Provide medical benefits to operatives	2.825	0.006	0.429	0.13	0.73
Ensure project team diversity in terms of race, gender and economic status	1.784	0.079	0.271	− 0.03	0.58
Conduct social impact assessment of project	4.315	0.000	0.586	0.31	0.86
Avoid price fixing in tendering/bidding	3.111	0.003	0.529	0.19	0.87
Ensure adequate planning and scheduling that complies with sustainable requirement	9.562	0.000	1.071	0.85	1.29
Disclosures towards anti-corruption measures	5.896	0.000	0.786	0.52	1.05
Provide a clear and adequate claim system for injured operatives	3.910	0.000	0.571	0.28	0.86
Provide decent accommodation to site operatives	4.993	0.000	0.686	0.41	0.96

(continued)

Table 9.3 (continued)

	t	Sig. (2-tailed)	Mean difference	95% Confidence interval of the difference	
				Lower	Upper
Prepare a supply chain management plan	6.939	0.000	0.843	0.60	1.09
Prepare a configuration management plan	5.925	0.000	0.757	0.50	1.01

design phase of the projects. The data also revealed that providing sufficient information to construction companies by the design team and client organisation is a main criterion for sustainable construction management practice. This is anticipated because inadequate information is the main reason for rework, poor client satisfaction, cost overruns, and delays in the construction sector [16]. The construction team required correct and complete information for proper decisions that increased productivity. The quality of decision-making is inherently limited by the quality of the information available.

The research found that an effective waste management plan is critical to sustainable construction management. A waste management plan is a strategic document to manage waste during the construction phase. It outlines the estimated types of waste and describes recycling and reuse methods for waste. The data revealed that sourcing material from sustainable sources is paramount to sustainable construction management. The design teams must specify materials and components that are obtained from sustainable sources. Most research on sustainable construction focuses on materials [12]. Additionally, renewable materials and components should be used in the projects to reduce reliance on the total primary energy supply for construction activities. The goal of sustainable construction management is to reduce the embodied energy consumed on-site during the material and production phases. Material and product passports must be prepared to aid the construction process and implementation.

Environmental impact assessments were found to be critical success factors for sustainable construction management. To ensure that the impact of construction processes and activities has a lower negative impact on the environment, it is necessary for the construction teams need to conduct environmental impact assessments. Such assessment will help in selecting alternative construction methods and developing measures to reduce their negative impact on the environment. In addition, the unique role of a construction manager in the greening construction process includes proper documentation for submittals and coordination of work by contractors and design teams for certification by the commissioning agents [23]. Planning

Table 9.4 Criteria success factor for sustainable construction management

Critical success factor	Lower	Upper	Std. Error mean	Kurtosis	Skewness	Std. Deviation	ACI
Avoid environmental pollution caused by the projects	85.00	92.80	1.98	1.50	− 1.47	16.56	88.80
The construction team should be given adequate information	84.60	92.00	1.88	0.10	− 1.07	15.78	88.20
The design team should be given adequate information	82.80	91.00	2.04	0.011	− 1.02	16.98	86.80
Conduct environmental impact assessments	81.40	90.00	2.16	− 0.17	− 0.97	18.06	85.80
Make use of a waste management plan	80.20	89.60	2.36	2.00	− 1.45	19.68	84.80
Use materials / components from sustainable sources	80.40	88.20	1.94	2.57	− 1.25	16.30	84.20
Make use of renewable material	77.60	87.00	2.36	1.55	− 1.26	19.72	82.20
Ensure adequate planning and scheduling that complies with sustainable requirement	77.00	85.80	2.24	3.666	− 1.34	18.74	81.40
Presence of method statement	76.40	85.40	2.24	− 0.83	− 0.52	18.78	80.80
Ensure proper health and safety management on sites	75.60	86.00	2.62	− 0.39	− 0.84	21.92	80.80
Engaged skilled workers	74.20	84.00	2.46	− 0.20	− 0.74	20.56	79.20
Design team should have a green/sustainable assessor	72.40	83.00	2.68	0.85	− 0.91	22.46	77.80

(continued)

Table 9.4 (continued)

Critical success factor	Lower	Upper	Std. Error mean	Kurtosis	Skewness	Std. Deviation	ACI
Make use of local materials	72.80	81.40	2.16	– 0.73	– 0.31	18.10	77.20
Respect labour rights and engagement	71.40	82.20	2.72	1.08	– 1.02	22.74	76.80
Prepare a supply chain management plan	72.00	81.80	2.42	2.21	– 1.13	20.32	76.80
Minimising traffic disruption to local neighbourhood	72.60	80.60	2.04	– 1.07	0.05	17.02	76.60
Conduct an indoor air quality assessment after the project is completed	70.60	80.80	2.54	1.31	– 0.90	21.24	75.80
Disclosures towards anti-corruption measures	70.40	81.00	2.66	0.70	– 0.79	22.30	75.80
Use value engineering/ management techniques	69.60	80.60	2.78	– 0.43	– 0.65	23.20	75.20
Prepare a configuration management plan	70.00	80.20	2.56	0.66	– 0.67	21.38	75.20
Green / sustainable consultants should be engaged	69.40	80.20	2.68	0.57	– 0.79	22.50	74.80
Develop project governance	70.80	78.80	2.02	– 1.05	0.23	16.92	74.80
Implement logistic management	70.00	79.00	2.26	– 0.33	– 0.27	18.94	74.60
Contractors should have green/sustainable certified staff	68.60	80.00	2.86	0.84	– 0.84	23.86	74.20

(continued)

Table 9.4 (continued)

Critical success factor	Lower	Upper	Std. Error mean	Kurtosis	Skewness	Std. Deviation	ACI
Provide descent accommodation to site operatives	68.20	79.20	2.74	0.47	– 0.71	22.98	73.80
Engage local operatives	68.60	78.40	2.46	– 0.76	– 0.19	20.64	73.40
Ensure that the construction company makes a profit	66.00	78.00	2.98	1.07	– 1.05	24.88	72.00
Use plants and equipment that use alternative energy source	66.40	77.60	2.80	0.38	– 0.78	23.44	72.00
Make use of building information modelling	67.60	75.80	2.06	– 0.59	– 0.06	17.20	71.80
Prevent anti-social behaviours among participants	65.60	77.80	3.08	– 0.02	– 0.80	25.82	71.80
Conduct social impact assessment of project	66.20	77.20	2.72	0.65	– 0.77	22.72	71.80
Provide clear and adequate claim system to injured operative	65.60	77.20	2.92	– 0.10	– 0.59	24.46	71.40
To resolve disputes, provide the proper claim management structure	64.40	76.80	3.10	0.67	– 0.10	25.86	70.60
Avoid price fixing in tendering/ bidding	63.80	77.40	3.40	0.17	– 0.82	28.44	70.60
Use building information modelling	65.40	74.60	2.26	0.10	– 0.32	18.88	70.00
Make political accountability a priority	62.60	75.20	3.16	0.27	– 0.80	26.52	68.80

(continued)

Table 9.4 (continued)

Critical success factor	Lower	Upper	Std. Error mean	Kurtosis	Skewness	Std. Deviation	ACI
Provide medical benefits to operatives	62.60	74.60	3.04	− 0.34	− 0.56	25.38	68.60
Offer corporate social responsibility	62.40	73.20	2.70	0.19	− 0.51	22.66	67.80
Use construction management procurement strategy	60.40	73.20	3.22	− 0.15	− 0.62	27.00	66.80
Use management contracting procurement strategy	60.20	73.00	3.22	0.13	− 0.70	26.96	66.60
Use design and build procurement strategy	60.00	72.60	3.20	− 0.11	− 0.60	26.72	66.20
Use industrialised building system	59.40	72.00	3.12	0.38	− 0.63	26.18	65.80
Ensure project team diversity in terms of race, gender and economic status	59.40	71.60	3.04	− 0.56	− 0.40	25.46	65.40
Use measurement contract strategy	57.80	70.80	3.22	− 0.03	− 0.55	27.00	64.20
Reduced parking footprint	56.80	69.60	3.20	0.07	− 0.60	26.74	63.20
Use conventional design, tender and construct procurement strategy	55.80	69.40	3.38	− 0.48	− 0.42	28.26	62.60
Use drone technology for site supervision and safety management	47.20	59.00	2.94	− 0.22	0.11	24.52	53.20
Make use of construction technology (e.g., 3D scanners)	46.40	59.20	3.20	− 0.38	− 0.17	26.82	52.80

and scheduling were found to be requirements for sustainable construction management. Plans (configuration, logistics, and supply chain) should be carefully monitored, and improvement remedies should be proposed and justified. This finding is like that of Hwang et al. [8] and Khun-anod and Limsawasd [11]. Their studies revealed that planning and scheduling problems are common in sustainable construction. For construction management practices to contribute to sustainable development, the sustainable construction manager needs to set sustainability as a deliverable in the planning and scheduling of the projects. Unpredictable demand and the shift towards contract work are making it harder for construction companies to maintain a continuous level of resource management. Therefore, construction companies must know whether the available resources and capacity match up with demand and what to do if resources are not adequate. However, most construction companies (e.g., resource planners) use spreadsheets for this purpose. However, in sustainable construction management practice, some organizations should improve their planning tools to incorporate digital techniques to simplify the process and optimise the organization's resource planning.

The research found that the method statement would contribute to sustainable construction management practices. A method statement describes how the construction work will be conducted. It also outlines health and safety plans, configuration plans, quality management plans, and work duration, which are essential for sustainable construction management. Thus, it would ensure that sustainable requirements are incorporated into the delivery process. Furthermore, preparing a sound health and safety plan is measured as a requirement for sustainable construction management. This is not surprising because construction health and safety problems are high (Oni et al. 2022, 2023). The health and safety management of construction workers is one unique aspect of social sustainability. However, many construction workers are not trained in the trades they work in. Many of the workers learn on the job. Most recent university and polytechnic graduates lack basic industrial experience for the jobs that have been assigned to them. However, this may lead to labour exploitation. Furthermore, it could lead to accidents, rework, and disputes, which may, in turn, increase claims and disputes. Since it is impractical for all the design and construction teams to hold sustainable /green practice certification, an independent sustainable assessor should be part of the design team. The independent sustainable construction manager is like professional architects, quantity surveyors, and engineers. This finding is impressive because, without a sustainable construction manager at the centre of the delivery, the tools, techniques, and methods would not produce the desired result. A competent sustainable construction manager has distinct responsibilities. Professional sustainable construction managers are generally focused on design requirements, while sustainable construction managers handle construction management from concept to completion. Sustainable construction managers serve as moderators among the project teams, and their unique skill set, and competencies influence the success of the projects.

The movement of people, plants, and equipment could lead to congestion on the sites. Hence, sustainable construction management must adopt an approach that controls traffic on and around the sites. The carbon footprint of a project can be reduced with effective traffic management plans. A contractor could also provide parking spaces designated for carpooling. Adequate parking and carpooling mean less vehicular movement, which will reduce greenhouse gas emissions, improve air quality, and increase worker savings. This is especially crucial for sites in cities where there are a lot of movement restrictions. It is not surprising that respect for labour rights and engagement were found to be critical factors for sustainable construction management. A sizeable portion of construction activities are performed by workers who are not paid what they deserve. The payment and condition of compensation to construction workers, especially trade and artisan workers, should receive strategic consideration. One of the major sectors that disrespects human dignity is the construction sector. It has been estimated that about 4.5 million construction workers are in forced labour globally [9], and anti-trafficking measures to prevent and prohibit labour trafficking in the sector are weak. Globally, most construction companies do not have policies to prevent labour exploitation or measures to ensure that wages are paid on time and fairly. Many emigrant workers died on the construction sites due to labour exploitation and many were engaged without valid documents, and many of these workers were underpaid.

A logistics management plan has been identified as an essential criterion for sustainable construction management. Logistics management is critical to procuring the specified resources in the correct form, time, and place. The problems associated with transportation and storage facilities have increased the importance of logistic management in the sustainable construction management framework. Sustainable construction management emphasizes green supply chain management. Green supply chain management integrates sustainable environmental processes into product design, material sourcing, and production to reduce CO_2 emissions. The consideration of digitalisation in supply chain management can help organisations attain the SDGs. In sustainable construction management, it is fundamental for contractors to submit a logistic management plan for approval by the sustainable construction manager. This will ensure that the overall management of construction resources is effectively managed to avoid delays resulting from acquisition, storage, transportation, and delivery. This will also ensure that projects consider sustainable practices beyond the design and site management functions of the projects. Variations are inevitable in construction projects. Configuration management involves a set of processes, techniques, and tools to ensure that the project meets its performance requirements [the final design configuration] and that changes to the agreed drawings and specifications are critically justified. Configuration management ensures that contractors make informed decisions, take the appropriate actions, and deliver all work and products as intended throughout the lifecycle. It also provided detailed information on the sustainability impact of the variations of projects.

Internationally, there has been a push for industrialised building systems, or IBS, to improve the sustainability of building methods and projects. Implementing IBS would improve quality performance while decreasing rework, time, and expense. Therefore, it is not surprising that IBS is a major requirement for sustainable construction management. Sustainable construction management aims to ensure that contracting companies make a significant contribution to the management of construction projects. To achieve this, the contractor organisations need to have in-house certified green/sustainable staff. This will aid in the clarifications and reconciliations of practices and claims with professional sustainable construction managers. Sustainable construction management aims to ensure that projects are completed to meet sustainable requirements upon completion. Therefore, sustainable construction managers will conduct indoor air quality assessments to ensure that the completed project meets the required indoor air quality standard.

The source of materials was found to be critical to sustainable construction management. Where practicable, materials should be sourced from or near the community where the projects are located. Sourcing materials locally would create jobs for the local community. It will reduce traffic congestion and energy and fuel consumption along the supply chain. A goal of sustainable construction management is to meet the needs of the local or specific construction market in ways that ensure construction companies complete projects that will ensure their long-term survival. Profit margin is a necessary factor in sustainable construction management practices. Eventually, low-profit margins are rampant in the construction sector globally when compared to other sectors. For construction companies to be sustainable, they need to earn a profit to ensure their existence and pay their staff fairly and on time. This finding is supported by the finding of Robichaud and Anantatmula [22] that contractors involved in sustainable construction require a higher profit margin. A sustainable profit margin is proportional to the number of tasks completed without exploiting clients or workers. Workers' wages and benefits should be based on their qualifications, expertise, and the nature of the work. Hence, it is not surprising that the respondents have agreed that construction companies need to earn a good profit margin to contribute to sustainable construction management.

The construction sector is corruption-laden. Among the major economic sectors, the construction sector appears to be the most corrupt. Unlike in other sectors, corruption in the construction sector occurs at all levels. This is compounded by the fragmented and divided interests in the construction sector. Therefore, sustainable construction managers must instill anti-corruption measures into the contract. The research revealed that value management would contribute to sustainable construction management practices. Value management is a practice that seeks the achievement of client value systems. The value management technique views decisions, materials, components and projects from the first principle. Value management ensures that the money invested in the project has value and drives away unnecessary costs. The involvement of stakeholders during the design phases is a basic requirement for managing sustainable construction. Value management is the most viable platform to include stakeholders, especially the client, during the concept and design stages of a project. Because it is about the user's or client's value system, value management is user- or client-focused. The primary function of a design or project is to meet the performance requirements that meet the users' or clients' requirements.

This research found that project governance is a requirement for sustainable construction management. Project governance is an auxiliary component that offers a comprehensive framework for project management, considering the client value system, to sustainable construction managers and other team members. It empowers sustainable construction managers to make authoritative decisions about incorporating sustainability elements into the management of the projects. The complexity and size of construction projects mean that plants and equipment are always required. Conventionally, the plants and equipment depend on fossil fuels. Sustainable construction managers emphasise the use of plants and equipment that use alternative energy sources on-site. Another implication of the SDGs is making sure that site operatives are well accommodated on sites. However, most of the accommodations provided for site workers are not suitable for human habitation. Most of the workers are cramped in the housing, often without adequate ventilation, lighting, or water [14]. Site accommodations are epicentres for the spread of diseases. Sustainable construction managers ensure that the housing for the construction workers is decent and comfortable and recognise the fact that recent construction will boost worker productivity and lower medical costs.

Governments and large construction clients expect considerable reductions in CO_2 emissions, as sustainability is high on the agenda of governments and corporate organisations. One solution to this issue is the use of BIM. Creating and effectively managing digital information about built facilities is known as building information modelling, or BIM. Adoption of BIM will enhance health and safety management, decrease claims, boost productivity, and decrease delays. Rework can be drastically eliminated by digitally replicating the construction phase of a project [16]. Many large contractors offer corporate social responsibility to residents of the communities where projects are located. Some contractors construct schools, roads, and boreholes and supply seeds to farmers in the communities where projects are located as part of their corporate social responsibility and social value. The social impact assessment and analysis of the social value of a project would allow construction companies to determine the extent to which a project would impact society's social fabric. It also provides information to the construction companies on the nature of the skillset available in the community. Hence, it is interesting to know that employing operatives from the communities would contribute to sustainable construction management.

When a construction project is required, a suitable procurement strategy is identified. There are various procurement options, but only one type is suitable for a particular client's requirements. While some of the strategies embrace the performance specification, others do not or do so to a lesser extent. For instance, a construction management strategy ensures that sustainability requirements are included in the project requirements. Provides an opportunity to incorporate changes into the design before site activities commence. Thus, it is not surprising that a construction management procurement strategy will contribute to sustainable construction management practices. A construction management strategy is a type of strategy in which a client engages the services of professional construction managers to oversee the project on their behalf [1]. The data revealed the importance of public accountability in construction management. Regardless of the type of client, ensuring public accountability among all the project stakeholders is very crucial. This ensures that project costs are not overtly inflated and that workers are paid on time and based on their competency. It is not surprising that the closest factor to this factor is price-fixing during tendering. Price fixing is endemic in the construction sector. Sustainable construction management aim to promote fairness and accountability by ensuring that tenderers do not collide to fix tender prices for a project.

Sustainable construction management aims to reduce litigation and claims on construction projects. It also seeks to minimise the length and intensity of negotiations needed to agree on transactions. When workers are injured or die due to work activities, speedy resolution of the disputes is essential to the victims and their families. It is essential for the other workers on site because it will affect their emotions and sentiments. Therefore, it is essential to provide a clear and adequate claim system for injured operatives. Another procurement strategy that would facilitate the incorporation of sustainability practices into total construction procurement is the management contracting procurement strategy, according to the data. Like the construction management strategy, the design and construction services are provided by separate organizations. While the strategy provides an opportunity to incorporate sustainability into construction management practice, such an opportunity may be reactive. The final cost may be significantly higher, and variations are usually greater due to lack of surety. However, with experienced teams, the strategy would be beneficial.

Apart from providing speedy claim management to the injured operatives, it was found that there was a need to provide a proper claim management structure to resolve disputes among the project team, as the data revealed. The construction sector is highly litigious. Disputes may arise for several reasons in construction contracts. In sustainable construction, it may be due to greenwashing. Sustainable construction management provides a collaborative guideline in the contract on how disputes are resolved among all parties to the contract. Sustainable construction management promotes the use of ADR practices rather than litigation. However, where litigation is identified as the most suitable means to resolve disputes and claims, it should be used. Because ADR is not necessarily the most proper in all cases. The data also revealed that another useful procurement strategy for sustainable construction management is measurement contracts. In a measurement contract, contractors are paid for the actual work performed plus an added agreed-upon fee to cover the contractor's profit and overhead expenses. However, the strategy is associated with high and multiple uncertainties. Therefore, the professional sustainable construction manager must conduct more thorough evaluations of the deliverables to ensure that the project meets sustainability standards. It was also found that sustainable construction management can be procured through conventional design, tendering, and construction strategies. This strategy allows for complete information before work can start. This may reduce rework, disputes, claims, delays, and cost overruns, which are essential for sustainability. However, professional sustainable construction managers should be included in the design team to ensure that sustainable elements and practices are incorporated into the design and process. However, unless the professional sustainable construction managers are the project's lead consultants, this does not guarantee that sustainable construction management is visible.

Construction projects involve various people from many countries. In most developed and developing countries, the site operatives are primarily foreigners, especially from underdeveloped and some developing countries with lower economic status.

However, gender discrimination and child labour are common in the sector, and site operatives are often exploited and abused. Furthermore, there is no genuine approach to ensuring that workers are employed based on diversity. Instead, it is based on profit maximization through mistreatment and abuse. Sustainable construction management aims to ensure that the diversity of contractors, subcontractors, operatives, suppliers, and other stakeholders is purposefully promoted and that it seeks to create employment for the disadvantaged in the projects. For instance, functions that do not require movement should be identified and assigned to those with movement restrictions. Some work should also be specifically dedicated to the extremely poor in the community.

Construction organisations should set aside a certain percentage of their workforce for individuals with disabilities. Sustainability drives are already paving the way for innovative technologies that enable construction projects to be completed safely and to a high standard. 3D scanners, robotics, and drones are becoming increasingly popular. The use of 3D scanners brings greater accuracy and efficiency to workflows and decreases rework. The use of robotics and drone technology for site supervision and safety management has been widely discussed among large contractors lately. Sustainable construction management seeks the application of technologies for tracking progress, conducting quality assurance checks, and investigating reworks to collect real-time data to support automated decision-making. Using technology to enhance communication and collaboration is not just between site operatives but also between different parties in a project team.

9.8 Conclusion and Implications

Research focusing on sustainable construction management holds immense value in supporting sustainable development goals and advancing the construction industry towards a more sustainable and resilient future. This research addresses a gap in the current literature on sustainable construction management by investigating its philosophy and the CSFs. While earlier studies have explored sustainable construction practices, there is limited research on how the theory can be effectively implemented in practice. Earlier attempts to evaluate the critical success factors have been fragmented and localized. This chapter presents evaluation criteria for the implementation of sustainable construction management from an international perspective. Theoretically, the findings support and contribute to the understanding of the fundamental theory of sustainable construction management. It demonstrates how sustainable construction can be profitable for the construction and design teams, clients, users, community, and policymakers. Sustainable construction management has drawn considerable attention over the past several years for a variety of reasons: interest in decarbonising the built environment; cost of production; durability; climate change; health; and lower operation and maintenance costs. Moving forward, sustainable construction requires sustainable construction management. Research in sustainable construction management can apply positivism, subjectivism, or their combination. It is imperative to draw upon a diversity of methodologies to provide an enriching understanding of the nature of sustainable construction research and practice. To put it differently, it is thereby necessary that the epistemology of research in sustainable construction tends towards multi-methodology. To implement sustainable construction management, critical success factors that combine the three elements of sustainability were evaluated using both objective and subjective ontologies. 48 critical success factors were revealed, measured, and clustered. As the empirical data unfolds, it exhibits a strong emphasis on project management structure, procurement plan, environmental ethics, engagement of professional sustainable construction managers, company policy, information management, health and safety plan, digital construction, and community engagement. The data would be useful for academic institutions that offer construction management, architecture, engineering, and quantity surveying programmes. This study advances knowledge in sustainable construction management and offers helpful advice for enhancing project success in the construction sector. It provides insightful information for both academics and practitioners. Below are the specific implications of this research:

Academic Implications:

- Theoretical Advancement: By finding and validating crucial success variables, this research adds to the theoretical foundation of sustainable construction management. It improves our present comprehension of the subtleties and complexity of sustainable construction projects.
- Research Agenda: It sets the stage for future academic investigations by highlighting the importance of specific success factors and areas that require further exploration. This can stimulate research in related fields.
- Interdisciplinary Insights: This study bridges the gap between construction management and sustainability disciplines by focusing on sustainable construction management, providing insights that can advance both fields of study.

Practitioner Implications:

- Enhanced Decision-Making: The discovered crucial success elements can be used by practitioners as a practical guide to help them decide wisely at various stages of sustainable construction projects, which will improve project outcomes.
- Risk Mitigation: Understanding these factors and components can make practitioners more prepared to foresee and reduce potential risks and difficulties related to sustainable building, leading to more durable and effective projects.
- Strategic Planning: These factors or components can be incorporated by organisations into their strategic planning procedures, enabling them to better match their goals with ethical behaviour and increase their competitive edge.

However, a major limitation of this research is the sample size. However, the statistics of goodness of fitness suggest that the data are normal. The profiles of the respondents suggest that the data and findings are generalizable. However, confirmatory research would be useful, and future research may replicate the research elsewhere.

Bibliography

1. Ahmed SM, Yaris C, Farooqui RU, Saqib M (2014) Key attributes and skills for curriculum improvement for undergraduate construction management programs. Int J Constr Educ Res 10(4):240–254
2. Arcadis (2021) International construction costs 2021: investing for a sustainable recovery. https://connect.arcadis.com/ICC-Report-download-confirm. www.arcadis.com/en/.
3. Banihashemi S, Hosseini MR, Golizadeh H, Sankaran S (2017) Critical success factors (CSFs) for integration of sustainability into construction project management practices in developing countries. Int J Project Manage 35(6):1103–1119
4. Brundtland Commission (1987) World commission on environment and development, our common future. Oxford University Press, Oxford and New York
5. Chegut A, Eichholtz P, Kok N (2019) The price of innovation: an analysis of the marginal cost of green buildings. J Environ Econ Manag 98:102248
6. Dodge Data and Analytics (2016) World green building and construction trends. Available at https://www.construction.com/toolkit/reports/world-green-building. 18 Jan 2022
7. He Z, Chen H (2021) Critical factors for practicing sustainable construction projects in environmentally fragile regions based on interpretive structural modeling and cross-impact matrix multiplication applied to classification: a case study in China. Sustain Cities Soc 74:103238
8. Hwang B-G, Leong LP, Huh Y-K (2013) Sustainable green construction management: schedule performance and improvement. Technol Econ Dev Econ 19(Supplement 1):S43–S57
9. ILO (2017) The ILO records that 18% of all forced labor cases are documented in the construction sector, pp 25–26
10. Ismael D, Shealy T (2018) Sustainable construction risk perceptions in the Kuwaiti construction industry. Sustainability 10(6):1854
11. Khun-anod K, Limsawasd C (2019) Pre-project planning process study of green building construction projects in Thailand. Eng J 23(6):67–81
12. Lima L, Trindade E, Alencar L, Alencar M, Silva L (2021) Sustainability in the construction industry: a systematic review of the literature. J Clean Prod 289:125730
13. Olanrewaju A, Tan SY, Abdul-Aziz AR (2018) Housing providers' insights on the benefits of sustainable affordable housing. Sustain Dev 26(6):847–858
14. Olanrewaju AA, Chong YS (2021) Post occupancy evaluation of green residential buildings, in the Greater Kuala Lumpur, Malaysia. J Housing Built Environ 36(2):825–857. https://doi.org/10.1007/s10901-021-09832-1
15. Olanrewaju A, Shari Z, Gou Z (eds) (2019) Greening affordable housing: an interactive approach. CRC Press
16. Olanrewaju AL (2022) An artificial neural network analysis of rework in sustainable buildings. In: IOP conference series: earth and environmental science (vol 1101, no 2). IOP Publishing, p 022003
17. Oni OZ, Olanrewaju A, Khor SC (2023) Review of critical success factors affecting Malaysia's construction industry's sustainable health and safety practices. Front Eng Built Environ 3(1):48–62
18. Oni OZ, Olanrewaju A, Khor SC (2023) Fuzzy synthetic evaluation of the factors affecting health and safety practices in Malaysia construction industry. J Eng Des Technol

19. Maqbool R, Amaechi IE (2022) A systematic managerial perspective on the environmentally sustainable construction practices of UK. Environ Sci Poll Res 1–18
20. Metz TR, Cheatham CW (2008) As green building moves forward, claims and disputes will follow. Constr Law Publ Contracts 57:56–59
21. Radosavljevic M, Bennett J (2012) Construction management strategies: a theory of construction management. Wiley
22. Robichaud LB, Anantatmula VS (2011) Greening project management practices for sustainable construction. J Manag Eng 27(1):48–57
23. Spiegel R, Meadows D (2010) Green building materials: a guide to product selection and specification. Wiley
24. Wuni IY, Shen GQ (2020) Critical success factors for modular integrated construction projects: a review. Build Res Inf 48(7):763–784

Chapter 10
PCA Framework for Sustainable Construction Management

Abstract This chapter introduces a Principal Component Analysis (PCA) framework for sustainable construction management, addressing the need to organise and interpret the complexity of critical success factors (CSFs) in global construction practices. Sustainable construction is increasingly prioritised, yet project outcomes often fall short due to fragmented tools and strategies. Through a survey of over 600 construction firms, 48 CSFs were identified and analysed using PCA to reduce dimensionality and group related variables. The analysis revealed nine key components: project management structure, procurement plan, environmental ethics, sustainable construction manager, company policy, information management, health and safety plan, digitalisation, and community engagement, explaining 80% of the total variance. The framework enhances systemic decision-making and provides a structured pathway for the successful implementation of sustainable construction across regions. It contributes to the philosophical and practical discourse by establishing sustainable construction management as a distinct discipline with dedicated methods and practices. The study's findings offer valuable guidance for policymakers, educators, and industry stakeholders to align construction practices with sustainability goals. Despite sample size limitations, the results are generalizable and pave the way for confirmatory and comparative studies. This chapter ultimately serves as a decision-making and pedagogical tool for promoting global sustainable development in the built environment.

Keywords Business case · Research paradigm · Pedagogy · Whole life appraisal · Greenwashing

10.1 Learning Outcomes

This chapter presents a Principal Component Analysis (PCA) framework based on the previous chapter. A popular dimensionality reduction method in several domains, such as signal processing, machine learning, and statistics, is principal component analysis. The PCA framework is employed to address the challenges associated

A. Olanrewaju, *Sustainable Construction Management: Research and Practice Companion*, Management in the Built Environment,
https://doi.org/10.1007/978-981-96-4910-5_10

with high-dimensional data and to enhance the interpretability and application of the analytical results presented in the previous chapter. The PCA aims to transform the original critical success factors into a new set of orthogonal principal components, that capture the largest variance in the data. The surge in demand for sustainable construction is outpacing the available supply, resulting in disappointing project outcomes. This predicament is primarily attributable to the inadequacies in the tools and strategies employed across all phases of sustainable construction projects, including design and construction. This paper reports research that develops an international framework for sustainable construction management based on the clustering of its critical success factors for systemic decision-making. The data were collected through a survey questionnaire. The questionnaires were administered to more than 600 construction companies based on the snowballing technique. The questionnaire makes up 48 critical success factors for sustainable construction management. Critical success elements for the creation of the framework were organised using the PCA. The factor analysis structured the 48 factors into nine components, which explained 80% of the total variance in the model, namely project management structure, procurement plan, environmental ethics, sustainable construction manager, company policy, information management, health and safety plan, digitalization, and community engagement. The salient finding from this research is that sustainable construction management practices are a distinct discipline with their practices, tools, methods, and strategies. Ontologically, even though the data are size-constrained, the results are generalizable. After studying this chapter, you would be to:

- Distinguish the interaction among the critical success factors
- Know the framework for the critical success factors
- To organized the critical success factors of implementing sustainable construction management
- Apply sustainable construction management framework
- Evaluate the application of the framework

10.2 Introduction

This chapter developed a decision-making framework based on the analytical results of the previous chapter. While the examinations of the CSFs are crucial, there is a need to develop a model to facilitate decision-making. It has also been proved that research on CSFs for sustainable construction management has not been analysed and categorised from an international or global perspective. This research addresses this gap by employing clustering techniques to categorize CSFs and subsequently developing a comprehensive framework. This framework aims to enhance the successful implementation of sustainable construction management practices across diverse countries and regions. By clustering the CSFs, commonalities and variations, can be identified, and will allow for a nuanced understanding of the factors that influence sustainable construction management outcomes globally. This approach not only facilitates the identification of region-specific challenges but also provides a structured framework for addressing them effectively. Clustering the CSFs will help to streamline and enhance decision-making on the deliverables of sustainable construction management. PCA was used as the clustering technique. Principal component analysis is a supervised machine learning of statistical procedure used to summarise the information content in large data tables by a smaller set of overview indices that can be more easily visualised and analyzed. To analyse the crucial success criteria for sustainable construction management, PCA offers a methodical and rigorous approach. It supports well-informed decision-making by assisting in the discovery of crucial drivers, interrelationship identification, simplification of complex information, and data visualization. The findings of this study will provide insights into the best practices and strategies that can be adopted to promote sustainable construction management practices globally. Also, the findings will help policymakers and industry practitioners make informed decisions for sustainable development.

10.3 Background and Theoretical Framework

In statistics and machine learning, dimensionality reduction techniques like PCA are used, in which a dataset is converted into a new coordinate system using mathematics. The principal components are the linear combinations of the original variables, and they are orthogonal to each other. This technique is used to simplify complex datasets by reducing the number of variables while keeping the most valuable information. This is achieved by finding a set of linearly uncorrelated variables, called principal components, that capture the largest amount of variance in the original data. The resulting principal components are sorted in descending order of variance, so the first few components explain most of the data's variance, while the later components account for decreasing amounts of variance. PCA is a powerful tool for exploratory data analysis, data visualization, and feature engineering to provide insights from complex datasets. Besides being used for dimensional reduction, it is also useful for data visualization, data pre-processing, and data compression. The justification for applying PCA to critical success factors (CSFs) for sustainable construction management lies in its ability to extract meaningful information from many CSFs to provide valuable insights for the stakeholders. For instance:

- Dimensionality reduction: CSFs for sustainable construction management frequently consider a wide range of elements, including regulatory compliance, social responsibility, environmental effect, and economic viability. By locating the most crucial variables (principal components) that account for the majority of the variance, PCA can help minimise the dimensionality of the data. The data's interpretation and analysis are made simpler as a result.

- Identification of key drivers: PCA makes it possible to decide the primary CSF drivers. The factors that have the greatest impact on sustainable construction management's effectiveness can be found by examining the loadings of the original variables on the primary components. With this knowledge, decisions can be made, and resources can be distributed to the most crucial sectors.
- Interrelationships among CSFs: PCA can reveal dependencies and correlations between the CSFs. It is possible to understand how various CSFs interact and affect one another by examining the correlations between variables. Designing integrated and comprehensive solutions for sustainable construction management benefits from having this information. Having this knowledge is helpful when designing integrated and comprehensive solutions for sustainable construction management.
- Enhanced decision-making: Evidence-based decision-making in sustainable construction management can benefit from the PCA's insights. Stakeholders can build targeted strategies to improve sustainability performance by prioritising actions, effectively allocating resources, and understanding the fundamental structure and significance of CSFs.

The Mathematical model for PCA is

$$Y = C1 * PC1 + C2 * PC2 + C3 * PC2 + C2 * PC2 \ldots + C10 * PC10 + C$$

where PCs are the components and Wis are the weights for each of the components. A simplified theoretical PCA model is displayed in the Fig. 10.1.

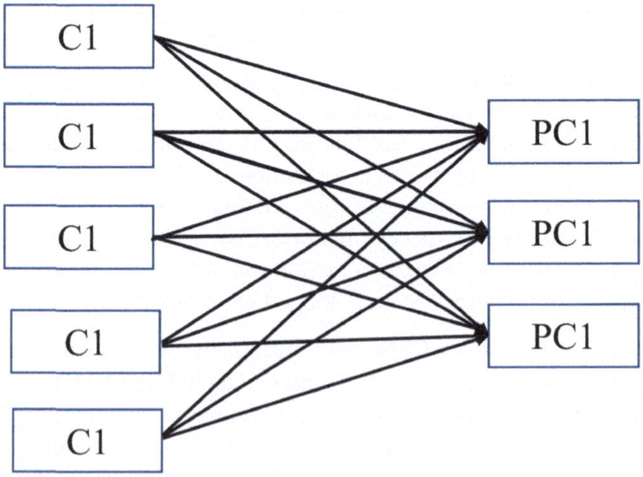

Fig. 10.1 Framework for principal component analysis

10.4 Results of the Factor Analysis on the Critical Success Factors for Sustainable Construction Management

The common method is for a factor name to be based on the name of the variable with the highest loading factor under a component. The alternative method is to base the name of the component on the similarity among the factors in a component. The second method, though sometimes challenging, is more realistic. Hence, in this chapter, the second method is adopted. For every component, an additional computation was made using a second-order factor analysis, which merged all of the determinants under a component. It takes multivariate decision-making tools to evaluate crucial success elements. The computation was done using the Varimax rotation approach and principal component analysis. The statistical determinant is 2.765E-28. The KMO (0.785) and Bartlett's test results (x2 (1081) = 3259.697, p = 0.00) show that the data are suitable for PCA. For useful decision-making, the 47 CSFs were grouped into 9 components by the results (Table 10.1). This was also demonstrated in Fig. 10.2, where the function essentially flattened out following the ninth component. The distribution of the crucial success elements for each component is shown in Table 10.2 78% of the total variance was explained by the nine components. Particularly for an exploratory study, the data are sufficient if the total variance explained is greater than 50% and each of the eigenvalues is greater than 1.

10.5 Discussion of Findings on the Critical Success Factors

10.5.1 Discussion of the Results of Factor Analysis

Altogether, a total of 48 critical factors were evaluated, but "develop project governance" was not clustered with any of the components despite several computations and treatments. Hence, it was removed from the framework. Following Olanrewaju and Idrus' [17] explanation, the components are named based on the similarity among the CSFs. The discussion is provided in the following subsections:

- **Component: Project management structure**

Project management structure refers to the framework or system that an organization uses to plan, execute, and manage projects. The structure typically includes roles, responsibilities, and reporting relationships for project team members, stakeholders, and other individuals involved in the project. A construction project's success depends on its project management structure, which offers a framework for organising, planning, and carrying out project activities. It guarantees open lines of communication, establishes roles and duties, and makes it easier for stakeholders to work together effectively. Tasks are assigned effectively, risks are recognised and handled,

Table 10.1 Total variance explained

Component	Initial eigenvalues			Extraction sums of squared loadings			Rotation sums of squared loadings		
	Total	% of Variance	Cumulative %	Total	% of variance	Cumulative %	total	% of Variance	Cumulative %
1	19.127	40.696	40.696	19.127	40.696	40.696	9.508	20.231	20.231
2	4.896	10.417	51.113	4.896	10.417	51.113	6.436	13.694	33.925
3	2.925	6.224	57.337	2.925	6.224	57.337	4.563	9.708	43.633
4	2.456	5.226	62.563	2.456	5.226	62.563	3.491	7.428	51.060
5	1.801	3.832	66.395	1.801	3.832	66.395	3.246	6.907	57.967
6	1.569	3.338	69.732	1.569	3.338	69.732	2.906	6.184	64.151
7	1.471	3.129	72.861	1.471	3.129	72.861	2.437	5.184	69.336
8	1.207	2.568	75.429	1.207	2.568	75.429	2.187	4.654	73.989
9	1.076	2.289	77.718	1.076	2.289	77.718	1.753	3.729	77.718
10	0.931	1.981	79.700						
11	0.827	1.759	81.459						
12	0.795	1.691	83.149						
13	0.766	1.630	84.779						
14	0.664	1.412	86.191						
15	0.626	1.332	87.524						
16	0.562	1.195	88.719						
17	0.471	1.002	89.720						
18	0.443	0.943	90.663						
19	0.420	0.893	91.556						
20	0.377	0.802	92.358						
21	0.359	0.764	93.123						

(continued)

Table 10.1 (continued)

Component	Initial eigenvalues			Extraction sums of squared loadings			Rotation sums of squared loadings		
	Total	% of Variance	Cumulative %	Total	% of variance	Cumulative %	total	% of Variance	Cumulative %
22	0.316	0.671	93.794						
23	0.295	0.628	94.423						
24	0.272	0.578	95.001						
25	0.247	0.526	95.527						
26	0.221	0.469	95.997						
27	0.202	0.430	96.427						
28	0.179	0.380	96.807						
29	0.176	0.374	97.181						
30	0.159	0.339	97.520						
31	0.150	0.319	97.839						
32	0.144	0.305	98.145						
33	0.134	0.285	98.430						
34	0.118	0.251	98.681						
35	0.099	0.211	98.892						
36	0.090	0.192	99.084						
37	0.076	0.163	99.246						
38	0.069	0.147	99.393						
39	0.058	0.124	99.517						
40	0.049	0.104	99.620						
41	0.044	0.093	99.714						

(continued)

Table 10.1 (continued)

Component	Initial eigenvalues			Extraction sums of squared loadings			Rotation sums of squared loadings		
	Total	% of Variance	Cumulative %	Total	% of variance	Cumulative %	total	% of Variance	Cumulative %
42	0.032	0.068	99.782						
43	0.028	0.060	99.842						
44	0.026	0.054	99.896						
45	0.024	0.051	99.947						
46	0.015	0.031	99.978						
47	0.010	0.022	100.000						

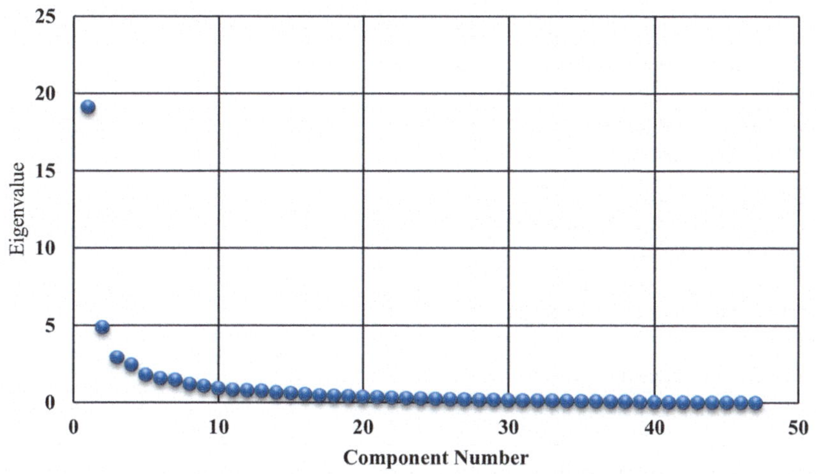

Fig. 10.2 Scree plot

and resources are maximised to complete the project within the allocated budget and time frame when there is a clear project management framework in place. The framework also makes it possible to track developments, solve problems, and adjust to shifting conditions. All of these things eventually result in the timely delivery of high-quality construction projects that satisfy stakeholders' demands and customer expectations. The first component has 14 critical success factors. The component has a validity of 0.708. A second-order factor analysis combined all the factors into one component. The communities range from 0.403 to 0.790. The findings revealed that sustainable construction managers should develop a defined structure to manage the project and client, control corruption, promote diversity and inclusion, and reduce anti-social activity and discrimination. Similarly, claim management should specify medical benefits for on-site workers as well as how injured employees' and workers' rights will be protected. The nature of the accommodations for the workers should be included in the structure. The structure should specify the expected profit margin for the project. Proper planning and scheduling that comply with sustainable requirements should be included in the structure. Each project should have information on the required construction technology. The application of construction technologies should be encouraged. Following the completion of a project, an indoor air quality assessment and an impact assessment of traffic disruption on the local community should be carried out.

Table 10.2 Rotated factor pattern for the selection criteria

Critical success factor	Component								
	1	2	3	4	5	6	7	8	9
Disclosures towards anti-corruption measures	0.86								
Provide clear and adequate claim system to injured operative	0.83								
Respect labour rights and engagement	0.80								
To resolve disputes, provide the proper claim management structure	0.78								
Ensure adequate planning and scheduling that comply with sustainable requirement	0.76								
Provide a medical benefit to operatives	0.76								
Prevent anti-social behaviours among participants	0.74								
Avoid price fixing in tendering/bidding	0.74								
Conduct social impact assessment of project	0.72								
Ensure project team diversity in terms of race, gender and economic status	0.66								
Provide descent accommodation to site operatives	0.66								
Minimising traffic disruption to local neighbourhood	0.58								
Ensure construction make profits	0.55								
Conduct indoor air quality assessment on completion of the project	0.49								
Use management contracting procurement strategy		0.86							
Use measurement contract strategy		0.86							
Use construction management procurement strategy		0.84							
Use conventional design, tender and construct procurement strategy		0.79							
Use design and build procurement strategy		0.78							
Use value engineering/management techniques		0.58							
Implement logistic management		0.55							
Prepare a configuration management plan		0.54							

(continued)

Table 10.2 (continued)

Critical success factor	Component								
	1	2	3	4	5	6	7	8	9
Prepare a supply chain management plan		0.50							
Make use waste management plan			0.78						
Use renewable material			0.77						
Conduct environmental impact assessment			0.71						
Use materials /components from sustainable sources			0.70						
Prevent environmental pollution associated with projects			0.64						
Contractors should have green/ sustainable certified staff				0.86					
Design team should have green/ sustainable assessor				0.75					
Green /sustainable consultants should be engaged				0.75					
Reduced parking footprint				0.53					
Offer corporate social responsibility					0.70				
Use plant and equipment that use alternative energy source					0.64				
Make political accountability a priority					0.55				
Make use of 3D scanners					0.55				
Use industrialised building system					0.53				
The construction team should be given adequate information						0.84			
The design team should be given adequate information						0.83			
Presence of method statement						0.55			
Use skilled workers							0.69		
Ensure proper health and safety management on sites							0.66		
Use an intelligent building system								0.71	
Use building information modelling								0.70	
Use drone technology for site supervision and safety management								0.47	
Use local materials									0.75
Engage local operatives									0.62

- **Component: Procurement plan**

The procurement plan is a document that is developed during the planning phase of the project and is reviewed and updated throughout the project's lifecycle as needed. It is a useful tool for making sure the procurement procedure for the project is successful, efficient, and achieves the project's objectives. The validity of this component is 0.698. Procurement is strategic for practising construction management, and even more so for sustainable construction management. Some procurement strategies ease sustainable construction management and some do not. Procurement strategies that discourage client participation in the development of client value systems and contribute to waste management are not suitable for practising sustainable construction. To ensure that a procurement strategy contributes to a sustainable project, value management must ensure that client value systems are made explicit to all project stakeholders. A shortage of resources is a major problem for sustainable construction projects. Therefore, logistics management should be included in the procurement plan.

The procurement plan should include supply chain management plans and configuration management plans. The procurement plan should outline the steps involved in designing, tendering, and hiring contractors for a project. In order to achieve sustainable results across environmental, social, and economic dimensions, risk management enabled sustainable construction management is be prioritised. Risk management enabled sustainable construction management refers to the strategic integration of risk management practices and concepts into the planning, execution, and monitoring of construction projects. By proactively detecting, evaluating, and managing "problem risk," this method can improve resilience, minimise negative effects, and maximise long-term value generation in building projects by affecting their sustainability performance. The risk allocation and contractual arrangements for each of the procurement strategies should be analysed appropriately. The risk connectivity of the risk should be identified and evaluated. Risk connectivity is the degree to which risks within a system or project are connected or related to one another. It represents the knowledge that risks are rarely isolated entities but are instead frequently linked together by convoluted interactions and relationships. Effective risk management requires an understanding of risk interconnection because it helps stakeholders recognise and evaluate the wider implications of risk interdependencies, foresee any knock-on consequences, and create comprehensive risk mitigation plans. Through the identification of interconnected hazards inside a project or system, organisations can improve their resilience, adaptability, and capacity to efficiently traverse convoluted difficulties.

- **Component: Environmental ethics**

This component deals with the ethical relationship between human beings and the natural environment. Because they encourage the ethical stewardship of natural resources and ecosystems, environmental ethics are essential for the success of building projects. Construction projects minimise environmental harm, alleviate pollution, and protect biodiversity by taking sustainability principles, sustainability impacts, and regulatory compliance into account. Project sustainability, resilience, and long-term viability are improved by ethical practices such waste reduction, the use of eco-friendly materials, and the application of green building techniques. Environmental ethics also promote goodwill among stakeholders, show corporate accountability, and improve the marketability and reputation of projects. In the end, including environmental ethics assures construction projects that are successful and sustainable while respecting the environment. Environmental ethics is a guiding our decisions and actions towards a more sustainable and ethical relationship with the natural world. It is concerned with understanding how human actions affect the environment and what our ethical responsibilities are towards the natural world. The validity of this component is high. About 50% of global waste is construction waste. In fact, to most people, sustainability, or "green" is about waste management. Correspondingly, construction production consumes significant energy. Construction companies develop waste management plans to control material usage, and some companies convert the waste to generate energy for use on-site. To reduce the amount of waste, construction companies should use renewable materials, use materials that are produced from sustainable sources, and provide effective waste and pollution control on their projects. In addition, the companies must conduct environmental impact assessments. Sustainable construction management ensures that sources of waste and pollution are found and mitigated throughout the production phases. Innovative means of generating and using energy should be encouraged. All forms of pollution, waste, emissions from vehicles, and noise should be controlled.

- **Component: Professional sustainable construction manager**

This component has a validity of 0.720. Regardless of the procurement strategy and the skills of the design team and construction team, there is a need for the client to engage an independent sustainable construction manager. To complement the independent sustainable construction manager, the contractor organisations should have at least one senior member in their team who is certified by a leading green or sustainable organization. It seems that if a project has a professional sustainable construction manager and the contractor has certified green/sustainable staff, there will be a reduction in carbon footprint on sites through an effective parking system. Contractors can reduce their carbon footprint by providing parking spaces designated for the use of the project buildings for the sole use of carpool vehicles on the sites. Successful construction projects require a professional sustainable construction manager because they have the know-how to incorporate sustainability principles into project planning, design, and execution. Coordinating many parties, putting green building concepts into practice, and making sure environmental standards are followed are all part of their job description. Sustainable construction managers increase project resilience, efficiency, and value by maximising resource utilisation, reducing environmental effects, and fostering social responsibility. They promote innovation, constant development, and the reduction of hazards related to resource shortages and climate change. In the end, their leadership promotes sustainable development, leaves a favourable legacy, and completes initiatives that satisfy the requirements of both the current and the next generation.

- **Component: Company Policy**

Many construction companies offer corporate social responsibility to the community where projects are found and uphold political accountability. Construction companies must use their corporate social responsibility tools to integrate social and environmental concerns into their business models. Some construction companies provide construction skills to members of communities where projects are found, especially in rural areas. Part of the responsibility involves public accountability. Using alternative energy sources is part of the environmental responsibilities of many construction businesses, which must be shown in the policy. The application of industrialised building systems would reduce the demand for energy and labour exploitation. Sustainable construction management emphasises using industrialised building systems, along with 3D scanners and plants and equipment that use alternative energy sources, to reduce energy consumption, rework, and waste on projects. 3D building scanners capture an exceedingly crucial level of detail that can help with the industrialised building systems and energy consumption of the construction project during design, construction, and operation.

Company policy is crucial for successful construction projects as it sets the framework for ethical conduct, safety protocols, and operational standards within the organization. By defining expectations, roles, and responsibilities, company policy ensures consistency, accountability, and compliance with regulatory requirements. It establishes guidelines for risk management, quality assurance, and stakeholder engagement, mitigating potential conflicts and liabilities. Moreover, company policy promotes a culture of transparency, integrity, and continuous improvement, fostering trust among employees, clients, and partners. Ultimately, adherence to company policy enhances project performance, reputation, and sustainability, driving success and long-term viability.

- **Component: Information management**

The purpose of an information management strategy is to ensure that an organization's information assets are effectively managed, secured, and used to support the organization's goals and aims. A well-designed and implemented information management strategy can help improve organisational efficiency, decision-making, and competitiveness. The validity of this component is 0.74. To reduce rework, variation, cost, and time overruns, there is a compelling need to provide adequate information to the project design team and construction team. The teams should focus on the latest and most up-to-date information when making decisions. Complete and correct information is needed for valid decision-making to take place. Construction projects that are successful depend on information management because it guarantees prompt access to reliable data, promotes collaboration, and aids in decision-making.

Information management lowers project risks, minimises errors, and improves stakeholder engagement by organising, storing, and distributing project-related data. It facilitates the management of budgets, resource allocation, and project progress tracking, encouraging accountability and efficiency. Furthermore, proactive issue detection, conflict resolution, and situational adaptation are made possible by efficient information management, which improves project resilience and results. In the end, it helps stakeholders make well-informed decisions, maximise performance, and complete building projects according to schedule, budget, and specifications. Lackk of information would lead to claims and disputes over the projects. Rework, deception, errors, and mistakes all contribute to massive waste, conflict, and social turpitude on a project. Therefore, there should be a clear information management strategy for a project on how information is generated, bought, stored, transferred, and shared among all the project members.

- **Component: Health and safety plan**

A health and safety plan are a document that outlines the steps and procedures that an organisation or individual takes to promote the health and safety of their workers, customers, or clients. The plan typically includes measures to prevent accidents, injuries, and illnesses in the workplace or other settings. The competency of the worker is critical in ensuring that workers adhere to the health and safety practises of organisations. According to research, accidents are more common among unskilled workers [14]. In the construction industry, health and safety are now clearly recognised as essential performance factors. As a result, making decisions that impact the health and safety of construction workers is crucial to sustainable construction management. According to the Occupational Safety and Health Administration (2020), sustainable organisations must safeguard the welfare, health, and safety of their employees. Despite the fact that sustainability and health and safety are not typically associated (ISO 2018), there is a strategic relationship between the two. A health and safety plan encourage sustainable construction management techniques by highlighting the value of competent construction workers and effective health and safety management. It places a high priority on employee well-being, ends risks and accidents, speeds up projects and saves money, inspires stakeholder confidence, guarantees regulatory compliance, and fosters a culture of ongoing improvement. In the end, this benefit both the well-being of the concerned parties and the general sustainability of construction projects.

- **Component: Digital construction**

The development of computers and other technology has led to a notable increase in the rate of digitization among construction companies. Prominent construction organisations engaged in sustainable projects leverage digitization and intelligent building system components to facilitate easy and convenient collaboration among design team members, regardless of their geographical locations. Writing, oral interviews, films, and graphics are commonly utilised as data supplements. Construction businesses are digitising the construction process in order to boost efficiency and profit margins by turning information into a digital format. High-speed bands and radio-frequency bands are becoming increasingly crucial to increase efficiency. Sanmargaraja et al. (2021) state that technologies such as the Internet of Things, the "Internet of everything," and smart technologies offer methods for gathering, organising, storing, processing, analysing, and dispersing vast amounts of data. Due to its ability to use technology to promote cooperation, expedite procedures, and improve project outcomes, digital construction is essential for project success. Digital construction allows precise project planning, visualisation, and coordination through the use of Building Information Modelling (BIM), virtual reality, and other digital tools. By enabling stakeholders to communicate, share data, and make decisions in real time, it lowers errors, delays, and expenses. In addition, digital construction facilitates quality control, safety procedures, and sustainability efforts, guaranteeing adherence to legal mandates and customer expectations. In the end, it improves project productivity, efficiency, and inventiveness, which propels success in the construction sector.

- **Component: Community engagement**

The needs of the local communities should be adequately addressed in the project's policies, environmental management plans, and social assessments. Community values should be respected. Those who may be impacted by the project, from the local community to those directly employed in construction, should be involved where necessary. Materials and components pose significant environmental and social risks to the SDGs. The sources, delivery, and use of materials should be carefully considered to contribute to the sustainability community. Construction projects that use regional resources and labour encourage community involvement in sustainable construction management techniques. The community is strengthened, skills are improved, economic gains are produced, cultural identity is preserved, sustainability is made clear, and cooperative ties are formed. Construction projects can generate stronger sustainability results and have a long-lasting beneficial effect on the community's resilience and well-being by appreciating and involving the local community. In order to build mutual understanding, trust, and collaboration between project stakeholders and the surrounding community, community engagement is crucial for construction projects to succeed. Communities are less likely to oppose projects and participate in conflict when their needs, interests, and concerns are taken into consideration through community engagement in the planning, design, and decision-making phases. It increases support and acceptability of the initiative by fostering social responsibility, accountability, and openness. Furthermore, community involvement fosters beneficial relationships, capacity building, and information sharing, which results in shared ownership and long-term advantages for all parties involved. In the end, it makes a favourable impact on construction projects' sustainability, success, and legacy.

10.6 Conclusion and Policy Implications

This research support and opens a discussion on the philosophy of sustainable construction management. Research in sustainable construction management can apply positivism, subjectivism, or their combination. This research also investigated the CSFs that contribute to the successful implementation of sustainable construction management practices across different countries and regions. A framework covering nine main tasks that a sustainable construction manager will perform on a construction project has been developed through factor analysis. They are mainly the tools, strategies, principles, techniques and practices needed to deliver sustainable construction projects. The insights provided by this research can guide policymakers and industry practitioners in making informed decisions to promote sustainable development. By implementing sustainable construction management practices, stakeholders can reduce the environmental impact of the construction industry, conserve natural resources, and promote social equity. However, the study also highlights the need for further research to explore the effectiveness of different strategies for implementing sustainable construction management practices in different contexts. All things considered, this study adds to the expanding corpus of information on sustainable construction management and lays the groundwork for more studies and applications in this field. This study has added to the expanding corpus of information regarding sustainable building management techniques. However, the sample size is likely a significant research restriction. However, the statistics of goodness of fitness suggest that the data are presentative, and the profiles of the respondents suggest that the data is generalizable. Yet, confirmatory research would be useful. Future research may be replicated. The data would be useful for academic institutions that offer construction management, architecture, civil engineering, environmental engineering, occupational health and safety, and quantity surveying programmes.

Bibliography

1. Ahmed SM, Yaris C, Farooqui RU, Saqib M (2014) Key attributes and skills for curriculum improvement for undergraduate construction management programs. Int J Constr Educ Res 10(4):240–254
2. Arcadis (2021) International construction costs 2021: investing for a sustainable recovery. https://connect.arcadis.com/ICC-Report-download-confirm. www.arcadis.com/en/
3. Banihashemi S, Hosseini MR, Golizadeh H, Sankaran S (2017) Critical success factors (CSFs) for integration of sustainability into construction project management practices in developing countries. Int J Project Manage 35(6):1103–1119
4. Brundtland Commission (1987) World commission on environment and development, our common future. Oxford University Press, Oxford and New York
5. Chegut A, Eichholtz P, Kok N (2019) The price of innovation: an analysis of the marginal cost of green buildings. J Environ Econ Manag 98:102248
6. Dodge Data and Analytics (2016) World green building and construction trends. Available at https://www.construction.com/toolkit/reports/world-green-building. 18 Jan 2022

7. He Z, Chen H (2021) Critical factors for practicing sustainable construction projects in environmentally fragile regions based on interpretive structural modeling and cross-impact matrix multiplication applied to classification: a case study in China. Sustain Cities Soc 74:103238
8. Hwang B-G, Leong LP, Huh Y-K (2013) Sustainable green construction management: schedule performance and improvement. Technol Econ Dev Econ 19(Supplement 1):S43–S57
9. ILO (2017) The ILO records that 18% of all forced labor cases are documented in the construction sector. ISO 45001:2018 Occupational health and safety management systems, pp 25–26
10. Ismael D, Shealy T (2018) Sustainable construction risk perceptions in the Kuwaiti construction industry. Sustainability 10(6):1854
11. Khun-anod K, Limsawasd C (2019) Pre-project planning process study of green building construction projects in Thailand. Eng J 23(6):67–81
12. Lima L, Trindade E, Alencar L, Alencar M, Silva L (2021) Sustainability in the construction industry: a systematic review of the literature. J Clean Prod 289:125730
13. Olanrewaju A, Tan SY, Abdul-Aziz AR (2018) Housing providers' insights on the benefits of sustainable affordable housing. Sustain Dev 26(6):847–858
14. Olanrewaju AA, Chong YS (2021) Post occupancy evaluation of green residential buildings, in the Greater Kuala Lumpur, Malaysia. J Housing Built Environ 36(2):825–857. https://doi.org/10.1007/s10901-021-09832-1
15. Olanrewaju A, Shari Z, Gou Z (eds) (2019) Greening affordable housing: an interactive approach. CRC Press
16. Olanrewaju AL (2022) An artificial neural network analysis of rework in sustainable buildings. In: IOP Conference series: earth and environmental science, vol 1101, no 2. IOP Publishing, p 022003
17. Olanrewaju A, Idrus A (2020) What is determining affordable housing shortages in the Greater Kuala Lumpur, Malaysia? Prop Manag 38(1):52–81
18. Oni OZ, Olanrewaju A, Khor SC (2023) Review of critical success factors affecting Malaysia's construction industry's sustainable health and safety practices. Front Eng Built Environ 3(1):48–62
19. Oni OZ, Olanrewaju A, Khor SC (2023) Fuzzy synthetic evaluation of the factors affecting health and safety practices in Malaysia construction industry. J Eng Des Technol
20. Maqbool R, Amaechi IE (2022). A systematic managerial perspective on the environmentally sustainable construction practices of UK. Environ Sci Pollut Res 1–18
21. Metz TR, Cheatham CW (2008) As green building moves forward, claims and disputes will follow. Constr Law Public Contracts 57:56–59
22. Radosavljevic M, Bennett J (2012) Construction management strategies: a theory of construction management. Wiley
23. Robichaud LB, Anantatmula VS (2011) Greening project management practices for sustainable construction. J Manag Eng 27(1):48–57
24. Sanmargaraja D, Olanrewaju A, Kean BK (2020) Implementation of artificial intelligence (AI) in the construction industry during the COVID-19 pandemic in Malaysia. Techies 10:19–22
25. Spiegel R, Meadows D (2010) Green building materials: a guide to product selection and specification. Wiley
26. Wuni IY, Shen GQ (2020) Critical success factors for modular integrated construction projects: a review. Build Res Inf 48(7):763–784

Chapter 11
Defects and Maintenance Management in Sustainable Buildings

Abstract As the demand for sustainable buildings rises, so does the urgency to address the persistent challenge of defects that compromise their environmental, economic, and social deliverables and performance. Despite their resource-efficient and occupant-friendly design, sustainable buildings are not immune to failures that increase costs, cause rework, and impact stakeholder confidence. This chapter investigates the nature, causes, and management of defects specific to sustainable buildings, revealing that even with green innovations, design and construction flaws remain prevalent across global contexts. Drawing on international data and case studies, the research examines how construction practices, material choices, and managerial strategies influence defect occurrences. A mixed-method approach was used to analyse defect patterns and propose a machine learning-based classification method to streamline repair and maintenance processes. The findings inform the development of a sustainable building pathology framework that guides industry stakeholders in diagnosing and mitigating defects proactively. This framework underscores the distinction between conventional and sustainable construction practices and emphasises the importance of quality control, skilled labour, integrated design, and long-term maintenance strategies. Ultimately, this chapter contributes to the advancement of defect-aware, sustainable construction management and offers practical insights for contractors, policymakers, educators, and developers aiming to enhance the longevity and performance of green buildings.

Keywords Building pathology · Maintenance-span · Building performance · Site operatives · Rework · Contractors · Green rating tools

11.1 Learning Outcomes

The demand for sustainable buildings has increased dramatically in recent times. Sustainable buildings are designed to be healthy for the environment, resource-efficient, and comfortable for their occupants. However, sustainable buildings are not immune to defects, which can compromise their sustainability performance and

A. Olanrewaju, *Sustainable Construction Management: Research and Practice Companion*, Management in the Built Environment,
https://doi.org/10.1007/978-981-96-4910-5_11

lead to increased costs, litigation, pollution, and negative impacts on occupants and stakeholders. This research explores the production and management strategies of defects in sustainable buildings. It examines the nature of sustainable building defects and explores the potential causes and remedies of these defects. At the end of the chapter, the reader will be able to:

- Explain defects in sustainable buildings.
- Explain the determinants of defects in sustainable buildings.
- Explain the construction management strategies for sustainable buildings.
- Explain the critical success factors for defect-sustainable construction management.
- Explain the opportunity cost of construction maintenance.

11.2 Introduction

Defects in buildings that result in quality problems are a topic of ongoing discussion in the construction industry, both generally and with regard to particular building project classes and types. According to Montague [19], improved quality management has the potential to save the UK construction sector up to £12 billion annually. In 2004, the US construction industry spent $1502 billion on building expenditures overall, while $75 billion was lost to defect costs [5]. The annual cost of failures in the Malaysian construction sector is almost equal to or close to 10% of the total annual production value [21]. This corresponds with studies of building projects in South Africa [22], Denmark [17], Australia [18, 23], Singapore, Spain, and Canada. Despite the efforts to reduce defects in buildings in the UK, USA, Sweden, Singapore, Australia, and Malaysia, the construction sectors have largely continued to experience the same level of defects [12, 15, 18, 20]. As a result, sustainable building was conceived to address the shortcomings inherent in conventional buildings. Despite limited available data, concerns about defects in sustainable buildings are steadily increasing. This may be linked to the unique characteristics of sustainable construction. Sustainable building practices typically involve the use of industrialised building systems, off-site prefabrication, locally sourced materials, and the integration of recycled or reused resources. They also often incorporate innovative—and sometimes relatively untested—materials and components, highlighting the sector's commitment to environmental stewardship and technological advancement. These force construction companies to meet a certain set of standards and competencies. However, these often lead to different defects in the buildings. Supporting this accession, Tollin [24] concurred that the risks in green building projects are due to defects and omissions by design professionals, contractors, and subcontractors.

Hwang et al. [15] also opined that the construction of green residential buildings faces various risks like those of traditional residential buildings. Bansal et al. (2017) found that the construction method selected for the construction of buildings has an impact on the performance of green buildings. There appears to be a practical as well as a research gap in understanding how and why defects are recurrent and persistent in sustainable buildings. Studies of defects in buildings have been idiosyncratic and compartmentalised [17, 20]. In other words, thorough research covering the whole construction project life cycle-from concept to operation-becomes necessary. To solve this research gap and problems, an ethnographic method is required. Ethnography demonstrates transparency and a close connection to the action taking place on the ground. It can also offer different interpretations, forecasts, and alterations of objective research questions and hypotheses. However quantitative techniques are needed as well, which can provide new dimensions to the processes and practises for stakeholders to manage defects in the buildings. Therefore, this research aims to use machine learning to automate the forecasting and classification of defects in sustainable buildings based on data collected during the design and construction processes. The conventional methods for this type of investigation are reactive, expensive, and time-consuming. The purpose of the research reported in this chapter is to examine how defects are produced and to ask the following questions: What are the management strategies to solve defect problems in sustainable buildings and restore defects in the buildings?

11.3 Background and Conceptual Framework

Quality is defined by the ISO [16] as the "inherent characteristic of an object related to a requirement," where a requirement is defined as a need or expectation that is stated explicitly, impliedly, or required [4]. Defects are effectively defined as deviations from an object's intrinsic features to satisfy certain criteria. Thus, the extent to which an object's intrinsic qualities satisfy project criteria might happen at any point during the development process. Defects, which are characterised as shortcomings in the building's functional performance, legal requirements, or user needs, are indicators of inferior quality. The ability of the buildings and services to function properly is hampered by defects. In addition to negatively impacting the happiness of tenants or clients, defects can result in disagreements and legal action between tenants, users, developers, and maintenance companies. Building defects may be fixed during the building process, during the defect liability period, or following the defect reliability period. Figure 11.1 displays the relationship between inferior quality, defects, rework, and maintenance [20].

11.4 Size, Nature, and Impact of Defect

Building defects are a widespread occurrence. According to Chan's [6] research, Hong Kong has a defect cost that falls between 5 and 20%. According to Heravi and Jafari [13], 7.4% of the project cost should be allocated to poor-quality costs in Iran. According to Simpeh et al. [22], the cost of rework to the original contract value was 5.12%. In Singapore, 24.94% of the increase in the construction schedule was

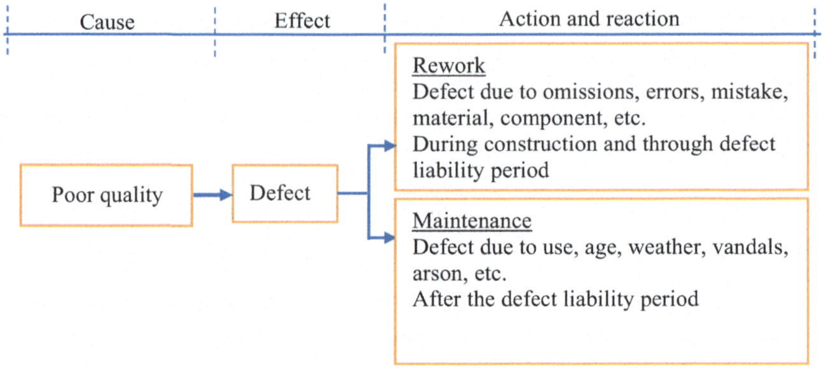

Fig. 11.1 Relationship between poor quality with rework and maintenance

due to rework [13]. According to data on the Israeli construction sector, non-quality expenses may account for as much as 14% of construction businesses' total revenue. According to Yap et al. [25], defects can cost anywhere from 3 to 6% in Malaysia and up to 10% in several other nations [11]. Current research indicates that the type, nature, size, and location of the project all impact on the defects in the construction.

11.5 Requirements of Sustainable Buildings

Despite a paucity of representative data, there is ongoing worry about faults in sustainable building (Olanrewaju et al. 2021) [3, 14]. There is a dearth of empirical study on the pathology of sustainable buildings, whereas there is a dearth of research on quality issues in conventional buildings. According to Olanrewaju's [20] early estimate, a recently finished sustainable building incurred defect costs equivalent to approximately 30% of the initial contract total.

Even though there are concerted efforts to reduce defects in buildings, the construction sector has largely contributed to experiencing similar-scale defects in buildings both locally and internationally. The construction industry is synonymous with poor quality and defective projects. From a micro perspective, defects are both recurrent and endemic at the project and company levels. Due to a lack of frameworks or hindsight, the construction industry continues to adopt conventional construction management practices for sustainable buildings. However, conventional and sustainable construction management approaches differ significantly in terms of their methods, practices, and priorities. These differences can lead to defects in buildings in several ways, as outlined in Table 11.1.

Sustainable construction management practices can help to reduce defects in buildings by using green materials, energy-efficient designs, and waste management practices, while conventional construction practices may lead to defects due to

Table 11.1 Causes of defects in conventional and sustainable management approaches

Factor	Explanation
Material selection	Low-cost materials are usually given priority in conventional construction management, even though they might not be the best sustainable or long-lasting choices. Contrarily, sustainable building emphasises the use of premium, green materials, which may initially cost more but offer durability and long-term cost benefits
Defects	While using green materials can guarantee a healthy living environment, using non-environmentally friendly products may pose health risks to inhabitants
Design	Conventional construction management frequently puts financial savings and time ahead of design quality, which can result in poorly designed buildings and structural flaws. The integration of design, which considers the building's energy efficiency, environmental effect, and occupant health and well-being, is given priority in sustainable construction management
Construction practices	Building faults may result from the prioritisation of speed and cost savings over quality and safety in conventional construction management techniques. Sustainable construction management places a strong emphasis on appropriate building methods, such as the use of premium materials, adequate site preparation, and efficient project management
Energy efficiency	Energy-efficient building systems are given priority in sustainable construction management because they can cut down on energy use and operational expenses. Energy efficiency could be disregarded by conventional construction management, which would increase energy costs and carbon emissions
Maintenance	To prolong the life of a building and avoid flaws, sustainable construction management places a strong emphasis on the necessity of continuous building maintenance and repair. Conventional construction management may put immediate financial savings ahead of ongoing maintenance requirements, which could result in increased maintenance expenses and structural flaws
Waste management	Conventional construction generally generates substantial amounts of waste, but sustainable construction approaches prioritise minimising, reusing, and recycling waste materials

the use of non-environmentally friendly materials, inefficient energy systems, waste generation, and a lack of holistic design considerations.

Furthermore, a variety of factors, including design, construction methods, materials utilised, and maintenance schedules, can affect the kinds of building problems that might arise in both conventional and sustainable buildings. Nonetheless, as the following list illustrates, there are some broad distinctions in the kinds of flaws that might be more prevalent in conventional buildings versus sustainable ones:

- Structural Defects: Because time and cost savings are prioritised over quality in construction processes, conventional buildings may be more vulnerable to structural flaws. Higher-quality, more resilient materials are often used in the construction of sustainable buildings, and correct construction methods are stressed to lower the risk of structural flaws.

- Moisture and Water Damage: Both conventional and sustainable buildings are susceptible to moisture and water damage, albeit the former may experience them more frequently because of the use of inferior materials that are more likely to leak and sustain water damage. High-quality, ecologically friendly materials that are less prone to absorb moisture and water damage are often used in sustainable structures.
- Thermal and Energy Efficiency: Conventional buildings may be less energy-efficient than sustainable ones, which could result in greater energy costs and uncomfortable temperatures for occupants. Energy-efficient technologies and materials are used in sustainable buildings, which can lower energy usage and increase occupant comfort.
- Indoor Air Quality: By utilising components and techniques that enhance indoor air quality, sustainable buildings place a high priority on the health and happiness of their occupants. Conventional buildings could contain materials that emit toxic compounds, which could result in poor interior air quality and possible health issues for residents.
- Maintenance and Longevity: To prolong the building's lifespan and avoid flaws, sustainable structures place a high priority on regular maintenance and repairs. Conventional buildings may put immediate financial savings ahead of ongoing maintenance requirements, which could result in increased maintenance expenses and structural flaws.

While structural flaws and moisture/water damage can affect conventional and sustainable buildings equally, sustainable buildings may be less vulnerable to these issues. They also place a higher priority on indoor air quality and energy efficiency, as well as continued upkeep and longevity to protect against flaws.

The generation of defects in sustainable construction is rising unabatedly along with the growth in both the supply and demand for sustainable. Regarding how and why flaws in sustainable buildings are significant and growing, both in absolute and relative terms, there exist research and practical gaps. Since there isn't as much research on this topic, including empirical studies, the gap is primarily due to a lack of awareness of the complexity of sustainable building nature and a lack of empirical foundation. This research's main goal is to investigate how flaws are made, produced, and managed in the context of construction management for sustainable buildings in order to address a number of related issues. This study fills a gap in the literature on construction management by emphasising problem-solving techniques instead of the usual prescriptive models that are used to account for site-specific scenarios.

A framework to improve management process decisions is presented in Fig. 11.2. The framework starts with observations made by maintenance organisations or occupants during defect repair or maintenance. After maintenance is completed, the person or entity who requested the maintenance should receive feedback or information from the repair. For construction and engineering services, the framework serves as a helpful model of a systematic maintenance management system.

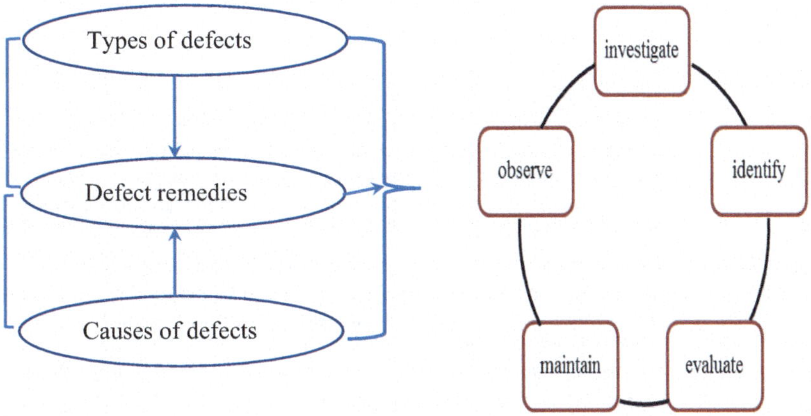

Fig. 11.2 A simplified conceptual framework for defect management during a building operation

11.6 Maintenance Economics

The economic theories and practices that are applied to the management and optimisation of maintenance tasks inside a company or across a range of assets and infrastructure are referred to as maintenance economics. It includes all aspects of analysis, decision-making, and resource allocation to guarantee that maintenance activities are economical, successful, and in line with overarching company goals. Table 11.2 lists the primary components of the maintenance economics.

Table 11.2 The major factors for the maintenance economics

Factor	Explanation
Cost analysis	Evaluating the expenses related to various maintenance tasks, such as predictive, corrective, preventive, and reliability-centred maintenance, is the focus of maintenance economics. Both direct costs-such as labour, supplies, and equipment-and indirect costs-such as lost output, downtime, and safety hazards-are considered in this study
Lifecycle costing	The entire lifespan costs of infrastructure and assets, including purchase, operation, and maintenance expenses as well as disposal costs at the end of useful life, are considered by maintenance economics. Organisations may optimise long-term value by making informed decisions on asset investment, maintenance plans, and replacement timing by taking the complete lifecycle costs into account
Cost–benefit analysis	Cost–benefit analyses are carried out in maintenance economics to assess the viability and feasibility of maintenance initiatives and expenditures. The anticipated advantages (such as increased productivity, longer asset life, decreased downtime, and enhanced reliability) are weighed against the costs of putting maintenance procedures into place in this analysis

(continued)

Table 11.2 (continued)

Factor	Explanation
Optimization strategies	The goal of maintenance economics is to maximise asset performance while minimising costs by optimising maintenance plans and schedules. To attain the ideal balance between dependability, availability, and cost, this may entail striking a balance between predictive maintenance (condition monitoring and diagnostics), reactive maintenance (repairing defects as they arise), and preventive maintenance (periodic inspections and servicing)
Risk [problem] management	The economic effects of asset malfunctions, breakdowns, and disruptions on overall corporate operations are considered by maintenance economics. Organisations can reduce financial losses and business impact by implementing risk mitigation strategies, allocating resources efficiently, and prioritising maintenance activities based on an assessment of the financial risks associated with various failure modes and downtime scenarios
Investment prioritization	Organisations can prioritise maintenance expenditures by using maintenance economics to assess the possible ROI and business performance impact of each investment. This entails determining which assets are essential, evaluating their performance and maintenance needs, and assigning resources to the highest-priority maintenance tasks that will yield the biggest financial returns
Continuous improvement	The field of maintenance economics places significant emphasis on the ongoing enhancement and optimisation of maintenance procedures, systems, and performance indicators. Organisations can find chances for efficiency gains, cost savings, and performance enhancement over time by analysing maintenance data, tracking KPIs, and benchmarking against industry standards and best practices

A methodical framework for organising maintenance tasks that strikes a compromise between financial concerns and asset dependability, safety, and operational effectiveness is offered by maintenance economics. Organisations may maximise the value and performance of their assets and infrastructure, optimise resource allocation, and reduce lifecycle costs by incorporating economic principles and analysis methodologies into maintenance decision-making.

11.7 Construction Maintenance Economics

The basic construction economics principle and theory have been discussed in Chap. 4 of this book. The fundamental aspect of economics is scarcity. Scarcity is probably the cornerstone of economics because it is not only human beings that require more than what they produce; animals also face scarcity problems because the available resources are not enough to meet all our demands. The problem of scarcity leads to the problem of choice. We are to choose between alternatives. Construction clients have to choose between maintenance or using the resource for alternative purposes. There are broadly 4 types of economic evaluation, namely cost minimisation, cost effectiveness cost utility and cost benefit analysis.

Burning fossil fuels to generate electricity generates about 30 billion metric tonnes of CO_2 into the atmosphere annually. When you make the choice to procure a conventional project, you leave your carbon footprint. However, you have a choice to lessen the carbon footprint by choosing sustainable construction. The direct cost of maintenance is the focused of most economic evaluations. The impact of poor maintenance and the indirect cost of maintenance are usually not evaluated in included in the investment appraisal largely because money worth is difficult to be ascribed to them. Maintenance plays pivotal role in the performance of construction projects. In the project life cycle, maintenance problem commences even before project is complete. In terms of ensuring durability and preventing breakdown, both maintenance and sustainability serves the same purpose. Maintenance decision are critical for sustainable projects, as clients need to make decision if to maintain the project along with when to maintain the projects considering both the direct and indirect cost on the one hand and opportunity of the maintenance on the other hand. Clients need to evaluate the essence of executing maintenance for a sustainable construction. Like sustainability, maintenance also aim to reduce or avoid causing harms to the project's users, workers and the general populace. The opportunity cost of different maintenance management strategies and interventions must be equally evaluated. The application of economic theories and techniques to the management and optimisation of maintenance processes in the construction sector is known as construction maintenance economics. Its fundamental objective is to assess the risks, costs, and advantages of maintaining facilities, infrastructure, and construction assets over the course of their lives. The key construction maintenance economics principles and tactics were listed in Table 11.3.

Table 11.3 The primary economics principles and strategies of the construction maintenance economics

Lifecycle cost analysis	The study of construction maintenance economics considers the entire cost of building assets over their entire life, including purchase, operation, maintenance, and disposal expenses. Through the process of cost analysis, organisations can arrive at well-informed judgements on asset allocation, maintenance planning, and resource allocation
Cost–benefit Analysis	Cost–benefit analysis is a key component of construction maintenance economics, which assesses the viability and economics of maintenance expenditures and operations. The projected advantages of maintenance, such as enhanced asset performance, longer service life, and decreased downtime, are compared in this analysis with the expenses associated with putting maintenance plans into action

(continued)

Table 11.3 (continued)

Optimization of maintenance strategies	The goal of construction maintenance economics is to maximise asset performance while minimising costs through the optimisation of maintenance plans and practices. To attain the best possible balance between availability, cost, and reliability, this may entail putting preventative maintenance plans into place as well as condition-based and reliability-centered maintenance methodologies
Risk management	The study of construction maintenance economics considers the financial effects that asset malfunctions, breakdowns, and interruptions have on construction projects and operations. Organisations can minimise financial losses and project delays by effectively allocating resources, prioritising maintenance activities, and implementing risk mitigation measures by evaluating the financial risks associated with construction asset failures and downtime
Investment prioritization	Organisations can prioritise maintenance expenditure by using construction maintenance economics, which considers the project performance impact and possible ROI. This entails determining which construction assets are essential, evaluating their performance and maintenance requirements, and assigning resources to the highest-priority maintenance tasks that have the biggest financial returns
Resource allocation	The field of construction maintenance economics focuses on the efficient allocation of resources, such as labour, materials, equipment, and technology, for maintenance tasks. Organisations may boost output, cut expenses, and improve project results by effectively allocating resources according to project requirements, asset condition, and maintenance priorities
Continuous improvement	The field of construction maintenance economics places significant emphasis on the ongoing enhancement and optimisation of maintenance procedures, frameworks, and performance indicators. Organisations can find opportunities for cost reductions, performance enhancement, and efficiency gains in construction maintenance activities by collecting KPIs, analysing maintenance data, and using lessons learned from past projects

Construction maintenance economics provides a systematic framework for managing maintenance activities in the construction industry, helping organizations optimize asset performance, minimize lifecycle costs, and maximize the value of construction assets and infrastructure over time.

11.8 Maintenance Management Enabled Sustainable Construction Projects

The phrase "maintenance management enabled sustainable construction projects" describes how to improve a construction project's sustainability performance by including efficient maintenance management techniques into the project's planning, design, construction, and operation. This method acknowledges that maintenance is essential to maintaining the long-term resilience, durability, and environmental efficiency of built assets. Maintenance management enabled sustainable construction projects involve the following activities: lifecycle approach, proactive maintenance planning, resilience and adaptability, environmental sustainability, resource efficiency, occupant health and comfort, optimised performance and productivity, and continuous improvement. The full justification of how maintenance management supports sustainable building projects is provided below.

1. Proactive Maintenance Planning: Proactive planning for the continuous care and upkeep of construction assets throughout their lifecycle is a key component of maintenance management. Sustainable building projects can reduce lifespan costs, maximise asset performance, and foresee future maintenance requirements by incorporating maintenance considerations into the planning and design stages of the project. Table 11.4 lists the tasks that are part of the proactive maintenance planning process.

Table 11.4 Activities in proactive maintenance planning

Activities	Explanation
Early integration in project lifecycle	The project lifetime incorporates maintenance considerations from the very opening of planning and design. In order to incorporate maintenance-friendly design elements, choose long-lasting materials, and specify equipment with easy access for maintenance duties, this entails working with architects, engineers, and maintenance specialists
Predictive maintenance strategies	Predictive maintenance techniques are used in sustainable construction projects to identify any equipment problems early on and take appropriate action. This could entail keeping an eye on asset status, spotting abnormalities, and planning preventive maintenance tasks based on performance patterns using sensors, data analytics, and predictive modelling
Lifecycle cost analysis	As part of maintenance planning, lifetime cost studies are carried out to assess the long-term expenses and advantages of various maintenance approaches. To determine the most cost-effective maintenance strategies for an asset over its lifetime, this analysis considers variables including energy efficiency, replacement cycles, and frequency of maintenance

(continued)

Table 11.4 (continued)

Activities	Explanation
Risk-based maintenance planning	The development of maintenance plans is predicated upon a comprehensive evaluation of the criticality, dependability, and performance needs of the asset. To reduce the chance of failure and the ensuing downtime, high-risk assets or components are given priority for more frequent inspections and preventive maintenance procedures
Maintenance scheduling and optimization	The goal of maintenance schedule optimisation is to maximise asset performance and dependability while minimising interference with ongoing operations. This could entail planning maintenance chores for off-peak times, making use of downtime for several repair jobs, and utilising sophisticated scheduling algorithms to maximise resource utilisation and reduce idle time
Performance monitoring and feedback	Setting up performance monitoring systems to measure asset performance, dependability, and maintenance efficacy over time is part of proactive maintenance planning. Regular monitoring of key performance indicators (KPIs) is conducted, and input from maintenance operations is utilised to enhance asset management tactics, optimise maintenance schedules, and promote ongoing enhancements
Training and skill development	In order to guarantee that maintenance staff members have the skills, resources, and know-how necessary to carry out their jobs efficiently, maintenance planning entails making investments in their education and training. To improve maintenance capabilities and efficiency, this may entail offering specialised training on equipment operation, maintenance methods, safety precautions, and cutting-edge technology
Lifecycle asset management	A lifecycle asset management strategy is used in sustainable building projects, considering the complete lifecycle of developed assets from acquisition to disposal. To maximise asset value and reduce total cost of ownership over time, this entails creating thorough asset management plans that cover maintenance, repair, replacement, and decommissioning activities

2. Lifecycle Approach: Construction projects are seen holistically by maintenance management, which considers the full lifecycle of created assets from conception to disposal. In order to maximise asset performance and lifetime, sustainable building projects use a lifecycle approach to maintenance, combining techniques for condition-based, predictive, and preventive maintenance. The explanations of the lifecycle approach's actions are provided in Table 11.5.
3. Environmental Sustainability: By utilising environmentally friendly materials and technology and following green maintenance practices, maintenance management helps sustainable construction projects to minimise their negative environmental effects. This covers tactics like water-saving fixtures, renewable energy sources, sustainable landscaping techniques, and energy-efficient HVAC and lighting systems. For an explanation of the actions involved in environmental sustainability, refer to Table 11.6.

Table 11.5 Activities in the lifecycle approach

Activities	Explanation
Inception phase	When choosing assets, establishing design guidelines, and choosing construction techniques, maintenance considerations are first considered at the project's initial stages. In order to ensure successful long-term maintenance, sustainable building projects give priority to materials and technology that are long-lasting, low-maintenance, and ecologically friendly
Design phase	Maintenance considerations and requirements are incorporated into the design phase of the process. This entails putting in place systems with redundancy and flexibility to meet future maintenance requirements, ensuring that equipment has easily accessible components for maintenance, and integrating sustainable design elements that lower lifecycle maintenance costs and environmental effect
Construction phase	Throughout the building phase, maintenance planning is ongoing, with a focus on making sure that construction activities do not jeopardise the needs for future maintenance. This could entail working with contractors to minimise disruption to building systems, recording construction specifics for later use, and performing quality control inspections to confirm that construction adheres to maintenance guidelines
Operational phase	After construction is finished, the facility enters the operational phase, during which routine maintenance tasks are integrated into facility management. Preventive maintenance programmes are used in sustainable building projects to handle regular maintenance chores including cleaning, lubrication, inspections, and small repairs in order to stop asset deterioration and extend service life
Predictive maintenance	Sustainable construction projects use predictive maintenance methods in addition to preventative maintenance to track the performance and condition of their assets in real time. In order to track equipment health, identify early indicators of degradation, and forecast possible breakdowns before they happen, proactive maintenance interventions may be necessary. These may involve the use of sensors, IoT devices, and data analytics
Condition-based maintenance	Maintenance activities are scheduled according to the real state of the assets, not on predetermined intervals, when condition-based maintenance techniques are used. As part of this process, diagnostic tools, performance measures, and asset health indicators are used to evaluate the state of the asset and rank maintenance tasks according to criticality, dependability, and performance goals
Optimization of maintenance strategies	Maintenance strategies are regularly optimised during the course of building projects to accommodate evolving operational conditions, developments in technology, and trends in asset performance. To increase efficiency and effectiveness, this may entail fine-tuning maintenance schedules, modifying maintenance intervals, and applying knowledge gained from prior maintenance experiences
End-of-Life considerations	Sustainable construction projects provide for ecologically responsible asset decommissioning and disposal at the end of the asset lifecycle. This entails evaluating the financial and environmental effects of various asset repurposing, recycling, or disposal choices in addition to adhering to legal obligations for waste management and environmental protection

Table 11.6 Activity in the environment sustainability

Activity	Explanation
Energy efficiency measures	Energy-efficient techniques are included into maintenance management to minimise environmental impact and lower energy use. This include replacing old fixtures with more energy-efficient models, installing lighting controls and automation to cut down on wasteful energy use, and routinely inspecting and maintaining lighting systems to guarantee optimal performance
HVAC system optimization	In order to increase indoor air quality and energy efficiency, sustainable construction projects place a high priority on the upkeep and optimisation of HVAC systems. This entails changing to energy-efficient HVAC equipment and technology, calibrating temperature and humidity controls, and routinely cleaning and replacing air filters
Water conservation practices	Water-saving techniques are put into place as part of maintenance management to reduce water usage and encourage conservation. To lessen dependency on potable water sources, this may entail fixing leaks, improving irrigation systems, installing low-flow fixtures and faucets, and putting in place water recycling and rainwater harvesting systems
Use of renewable energy sources	Renewable energy sources are incorporated into maintenance plans for sustainable construction projects in order to minimise greenhouse gas emissions and lessen dependency on fossil fuels. This could entail setting up geothermal, wind, or solar energy systems, or solar panels, to provide clean, renewable energy on-site, minimising environmental effect and fostering sustainability
Sustainable landscaping practices	Implementing sustainable landscaping techniques is part of maintenance management, which reduces environmental impact and increases biodiversity. Utilising native plants that need less water and upkeep, using organic pesticides and fertilisers to reduce chemical runoff, and installing green wall and roof systems to enhance air quality, lessen the impact of the urban heat island, and encourage habitat restoration are a few examples of how to achieve this activity
Waste reduction and recycling	In order to reduce waste creation and increase resource efficiency, maintenance management integrates recycling and waste reduction techniques. This entails appropriate waste segregation, recycling building and maintenance materials, and putting waste management plans into practice in order to keep waste out of landfills and advance the ideas of the circular economy
Lifecycle assessment	In order to assess the environmental impact of maintenance activities and pinpoint areas for improvement, sustainable building projects carry out lifetime evaluations. This entails calculating the environmental impact of maintenance activities, evaluating the life cycle effects of various maintenance techniques, and pinpointing areas where maintenance practices and strategies might be modified to improve environmental performance
Compliance with environmental regulations	Maintenance management complies with regulations governing maintenance activities and environmental standards. Maintaining environmental permits, licences, and certificates; implementing pollution prevention measures; and carrying out regular environmental audits and inspections are all necessary to guarantee compliance with pertinent laws and regulations

4. Resource Efficiency: Resource efficiency is given top priority in sustainable construction projects, which optimise maintenance tasks to reduce waste production, energy use, and resource depletion. By efficient maintenance planning and execution, maintenance management solutions aim to increase the effectiveness of maintenance operations, decrease downtime, and extend the service life of construction assets. Table 11.7 contains the activity in the resource efficiency.

Table 11.7 Activity in the resource efficiency

Activity	Explanation
Waste minimization	Maintenance management solutions put an emphasis on minimising waste by putting in place procedures for reducing, reusing, and recycling resources and materials. This include using construction surplus materials for maintenance repairs, streamlining inventory management to avoid overstocking, and putting in place waste segregation and recycling initiatives to keep items out of landfills
Energy conservation	Through the application of energy-efficient techniques and technology, sustainable construction projects concentrate on energy conservation in maintenance activities. To cut down on energy usage and expenses, this involves upgrading to energy-efficient equipment and lighting systems, conducting energy audits to find potential for energy savings, and optimising equipment operation schedules to minimise energy consumption during off-peak hours
Water management	Water management techniques are part of maintenance management, which maximises water use and reduces waste. To reduce water usage in landscaping and outdoor spaces, this may entail installing water-efficient fixtures and appliances, fixing leaks as soon as possible to stop water loss, and optimising irrigation systems
Material efficiency	By choosing strong, long-lasting materials that require little upkeep and replacement, maintenance management systems place a high priority on material efficiency. In order to minimise resource consumption and environmental impact, this involves utilising materials with low embodied energy and carbon footprint and specifying premium products with low environmental impact, such as sustainable or recycled materials
Optimized equipment utilization	The goal of maintenance management is to reduce resource consumption and maximise production by optimising the use of equipment. This entails putting preventative maintenance programmes into place to stop equipment breakdowns and failures, properly scheduling maintenance tasks to minimise downtime and maximise equipment availability, and optimising equipment usage patterns to lessen wear and tear and extend asset lifespan
Lifecycle cost considerations	When making maintenance decisions, sustainable building projects take lifecycle costs into account to maximise resource efficiency over the course of the built asset's existence. In order to determine which maintenance strategy is the most economical and resource-efficient, lifecycle cost assessments must be performed to assess the total cost of ownership of various maintenance strategies, including initial investment, operation, maintenance, and disposal expenses

(continued)

Table 11.7 (continued)

Activity	Explanation
Training and skills development	Enhancing the knowledge and skills of maintenance staff in resource-efficient maintenance techniques is a key component of maintenance management, which involves training and skill development. Assisting maintenance teams in operating and maintaining equipment in an effective and sustainable manner may entail offering training on energy saving tactics, waste minimization initiatives, and equipment operation and maintenance techniques
Continuous improvement	Sustainable construction projects place a strong emphasis on maintenance procedures that are continuously improved in order to gradually increase resource efficiency. Monitoring KPIs pertaining to resource consumption, waste generation, and energy usage is part of this process. Data analysis, performance benchmarking, and feedback mechanisms are used to discover potential for optimisation and efficiency increases

5. Resilience and Adaptability: Maintaining well-maintained assets that can endure shifting environmental conditions, natural disasters, and operational pressures is how maintenance management increases the resilience and flexibility of sustainable construction projects. The identification of vulnerabilities and the implementation of proactive actions to mitigate risks are facilitated by routine maintenance inspections, condition monitoring, and asset management techniques. The specific actions related to resilience and adaptation are listed in Table 11.8.

Table 11.8 Activities in resilience and adaptability

Activity	Explanation
Risk identification and assessment	Comprehensive risk assessments are carried out as part of maintenance management to find potential risks and weaknesses that might affect construction assets. This entails evaluating the risks associated with operational pressures (such heavy usage or equipment breakdowns), environmental conditions (like temperature changes and humidity levels), and natural disasters (like hurricanes, floods, and earthquakes)
Vulnerability analysis	In order to determine how vulnerable an asset is to different risks and threats; sustainable construction projects examine asset vulnerabilities. In order to pinpoint weak places and probable failure mechanisms, this entails assessing the structural integrity, material durability, and performance capabilities of construction assets under various stressors and environmental circumstances
Proactive maintenance interventions	Proactive maintenance actions are given priority in maintenance management in order to reduce risks and improve asset resilience. To identify early indicators of deterioration or damage, this entails doing routine maintenance inspections, condition monitoring, and structural assessments. It also entails putting in place prompt repairs, reinforcements, or retrofits to address weak points and avert catastrophic failures
Asset management practices	Sustainable construction projects implement asset management practices to enhance resilience and adaptability. This involves maintaining comprehensive asset inventories, documenting asset histories, and establishing asset performance benchmarks to track performance trends, identify emerging issues, and prioritize maintenance actions based on criticality and risk

(continued)

Table 11.8 (continued)

Activity	Explanation
Emergency preparedness planning	Establishing emergency response protocols, educating maintenance staff on emergency procedures, and collaborating with pertinent parties (such as emergency services, contractors, and suppliers) to guarantee a coordinated and prompt response to emergencies are all examples of emergency preparedness and response strategies that are developed as part of maintenance management in order to effectively manage and mitigate risks during crisis situations
Resilient design and construction	Resilient design and construction methods are used in sustainable construction projects to increase asset longevity and resistance to environmental pressures and hazards. In order to increase structural integrity and lessen vulnerability to outside threats, this involves adopting resilient building materials and techniques, such as reinforced concrete, hurricane-resistant windows, and earthquake-resistant structures
Monitoring and surveillance systems	Monitoring and surveillance systems are used by maintenance management to continuously track the performance of assets and the state of the environment in real time. This entails setting up cameras, sensors, and monitoring equipment to gather information on environmental factors, structural integrity, and asset health. This will enable early anomaly detection and proactive handling of possible risks and hazards
Adaptive maintenance strategies	Adaptive maintenance techniques that can react fast to shifting circumstances and new hazards are used in sustainable construction projects. This is adjusting priorities in response to new threats, keeping maintenance plans and schedules flexible to account for unforeseen circumstances, and modifying maintenance interventions in real-time to efficiently handle urgent needs and reduce risks

6. Occupant Health and Comfort: In sustainable construction projects, routine HVAC maintenance, and cleaning of ventilation systems and systems are all done to maintain indoor environmental quality (IEQ) requirements and prioritise the health and comfort of occupants. By lowering the possibility of mould formation, indoor air pollution, and other health risks, effective maintenance management helps to create a healthier indoor environment. Table 11.9 lists the actions that pertain to the comfort and well-being of tenants.
7. Optimized Performance and Productivity: Through ensuring that assets run effectively and dependably, maintenance management helps sustainable construction projects achieve maximum performance and productivity. Higher levels of satisfaction and productivity are supported by well-maintained infrastructure and buildings, which offer occupants a secure, pleasant, and productive environment. Table 11.10 contains the activity that are involved in the optimised performance and productivity stage.
8. Continuous Improvement: In order to improve maintenance practices over time, sustainable construction projects adopt a culture of continuous improvement in maintenance management. This is achieved by utilising data-driven insights, feedback systems, and performance indicators. Higher levels of operational efficiency and sustainability performance can be attained by projects through the

Table 11.9 Activities in the health and comfort of occupants

Activity	Explanation
Regular HVAC maintenance	HVAC systems are regularly maintained in sustainable building projects to guarantee peak efficiency and indoor air quality. To maintain appropriate indoor conditions, this entails changing and cleaning air filters, checking ducting for leaks or obstructions, and calibrating temperature and humidity controls
Ventilation system cleaning	Cleaning and maintaining ventilation systems on a regular basis to get rid of dust, allergens, and other impurities that might lower indoor air quality is part of maintenance management. To guarantee appropriate ventilation and airflow throughout the building, this entails cleaning exhaust fans, ductwork, and air vents to prevent the accumulation of pollutants
Pest control measures	In order to avoid infestations and reduce exposure to allergies and pests, sustainable construction projects use pest control techniques. This involves employing environmentally friendly pest control techniques to prevent pests and safeguard the health of occupants, sealing entry points, and placing screens on windows and doors
Indoor air quality monitoring	To make sure that indoor air quality regulations are fulfilled, maintenance management involves keeping an eye on variables including temperature, humidity, carbon dioxide concentrations, and volatile organic compounds. Deploying IAQ sensors and monitoring equipment to continuously check the quality of the air and spot possible problems that can compromise occupant comfort and health may be necessary for this activity
Mold prevention and remediation	Measures to stop mould growth and moisture-related problems that can affect indoor air quality and occupant health are implemented in sustainable construction projects. This entails keeping an eye out for any symptoms of moisture infiltration or leaks, keeping up with ventilation and drainage systems, and acting quickly to stop any mould growth with cleaning and remediation techniques
Daylighting and natural ventilation	Strategies for natural ventilation and daylighting are supported by maintenance management in order to improve occupant comfort and health. To enhance well-being and productivity, this entails keeping windows, skylights, and other apertures open to maximise natural light and airflow, minimising the need for artificial lighting and mechanical ventilation systems, and encouraging a connection to the outdoors
Noise control measures	Noise management techniques are used in sustainable construction projects to reduce noise pollution and give residents a calmer, cosier indoor environment. To limit noise disruptions from mechanical equipment and outside sources, this entails keeping up with sound-insulating materials, caulking gaps and cracks to stop sound transmission, and adding acoustical treatments and noise-absorbing finishes
Occupant education and engagement	As part of maintenance management, tenants are informed about the value of a healthy interior environment and are encouraged to adopt habits that support their well-being. To improve overall occupant well-being, this may entail educating people on how to properly operate HVAC systems, ventilation controls, and indoor pollution sources. It may also entail promoting routine cleaning and maintenance of personal areas

Table 11.10 Activity in the optimised performance and productivity stage

Activity	Explanation
Enhanced equipment reliability	Maintenance management employs routine inspections, preventative maintenance, and prompt repairs to guarantee the dependable and efficient operation of construction assets, machinery, and systems. By lowering the possibility of unplanned malfunctions or downtime, this promotes smooth construction operations and minimises delays to project timelines
Maximized asset utilization	By keeping resources and assets in top shape for the duration of their lives, sustainable construction projects optimise their utilisation. This entails making certain that machines and equipment are on hand and ready to work when needed, cutting down on downtime, and increasing output on building sites
Improved operational efficiency	By locating and fixing inefficiencies, bottlenecks, and performance gaps in construction processes and workflows, maintenance management improves operational efficiency. To reduce waste and enhance resource usage, this may entail optimising equipment settings, expediting maintenance processes, and applying lean construction concepts
Reduced downtime and delays	There are fewer instances of downtime and delays caused by equipment failures, breakdowns, or maintenance concerns in well-maintained buildings and infrastructure. This increases project productivity and lowers the possibility of expensive delays or disturbances that could affect project budgets and schedules
Enhanced safety and risk mitigation	By making sure that construction assets adhere to legal requirements and safety standards, maintenance management places a priority on worker safety. In order to reduce safety risks and provide a secure working environment for construction workers and occupants, this entails carrying out safety inspections, addressing potential hazards and risks, and putting preventive measures into place
Optimized energy and resource consumption	By sustaining energy-efficient building systems and putting resource-saving measures into place, sustainable construction projects maximise the use of energy and resources. This entails planning equipment operations for off-peak times, optimising HVAC systems' energy settings, and putting in place water-saving strategies to cut down on usage and utility expenses
Improved indoor environmental quality	Maintaining well-maintained buildings free from indoor air pollutants, allergens, and toxins is one way that maintenance management helps to improve indoor environmental quality. As a result, people can live in a healthier, cosier interior environment, which lowers absenteeism, increases focus and attention, and boosts general productivity and pleasure
Long-term asset value	By keeping assets in top condition and extending their useful lives, sustainable building projects protect their long-term worth. This supports the overall sustainability and resilience of built environments by safeguarding the investment made in building projects and ensuring that assets continue to deliver value and performance over time

analysis of maintenance data, trend identification, and application of lessons learned. The continual improvement activity is found in Table 11.11.

Table 11.11 Activity in the continuous improvement

Activity	Explanation
Data-driven decision making	Data-driven insights are used in sustainable building projects to guide maintenance decisions and propel ongoing improvement initiatives. In order to find patterns, trends, and opportunities for improvement, this entails gathering and evaluating maintenance data, such as equipment performance indicators, failure rates, downtime statistics, and maintenance expenditures
Root cause analysis	Root cause analysis is a technique used by maintenance management to determine the fundamental reasons for equipment malfunctions, breakdowns, and poor performance. Projects can address underlying issues and stop the recurrence of similar issues in the future by implementing targeted solutions and preventive measures after recognising the basic causes of maintenance problems
Performance metrics and KPIs	KPI and performance metrics are established in sustainable building projects to assess the efficacy and efficiency of maintenance procedures. Metrics that offer crucial insights into maintenance performance and aid in tracking advancement over time, such as mean time between failures, mean time to repair, equipment uptime, maintenance costs as a percentage of asset value, and overall equipment effectiveness, may be included in this activity
Feedback mechanisms	In order to obtain feedback from operators, maintenance staff, and other stakeholders participating in maintenance activities, maintenance management contains feedback methods. The maintenance team's culture of continual learning and progress is fostered by this input, which also helps identify areas for improvement and address problems and challenges
Benchmarking and best practices	In order to evaluate maintenance performance to industry benchmarks and best practices, sustainable construction projects participate in benchmarking efforts. Projects can find areas for development and strength by comparing performance metrics and practices to peers and industry norms. This allows them to put strategies into place to reach higher performance and efficiency levels
Training and skills development	Continuous improvement efforts include investing in training and Maintenance staff can improve their knowledge, skills, and competences by developing their skills. To encourage an environment of responsibility, cooperation, and creativity among the maintenance team members, this may entail offering leadership development courses in addition to technical instruction on equipment maintenance, safety protocols, and best practices
Technology adoption	In order to automate data gathering, optimise maintenance procedures, and enhance decision-making, sustainable construction projects make use of technological solutions including computerised maintenance management systems, predictive maintenance tools, and Internet of Things-enabled sensors. Projects can improve the efficiency, accuracy, and efficacy of maintenance by implementing cutting-edge technologies, which will boost operational results and sustainability performance
Continuous review and adaptation	Maintenance procedures must be regularly reviewed, evaluated, and adjusted as part of the continuous improvement process in response to shifting demands, goals, and conditions. Maintaining alignment between maintenance practices and project goals and objectives is ensured by regular evaluations of maintenance performance, evaluation of the impact of improvement measures, and strategy adaptation as needed in sustainable building projects

Maintenance management make certain the long-term viability, efficiency, and resilience of constructed assets. This is crucial in helping sustainable construction projects to meet their sustainability objectives. Sustainable construction projects offer long-term benefits to communities and stakeholders by minimising environmental impact, optimising asset performance, and incorporating maintenance considerations into project planning, design, and operation.

11.9 Opportunity Cost of Maintenance

The potential gains, alternative cost or opportunities lost when resources-like time, money, or effort-are devoted to maintenance tasks rather than being put to better use are known as the opportunity cost of maintenance. Stated differently, it denotes the worth of the next best option that could have been investigated had maintenance resources been allocated in a different manner. Opportunity costs ensure efficient use of resources. The principle is central to maintenance economies. The opportunity cost of building maintenance is the value that is derived if the money that is supposed to be invested in building maintenance is invested for, say, renovation or to upgrade the building of a facility, or rather, instead of maintaining the building, the money is spent on a new project while the other building is allowed to deteriorate or decay, which may eventually lead to demolition. It is negative if the investment is not profitable. In the same vein, it is positive if the maintenance is profitable. Table 11.12 contains the primary in maintenance cost opportunity.

Table 11.12 Main concept in the opportunity cost in maintenance of construction

Concept	Explanation
Lost revenue potential	Organisations risk missing out on chances to make more money or profits from other uses of those resources if they devote those resources to maintenance tasks. For instance, money used for equipment upkeep and repairs may have gone into expansion plans or projects that would have brought in money
Delayed innovation and upgrades	When resources are devoted to maintenance, innovation, R&D, or technological advancements that could improve competitiveness and market position may have to wait. The potential gains from investing in new technology, enhanced products, or innovative processes are included in the opportunity cost
Reduced productivity and efficiency	Maintenance tasks could take time and resources away from projects aimed at improving productivity, streamlining procedures, or increasing efficiency. The potential increases in output, productivity, or efficiency that could have been attained if funds had been dedicated to projects meant to enhance operational performance are included in the opportunity cost

(continued)

Table 11.12 (continued)

Concept	Explanation
Missed growth opportunities	Organisations risk missing out on chances for growth, expansion, or diversification into new product lines or markets by concentrating only on maintenance tasks. The possible gains from seizing expansion opportunities, such a larger market share, fresh revenue sources, or economies of scale, are included in the opportunity cost
Competitive disadvantage	If maintenance is neglected, there could be a decline in quality, dependability, or customer happiness, which would take away from a company's competitive edge. The possible advantages of preserving a solid reputation, devoted clientele, and a dominant position in the market are included in the opportunity cost
Long-term sustainability	Ignoring maintenance can cause an asset's value to decline, raise the likelihood of a failure, and increase future repair or replacement expenses. The potential savings and advantages of preventive maintenance procedures, which save expensive malfunctions, increase the lifespan of assets, and guarantee long-term sustainability, are included in the opportunity cost
Risk of regulatory non-compliance	In many industries, neglecting to carry out necessary maintenance tasks might lead to fines, penalties, or legal ramifications. The potential costs and repercussions of non-compliance, such as harm to one's reputation, legal fees, and missed commercial prospects, are included in the opportunity cost
Employee morale and satisfaction	Ignoring maintenance can result in more downtime, malfunctioning equipment, and disruptions to the workplace, all of which lower employee satisfaction and morale. The potential advantages of keeping a secure, dependable, and efficient workplace, such as increased worker productivity, engagement, and retention, are included in the opportunity cost

11.10 The Opportunity Cost of Maintenance of Sustainable Construction Projects

When resources are dedicated to maintenance activities instead of other uses that could support sustainability objectives, there are potential benefits or opportunities that may be lost. This is known as the opportunity cost of maintenance in sustainable construction projects. When resources, including time, money, or effort, are allocated to preserving green building elements and sustainable infrastructure rather than pursuing other sustainability projects or investments, opportunity costs may occur in the context of sustainable construction. Table 11.13 explains the ideas in the opportunity cost of maintenance of sustainable construction projects.

Table 11.13 The concepts in the opportunity cost of maintenance of sustainable construction projects

Concept	Explanation
Investment in renewable energy	The prospective expense of investing in more renewable energy capacity or energy efficiency initiatives could outweigh the benefits of regularly maintaining sustainable features like wind turbines or solar panels. This could entail introducing energy-saving technologies to improve overall sustainability performance or growing renewable energy systems to further lessen dependency on non-renewable sources
Green building certifications	The opportunity cost of investing in additional sustainability certifications or accreditations that could further improve the project's reputation, marketability, or environmental effect may outweigh the resources committed to maintaining green building certifications. To reach greater degrees of sustainable recognition, this can entail obtaining certifications in the areas of indoor air quality, materials sourcing, and water efficiency
Natural habitat restoration	Investing in natural habitat restoration projects that improve biodiversity, ecosystem services, and environmental resilience may be a better use of resources than concentrating only on upkeep of green infrastructure features like rain gardens or vegetated roofs. To lessen the negative ecological effects of building and encourage long-term sustainability, this may entail funding programmes for habitat conservation, reforestation, or the improvement of wildlife habitat
Community engagement and education	The opportunity cost may be the potential gains from funding community engagement and education initiatives that support social justice, behaviour change, and sustainability awareness while funds are allocated to maintaining sustainable elements that benefit the structure or infrastructure itself. To enable local stakeholders to embrace sustainable practices and support more general sustainability goals, this may entail providing funding for community workshops, outreach initiatives, or educational campaigns
Resilience planning and adaptation	The opportunity cost could be the potential gains from investing in resilience planning and adaptation strategies that address the effects of climate change, natural disasters, and other environmental risks, as opposed to concentrating only on preserving the current sustainable infrastructure. This could entail providing funding for infrastructure improvements, disaster preparedness programmes, or studies of climate resilience that improve the project's capacity to face obstacles and uncertainties in the future
Investment in sustainable innovation	The potential benefits of funding research and development of cutting-edge sustainability technologies, materials, or practices may outweigh the opportunity cost of focusing only on maintaining the sustainable characteristics that are already in place. This could entail providing funds for pilot programmes, initiatives for demonstration, or joint research projects with the goal of developing sustainable construction techniques, lowering environmental effects, and encouraging innovation in the sector

(continued)

Table 11.13 (continued)

Concept	Explanation
Social equity and inclusive design	Although resources are allocated to uphold environmentally or building occupant-beneficial sustainable features, there may be a missed opportunity cost associated with not investing in social equity and inclusive design strategies that tackle socio-economic disparities, accessibility obstacles, and community needs. Allocating resources to projects that prioritise universal design principles, affordable housing, or fair access to natural spaces and facilities could be one way to enhance project sustainability and promote social inclusion

11.11 Summary

From the foregoing, studies have examined defects in buildings, yet empirical data on defects in sustainable construction is nascent and even more so for forecasting and classification of defects in sustainable buildings. Therefore, most construction companies devote many human resources and economic costs to managing defects. This research aims to explore defects in sustainable buildings and to utilise the findings to develop a sustainable building pathology framework to pave the way to systemic sustainable construction management strategies. Sustainable construction management prioritizes high-quality, environmentally friendly materials, integrated design, proper construction techniques, energy efficiency, and ongoing maintenance. Conventional construction management may prioritize speed, cost savings, and short-term solutions, leading to defects in buildings over time. Generally, the types of building defects that can occur in conventional and sustainable buildings differ based on the materials used, the construction techniques employed, and the design priorities. While sustainable buildings are designed to minimize defects, proper design, construction, and maintenance practices are necessary to ensure their longevity and durability. This research proposes a mixed-method approach to examine defects in sustainable buildings and to set forth a machine learning approach for investigating whether a specific defect can be autonomously categorised into one of the categories of repair tasks. The defects in the design, construction and maintenance of sustainable buildings continue to reduce the fundamental essence of sustainable buildings. This research yielded a prototype decision-making framework for construction companies, government, industry, academic institutions, and local governments to reduce defects in the construction of sustainable buildings.

Bibliography

1. Abdul-Rahman H, Wang C, Wood LC, Khoo YM (2014) Defects in affordable housing projects in Klang Valley, Malaysia. J Perform Constr Facil 28(2):272–285
2. Ahzahar N, Karim NA, Hassan SH, Eman J (2011) A study of contribution factors to building failures and defects in construction industry. Procedia Eng 20:249–255
3. Al-Tmeemy SM, Abdul-Rahman H, Harun Z (2012) Contractors' perception of the use of costs of quality system in Malaysian building construction projects. Int J Project Manage 30(7):827–838
4. BSI (2015) BSI Standards Publication: BS EN ISO 9000:2015 Quality management systems Fundamentals and vocabulary. ISBN 978 0 580 78878 9
5. Bureau of Economic Analysis (BEA) (2006) Gross domestic product by industry in current dollars. http://bea.gov/bea/dn2/gdpbyind_data.htm
6. Chan DW (2019) Sustainable building maintenance for safer and healthier cities: effective strategies for implementing the mandatory building inspection scheme (MBIS) in Hong Kong. J Build Eng 24:100737
7. Cohen L, Manion L, Marrison K (2017) Research methods in education, 7th edn. Routledge Publisher, New York
8. Clean Malaysia (2018) Green buildings are 'on the up'. Available at: https://cleanmalaysia.com/2018/02/16/green-buildings-are-on-the-up/. Accessed on 21 Dec 2020
9. Construction Industry Development Board Malaysia (CIDB) (2018a) Built green: an overview of sustainability green building rating tools in Malaysia
10. Construction Industry Development Board Malaysia (CIDB) (2016) MyCRest (Malaysian Carbon Reduction & Environmental Sustainability Tool, Version l) a reference guide for operation and maintenance Stage. www.cidb.gov.my
11. Faqih F, Zayed T, Soliman E (2020) Factors and defects analysis of physical and environmental condition of buildings. J Build Pathol Rehabil 5:1–15
12. Forcada N, Macarulla M, Gangolells M, Casals M (2016) Handover defects: comparison of construction and post-handover housing defects. Build Res Inf 44(3):279–288
13. Heravi G, Jafari A (2014) Cost of quality evaluation in mass-housing projects in developing countries. J Constr Eng Manag 140(5):04014004
14. Haroon R (2022) 12:10am. https://www.nst.com.my/opinion/columnists/2022/01/761430/we-have-tolerated-corporate-greenwashing-too-long. https://www.nst.com.my/opinion/columnists/2022/01/761430/we-have-tolerated-corporate-greenwashing-too-long
15. Hwang BG, Shan M, Phua H, Chi S (2017) An exploratory analysis of risks in green residential building construction projects: the case of Singapore. Sustainability 9(7):1116
16. ISO 9000: 2015 Quality management systems
17. Koch C, Schultz SC (2019) The production of defects in construction—an agency dissonance. Constr Manag Econ 37(9):499–512
18. Love PE, Smith J, Ackermann F, Irani Z (2019) Making sense of rework and its unintended consequence in projects: the emergence of uncomfortable knowledge. Int J Project Manage 37(3):501–516
19. Montague A (2018) Defects cost more than profits: CIOB launches urgent course on quality. Global Construction Review. Available at http://www.globalconstructionreview.com/news/defects-cost-more-profits-ciob-launches-urgent-cou. Accessed on 17 Feb 2020
20. Olanrewaju A, Lee HJA (2022) Analysis of the poor-quality in building elements: providers' perspectives. Front Eng Built Environ 2(2):81–94
21. Othman NL, Jaafar M, Harun WMW, Ibrahim F (2015) A case study on moisture problems and building defects. Procedia Soc Behav Sci 170:27–36

22. Simpeh EK, Ndihokubwayo R, Love PE, Thwala WD (2015) A rework probability model: a quantitative assessment of rework occurrence in construction projects. Int J Constr Manag 15(2):109–116
23. Sandanayake M, Yang W, Chhibba N, Vrcelj Z (2021) Residential building defects investigation and mitigation–a comparative review in Victoria, Australia, for understanding the way forward. Eng Constr Archit Manag
24. Tollin HM (2011) Green building risks: it's not easy being green. Environ Claims J 23(3–4):199–213
25. Yap JBH, Low PL, Wang C (2017) Rework in Malaysian building construction: impacts, causes and potential solutions. J Eng Des Technol 15(5):591–618

Chapter 12
Emerging Technologies in the Built Environment

Abstract This chapter explores the transformative role of emerging technologies in shaping a more sustainable and resilient built environment. As global pressures for environmental responsibility and resource efficiency intensify, the construction industry is undergoing a digital revolution that enhances productivity, reduces environmental impact, and fosters collaboration across project stakeholders. The chapter delves into the theoretical and practical applications of digital construction technologies, including digital twins, smart systems, and automation and examines their role in advancing sustainable construction practices throughout the project lifecycle. Drawing on international survey data, the chapter provides insight into the current adoption levels of digital technologies, identifies constraints hindering wider implementation, and proposes mitigation strategies to overcome these barriers. It underscores that technology complements rather than replaces human input, especially in tasks that are hazardous, time-consuming, or labour-intensive. The growing relevance of digital innovation is reflected in industry-wide recognition, such as the UK's annual Digital Construction Awards. Ultimately, the chapter offers a practical framework for prioritizing digital tools in construction, aligning technological adoption with sustainability goals. Readers will gain a comprehensive understanding of how emerging technologies support sustainable construction management and contribute to a greener, more efficient, and future-ready built environment.

Keywords Digital construction · Project benchmark

12.1 Learning Outcomes

The construction sustainability movement is largely about digital space; e.g., the digital approach is extremely crucial. Digital construction is part of the revolution in computing that has been taking over the last decades. Digital tools can help reduce energy consumption, accidents, and improve collaboration and communication, etc. in the built environment. The devices collect personalized data about construction. It is crucial to stress that construction technologies that do and will not replace humans,

A. Olanrewaju, *Sustainable Construction Management: Research and Practice Companion*, Management in the Built Environment,
https://doi.org/10.1007/978-981-96-4910-5_12

but they are needed to perform tasks that are expensive to perform by humans or that are difficult to perform by humans. Some construction technologies are also used during periods when it is not comfortable or convenient to work, like during the night or when the weather is harsh. Research also revealed that technological innovation could reduce capital costs for net-zero technologies faster than expected. However, this chapter also presents the preliminary results of the survey on the application of digital twins in the construction sector from international perspectives. Primarily, the aim is to offer practical insight into the theoretical basis in the earlier chapters. The data were collected based on a survey questionnaire that was administered over one month. After studying this chapter, you will be able to:

- Explain digital construction technologies.
- Explain the application of digital technologies for sustainable construction.
- Explain the constraining factors for the application of digital construction technologies in sustainable construction management.
- Explain the techniques to mitigate the constraints for the applications of digital construction technologies.
- Prioritise construction technologies.

12.2 Introduction

In the ever-evolving landscape of construction, the integration of emerging technologies has become a catalyst for transformative change. As the global significance for sustainable practices intensifies, the construction sector stands at the forefront of a technological revolution that is reshaping the way construction projects are conceived, designed, constructed, and operated in built environments. This paradigm shift is driven by a commitment to environmental protection, resource efficiency, and the creation of construction projects that not only withstand the test of time but also contribute positively to the world we inhabit. Emerging technologies in sustainable construction represent a convergence of innovation and environmental responsibility. These advanced solutions go beyond conventional construction methods to harmonise with the ecosystem, consume fewer resources, generate less waste and are resilient in the face of climate challenges. From advanced materials and construction methodologies to smart systems that optimize energy usage, the trajectory of sustainable construction is being charted by technologies that redefine what is possible in the built environment. The application of digital technologies in construction is increasing so much that there is an annual award for it starting in 2022 in the UK. The Digital Construction Awards celebrate best practices and reward innovation in the application of digital technology in the built environment sector. The award features a total of 14 categories, which are: Digital Construction Project of the Year, Digital Contractor of the Year, Digital Consultancy of the Year, Digital Multi-disciplinary Consultancy of the Year, Digital Collaboration of the Year, Digital Rising Star of the Year, Digital Construction Champion of the Year, Best Application of Technology, Product Innovation of the Year, Best Use of Data on a Project, Delivering Sustainability with Digital Innovation, Digital Innovation in Asset Management, Digital Innovation in Health, Safety, and Wellbeing, and Digital Innovation in Productivity. In 2023, there were more than 140 entries, and nearly 450 digital construction professionals attended the gala dinner.

12.3 Construction Technologies

By using a variety of data sources, construction companies can assess, enhance, and maximise their projects through the application of technologies. Automating data collecting and report generation can aid in time-saving data management and integration of data sets. For example, through the utilization of technologies, construction companies may build and exchange private databases that gather and compute carbon factor data, which can then be merged with their 3D models. By adding carbon data to the current 3D models, the company will be able to use the 5D-plus model for its project as a sole source of truth and create a structured database for data analytics. By using technologies, many construction firms have made site crews the focal point of their value-creation initiatives. The drives have been shown to boost teamwork, increase site connectivity by over 50%, and produce quantifiable, repeatable efficiency gains across a range of processes, such as cutting the time it takes to process site data from hours to minutes. Both new and existing constructions can benefit from digital construction. Building fabric components, both new and old, can be added to a digital asset register by construction businesses using digital construction. After the project is finished, an asset information model can be created for FM reasons. Facilities management organisations use digital technologies to enforce adherence to manufacturer specifications, which unites responsible parties and fosters ownership. It also lowers the level of resources needed to implement a robust fire safety process and continuously provides end users with validated safety information to help them operate the building safely. Technology is used by clients for road maintenance contractors. With the use of construction technology, paper is eliminated from the process and replaced with a more intelligent, user-friendly, and certain 3D approach. This results in a 500% reduction in the time it takes to order supplies. The combination of these technologies has the potential to produce environmentally friendly buildings that can support the health of their occupants as well as the earth.

12.4 Digitalization

Digitalization is the process of transforming traditional corporate processes, activities, and models using digital technologies. It entails making the most of digital tools, technology, and data to improve an organization's or an industry's overall performance, effectiveness, and efficiency in a variety of areas. Beyond just using digital tools, digitalization entails a comprehensive integration of digital technologies into every facet of an organization's operations and functions. The process of digitalization is revolutionary and can affect many sectors and industries, such as business, manufacturing, healthcare, education, and construction. To enhance project planning, teamwork, and overall project delivery, digitalization in the construction sector entails implementing technologies including digital twins, BIM, VR, AR, construction management software, and other digital tools. Using technology to spur innovation, increase operational effectiveness, and adjust to the rapidly changing digital environment is the aim of digitalization.

In sustainable construction, digitalization is the strategic application of digital technology and procedures to improve the environmental, social, and economic sustainability of construction projects. The aim is to integrate digital technologies with sustainable methods to produce built environments that are more efficient, socially conscious, and ecologically friendly. Building information modelling, energy modelling and simulation, IOT and smart building technologies, digital twins, supply chain digitalization, collaboration platforms, recycling technologies, drones and remote monitoring, augmented reality and virtual reality, data analytics and machine learning, and digital platforms for stakeholder engagement are just á few of the aspects of digitalization in sustainable construction that are involved.

Building Information Modeling

The structural and operational features of a building or infrastructure are represented digitally using BIM or digital twins. It enables stakeholders to work together to design, visualise, and simulate construction projects, it maximises productivity and minimises waste. Sustainable construction practices are advanced using BIM. Beyond traditional two-dimensional drawings, it creates a three-dimensional model that incorporates data and information about many project components. As listed in Table 12.1 are the uses of BIM in the sustainable construction management.

Stakeholders can improve the sustainability, accuracy, and efficiency of building projects by integrating BIM into sustainable construction processes. Sustainable design, building, and operation concepts are aligned with BIM's collaborative and data-driven approach.

Energy Modeling and Simulation

Energy modelling and simulation to evaluate building energy performance are made possible by digital tools. This supports the use of renewable energy sources, sustainable material selection, and energy-efficient design optimisation. Because they enable stakeholders to evaluate and optimise a building's energy efficiency over its entire life cycle, energy modelling and simulation are essential to sustainable construction. In these procedures, a building's energy usage and efficiency are modelled, examined, and simulated using sophisticated computing tools. Table 12.2 contains a description of how energy modelling and simulation are used in sustainable construction.

Table 12.1 Application of BIM in supporting sustainable construction

Application	Explanation
Design optimization	Using BIM, designers and architects may produce precise and in-depth 3D models of buildings. This makes it possible to investigate several design possibilities to maximise daylighting, energy efficiency, and sustainability in general
Energy performance analysis	Energy simulations and assessments can be carried out using BIM technologies. Using energy modelling throughout the design stage, interested parties can evaluate and enhance a building's energy efficiency, considering the effects of various materials, layouts, and systems
Materials selection	Information about the environmental impact of building materials can be stored in BIM databases. Data on recycled content, embodied carbon, and other sustainability measures are included in this. With knowledge, designers can select materials that have less of an impact on the environment
Life cycle assessment	By offering a framework for examining a building's environmental effects throughout its whole life, BIM aids in life cycle assessment. This includes obtaining raw materials, manufacturing, constructing, operating, and maintaining the system, as well as any necessary recycling or eventual demolition
Collaboration and coordination	Collaboration between project stakeholders, such as architects, engineers, contractors, and owners, is eased by BIM. By encouraging coordination and communication throughout the project lifespan, this partnership can produce more long-lasting results
Waste reduction	By easing accurate quantity takeoffs and improved material management, BIM may reduce construction waste. BIM supports sustainable building methods by maximising resource utilisation and cutting waste
Construction sequencing and planning	Building process simulation and visualisation are made possible by BIM. This helps to minimise disturbances to the surrounding environment, optimise resource utilisation, cut down on building time, and organise and sequence construction activities
Facility management and operation	Facility operation and management can benefit from the use of BIM data. This entails keeping precise records of the systems, parts, and maintenance plans of the facility. Long-term sustainability can be improved by using BIM to increase building operations' efficiency
Renewable energy integration	The incorporation of renewable energy systems into architectural designs is facilitated by BIM. Stakeholders can decide whether to deploy solar panels, wind turbines, or other sustainable energy solutions by modelling and evaluating the performance of renewable energy installations
Green building certifications	The documentation and analysis needed to obtain certification for green buildings can be facilitated using BIM. The information required for certification processes is provided by the data-rich models, guaranteeing adherence to sustainability standards

Table 12.2 Application of energy modeling and simulation in supporting sustainable construction

Application	Explanation
Design optimization	During the design process, energy modelling is utilised to assess how unique design decisions would affect energy performance. To maximise the energy efficiency of the structure, architects and engineers can simulate several scenarios, such as changing the amount of insulation, where to put windows, and how to shade the space
Performance prediction	With the aid of simulation tools, one can precisely forecast how energy-efficient a building will be under various operating scenarios. Predicting heating and cooling loads, lighting needs, and total energy consumption based on variables like temperature, occupancy patterns, and equipment use are all included in this
Renewable energy integration	When evaluating the viability and efficiency of incorporating renewable energy sources, such as solar or wind power, energy modelling comes in handy. The economic feasibility, environmental impact, and prospective energy generation of various renewable energy systems can all be assessed through simulations
HVAC system optimization	HVAC system optimisation is made possible by simulations. Through an examination of various HVAC setups and management techniques, interested parties can determine the most energy-efficient options that satisfy air quality and comfort standards
Daylighting analysis	To evaluate a building's natural light availability, energy modelling incorporates daylighting models. This reduces the need for artificial illumination and increases overall energy efficiency by optimising window placement and size
Life cycle assessment	Life cycle assessment procedures incorporate energy modelling to estimate a building's environmental impact over its whole life. This entails evaluating the long-term energy consumption during operation as well as considering the embodied energy in building materials
Code compliance	Energy codes and standards compliance in buildings is aided by simulation tools. Stakeholders can fulfil or surpass mandatory performance criteria by evaluating a design about energy efficiency regulations
Operational cost predictions	Simulations offer valuable insights into the operational expenses linked to energy usage. When making judgements about purchasing energy-efficient systems and technologies, owners and operators can use this information to make well-informed choices
Occupant comfort and productivity	Simulations can evaluate indoor environmental quality, including thermal comfort and daylight availability, in addition to energy economy. Design decisions that improve occupant comfort and well-being are a part of sustainable building methods
Continuous monitoring and feedback	Continuous feedback on the energy performance of a building can be obtained using real-time monitoring instruments that are integrated with building management systems. This promotes long-term sustainability by enabling modifications and advancements throughout time

Achieving sustainability targets in construction requires the use of energy modelling and simulation technologies. Stakeholders may minimise energy usage, increase overall performance, and lessen their influence on the environment by using these tools to make educated decisions.

IoT and Smart Building Technologies

Building sensors, devices, and system integration are made easier by the IoT. Real-time data is used by smart building technology to optimise energy use, track indoor air quality, and improve overall building efficiency. A key factor in increasing sustainability in the construction sector is the use of smart building technology and the IoT. These technologies improve construction performance, efficiency, and environmental effects by using sensors, data analytics, and networked equipment. Table 12.3 lists outlines how IoT and smart building technologies can be used in sustainable building practices.

Table 12.3 Application of IoT and smart building technologies in supporting sustainable construction

Benefit	Explanation
Energy management	Real-time monitoring of energy consumption by IoT sensors and metres yields insights into usage trends. By using this data, energy efficiency may be maximised, problem areas can be found, and overall energy usage can be decreased
Occupancy sensing	IoT sensors identify a building's occupancy levels in various sections. By using this data, energy waste in vacant spaces is decreased as lighting, HVAC, and air conditioning systems are automatically adjusted based on actual occupancy
HVAC optimization	HVAC systems with IoT connectivity modify airflow and temperature according to occupancy and current conditions. Energy savings, increased comfort, and a smaller environmental effect result from this optimisation
Lighting control	IoT-enabled lighting systems change colour temperature and brightness in response to occupancy and natural light levels. This promotes a better interior atmosphere in addition to increasing energy efficiency
Renewable energy integration	Renewable sourced energy, like solar and wind power, can be integrated into a building's energy infrastructure thanks to the Internet of Things technologies. Distributing and using renewable energy more efficiently is made possible by smart grids
Predictive maintenance	Data on the functionality and condition of equipment is gathered by sensors. To reduce downtime and increase the lifespan of building systems, predictive analytics is used to analyse this data and forecast maintenance requirements
Water management	IoT sensors aid in optimising water consumption, cutting waste, and improving water efficiency in buildings by monitoring water usage and spotting leaks
Waste management	IoT-enabled waste cans make waste collection more effective by alerting users when they are full. This minimises fuel use and emissions by cutting down on pointless rubbish collection visits

(continued)

Table 12.3 (continued)

Benefit	Explanation
Indoor air quality monitoring	Sensors track elements of air quality, like contaminants and CO_2 concentrations. By adjusting ventilation systems using this data, indoor air quality and occupant health are guaranteed to be at their best
Security and access control	Security systems built on the Internet of Things include alarm, surveillance, and access control features. By using HVAC controls and adaptive lighting, these systems decrease energy usage and improve security
Occupant engagement	Interfaces for smart building systems often let residents monitor and control their surroundings. Building users are encouraged to adopt sustainable behaviours and become more conscious because of this interaction
Data analytics for continuous improvement	Big Data Analytics: To find trends, patterns, and areas for improvement, the enormous volume of data produced by IoT devices is analysed. This data-driven strategy improves overall sustainability and allows for further optimisation

Digital Twins

Digital twins are digital copies of actual construction. Digital twins can be used to improve sustainability, maximise resource utilisation, and model the environmental effects of building projects. Using digital twins in sustainable construction is revolutionary as they offer a digital duplicate of tangible assets, procedures, and frameworks. By simulating, tracking, and optimising many parts of construction projects, this technology helps stakeholders increase sustainability. The use of digital twins to aid sustainable construction is described in Table 12.4.

Table 12.4 Application of digital twins in supporting sustainable construction

Design and planning benefit	Explanation
	Buildings and infrastructure can be virtually prototyped thanks to digital twins. Before actual building starts, stakeholders can investigate various design possibilities, evaluate their effects on the environment, and maximise sustainability elements
Energy efficiency	Building energy performance is simulated and analysed using digital twins that integrate energy modelling. This makes it easier to spot areas where energy efficiency can be increased, such as through insulation, HVAC system optimisation, and the incorporation of renewable energy sources
Resource optimization	Digital twins monitor how building materials are used and how they affect the environment. This enables effective resource utilisation throughout the project, reduces waste, and helps the selection of sustainable materials
Construction process optimization	Digital twins can be used to model construction processes, which enables stakeholders to find possible bottlenecks, streamline workflows, and shorten construction schedules. This reduces environmental disturbance on-site and saves resources

(continued)

Table 12.4 (continued)

Design and planning benefit	Explanation
Life cycle assessment	Digital twins help in life cycle assessment by simulating how building materials and procedures affect the environment. This thorough examination considers a building's whole life cycle, from construction through use and final decommissioning
Operational efficiency	Through the provision of real-time data on building systems, equipment performance, and tenant behaviour, digital twins facilitate effective facility management. Proactive maintenance is made possible by this ongoing monitoring, which maximises operational sustainability and efficiency
Predictive maintenance	Digital twins incorporate sensors to keep an eye on the state of architectural elements. Based on this data, predictive analytics can perform preventative maintenance, minimising the need for replacements, cutting downtime, and prolonging the life of the equipment
Occupant comfort and well-being	Air quality, illumination, and thermal comfort are just a few of the elements of indoor environmental quality that digital twins evaluate and optimise. This improves occupant satisfaction, productivity, and well-being
Renewable energy integration	Digital twins replicate how renewable energy sources—like solar and wind turbines—would be integrated into the building's energy infrastructure. This guarantees the best arrangement and positioning for optimum energy production
Waste reduction	By simulating material movements, digital twins facilitate effective waste management in construction. This facilitates the identification of options for recycling, appropriate disposal techniques, and waste reduction
Adaptive and resilient design	Scenario modelling is made possible by digital twins for resilient and adaptable design. To improve building resilience, this involves modelling the effects of climate change, extreme weather, and other external elements
Continuous improvement	Large volumes of data are produced by digital twins, which give stakeholders the ability to make defensible choices based on current information. Throughout the building's life cycle, this data-driven strategy supports ongoing improvements in sustainability performance

Through the integration of digital twin technology with sustainable construction methods, stakeholders may design and construct infrastructure and structures that are not only ecologically conscious but also flexible, effective, and resistant to changing circumstances. In the construction sector, digital twins offer a dynamic and data-rich framework for accomplishing and upholding sustainability targets.

Supply Chain Digitalization

Construction material tracking and optimisation are made possible using digital technologies to supply chain management. In addition to ensuring the use of sustainable and ethically sourced resources, this lowers waste and increases transparency. Digital technologies are integrated into the construction supply chain to improve its sustainability, efficiency, and transparency. This process is known as supply chain digitalization in sustainable construction. To streamline operations, cut waste, and guarantee ethical material sourcing, this transformation makes use of digital tools and platforms. Table 12.5 shows how the applications of supply chain digitalization that contribute to sustainable construction.

Table 12.5 Application of supply chain digitalization in supporting sustainable construction

Benefit	Explanation
Transparent supply chain	A traceable and transparent supply chain is made possible by blockchain. From extraction to delivery, every stage of the construction material's path may be tracked and validated, guaranteeing accountability and transparency
Sustainable material sourcing	Digital platforms can help construction projects validate and confirm the sustainable qualities of their supplies. Information about carbon footprint, recycled content, and compliance with environmental regulations are all included
Data-driven decision-making	The process of digitalization makes it possible to analyse enormous volumes of data about the sourcing, usage, and transportation of materials. This data-driven methodology aids in locating chances for increased productivity and long-term decision-making
Real-time inventory management	Inventory management is enhanced by real-time building material tracking with RFID tags and Internet of Things sensors. As a result, there is less overordering, less surplus inventory, and less environmental damage from needless material manufacture
Collaboration platforms	Cloud-based solutions make it easier for different supply chain stakeholders to collaborate. To promote cooperation and communication for sustainable practices, this includes manufacturers, suppliers, contractors, and project owners
Supplier assessment and auditing	The evaluation and auditing of suppliers according to sustainability standards is made possible by digital tools. This guarantees that suppliers run their businesses by moral and environmental norms
Automated procurement processes	Automated procurement systems ensure that sustainable factors are considered when selecting materials, hence streamlining the purchasing process. In addition to improving efficiency and supporting sustainable buying methods, this lessens paperwork
Waste reduction and recycling	Waste management and tracking are aided by digital tools. Construction teams can apply tactics to minimise waste, encourage recycling, and lessen environmental impact by keeping an eye on waste generation and disposal

(continued)

Table 12.5 (continued)

Benefit	Explanation
Carbon footprint assessment	The carbon footprint of material transportation in construction projects can be evaluated with digital techniques. Selecting vendors and logistics partners with less of an influence on the environment is made easier by this information
Systemic logistics	Transportation routes are optimised via digital platforms, which lower pollutants and fuel usage. This is in line with sustainable transportation principles and reduces operating costs as well
Ethical labour practices	Digitalization of the supply chain includes labour management, and guaranteeing moral work practices. Fair treatment and adherence to labour norms can be ensured using digital tools to monitor and verify working conditions
Continuous improvement	Analytics on supply chain performance are made possible by digitalization. This information can be used in construction projects to pinpoint problem areas, establish sustainable objectives, and track advancement over time

The construction sector may transition to sustainable methods by using digital supply chain processes. Digital solutions enable stakeholders to optimise processes, make well-informed decisions, and establish a transparent, ethical, and responsible supply chain that takes environmental and ethical concerns into account.

Collaboration Platforms

Digital collaboration platforms facilitate instantaneous contact and information exchange amongst stakeholders involved in a project. This raises overall project efficiency, decreases errors, and increases coordination. The utilisation of digital twins in sustainable construction is revolutionary as they offer a digital duplicate of tangible assets, procedures, and frameworks. By simulating, tracking, and optimising many parts of construction projects, this technology helps stakeholders increase sustainability. The following in Table 12.6 are some ways that collaboration platforms might help with sustainable building.

Through the integration of digital twin technology with sustainable construction methods, stakeholders may design and construct infrastructure and structures that are not only ecologically conscious but also flexible, effective, and resistant to changing circumstances. In the building business, digital twins offer a dynamic and data-rich framework for accomplishing and upholding sustainability targets.

Recycling Technologies

Waste management procedures are optimised using digital technologies. This entails monitoring and controlling waste generated during construction, encouraging recycling, and lessening the environmental effects of waste disposal. Recycling methods are essential to waste reduction, resource conservation, and the advancement of a circular economy in sustainable construction. By concentrating on recovering and recycling materials from waste generated during construction and demolition, these

Table 12.6 Application of collaboration platforms in supporting sustainable construction

Benefit	Explanation
Design and planning	Buildings and infrastructure can be virtually prototyped thanks to digital twins. Before actual building starts, stakeholders can investigate various design possibilities, evaluate their effects on the environment, and maximise sustainability elements
Energy efficiency	Energy modelling is used in digital twins to simulate and assess building energy performance. This facilitates the identification of areas that can be improved for energy efficiency, such as insulation, HVAC system optimisation, and the incorporation of renewable energy sources
Resource optimization	Digital twins monitor how building materials are used and how they affect the environment. This enables effective resource utilisation throughout the project, reduces waste, and helps the selection of sustainable materials
Construction process optimization	Stakeholders can discover possible bottlenecks, streamline workflows, and shorten building times by using digital twins to model construction processes. This minimises environmental disturbance on-site and results in resource savings
Life cycle assessment	Digital twins help in Life Cycle Assessment by simulating how building materials and procedures affect the environment. This thorough examination considers a building's whole life cycle, from construction through use and final decommissioning
Operational efficiency	Through the provision of real-time data on building systems, equipment performance, and tenant behaviour, digital twins facilitate effective facility management. Proactive maintenance is made possible by this ongoing monitoring, which maximises operational sustainability and efficiency
Predictive maintenance	Sensors are integrated with digital twins to track the state of building elements. Based on this data, predictive analytics provide proactive maintenance, which lowers downtime, prolongs the life of equipment, and minimises the need for replacements
Occupant comfort and well-being	Indoor environmental quality elements including air quality, illumination, and thermal comfort are evaluated and optimised by digital twins. Increased productivity, satisfaction, and well-being are the results of this
Renewable energy integration	Digital twins replicate how renewable energy sources—like solar and wind turbines—would be integrated into the building's energy infrastructure. This guarantees the best arrangement and positioning for optimum energy production
Waste reduction	By simulating material movements, digital twins facilitate effective construction waste management. This makes it easier to spot opportunities for recycling, waste minimization, and proper disposal techniques

(continued)

Table 12.6 (continued)

Benefit	Explanation
Adaptive and resilient design	Scenario modelling is made possible by digital twins for resilient and adaptable design. To improve building resilience, this involves modelling the effects of climate change, extreme weather, and other external elements
Continuous improvement	Large volumes of data are produced by digital twins, which give stakeholders the ability to make defensible choices based on current information. Throughout the building's life cycle, this data-driven strategy supports ongoing improvements in sustainability performance

technologies help to conserve the environment, reduce economic loss. The recycling technologies used in sustainable construction are listed in Table 12.7.

Recycling technology use in sustainable construction is consistent with the values of resource conservation, waste minimization, and environmental responsibility. By reducing the demand for additional raw materials, closing the loop on construction

Table 12.7 Application of recycling technologies in supporting sustainable construction

Application	Explanation
Concrete recycling	To create aggregates that can be utilised in place of natural aggregates in newly mixed concrete, crushed concrete must first be demolished. As a result, the need for virgin resources is decreased, and the environmental impact of producing concrete is reduced
Asphalt recycling	Reclaimed asphalt pavement is recycled by mixing it into fresh asphalt mixtures. When RAP is used, less new asphalt needs to be produced because it is created by grinding and removing old asphalt surfaces
Wood recycling	Using wood recycling methods, waste wood from building sites is converted into useful commodities. To create wood chips, mulch, or recovered wood products, wood waste can be chipped, ground, or milled
Metal recycling	Metal parts, including aluminium and steel, can be recycled through remelting and processing. This reduces the need for raw metals, conserves energy, and lessens the harm that metal extraction causes to the environment
Plastic recycling	Plastic waste is transformed into building materials, such as plastic tiles, timber, or insulation, via technologies for recycling plastic. By doing this, the environmental impact of producing plastic is lessened and plastic is diverted from landfills
Gypsum recycling	Recycling is an option for gypsum waste from building sites, such as plasterboard and plasterboard. Gypsum is separated and processed using technologies such that it can be used again in agriculture or new building materials
Glass recycling	Crushing and processing used glass debris into fine or coarse aggregates is known as glass recycling in the building industry. By using these glass aggregates in concrete, asphalt, or landscaping applications, less additional raw materials will be required

(continued)

Table 12.7 (continued)

Application	Explanation
Brick and masonry recycling	It is possible to crush and repurpose demolished brick and masonry components into aggregates for future construction projects. This reduces the environmental effect of brick manufacture and conserves resources
Paint recycling	Paint recycling technologies entail using leftover paint by sorting and processing it. This promotes the production of recycled paint products and lessens the amount of hazardous paint waste that is disposed of
Construction and demolition waste sorting	Innovative sorting technologies help separate various materials in building and demolition waste streams. Examples of these technologies include automated conveyor belts and optical sorting systems. This lowers contamination and makes recycling more effective
Mobile recycling units	On-site recycling of construction debris is made possible by mobile recycling equipment that comes with crushers, screeners, and sorting devices. While building projects, this reduces transportation expenses and encourages sustainable methods
3D printing with recycled materials	Construction components can be produced using 3D printing technology using recyclable materials like concrete or plastic. This creative method encourages resource efficiency and minimises waste

materials, and lessening the environmental impact of construction activities, these technologies support the circular economy.

Drones and Remote Monitoring

An automated way to examine construction sites, track developments, and evaluate environmental conditions is made possible by drones and remote monitoring systems. This can lessen the environmental impact of construction activities, increase safety, and eliminate the need for on-site visits. In sustainable construction, the use of drones and remote monitoring technologies is growing. These tools provide creative ways to boost productivity, lessen environmental effects, and increase project management in general. Table 12.8 shows some applications that drones and remote monitoring might support environmentally friendly building practices.

Drones and remote monitoring technology can be integrated to improve efficiency, minimise environmental impact, and manage resources more effectively in sustainable construction methods. These technologies help to make the design, carrying out, and monitoring of building projects more sustainable and ethical.

Augmented Reality and Virtual Reality

Construction project visualisation and comprehension are improved by AR and VR technologies. This can help with training, stakeholder participation, and design reviews, all of which improve decision-making and project results. Technologies like VR and AR are transforming the construction sector by providing innovative instruments to improve sustainable practices. Some ways that R and VR can support environmentally friendly construction practices as listed in Table 12.9.

Table 12.8 Application of drones and remote monitoring in supporting sustainable construction

Application	Explanation
Site inspection and surveying	High-resolution aerial imaging is made possible by drones fitted with cameras and sensors, enabling thorough site inspections. As a result, fewer ground-based surveys are needed, reducing the negative environmental effects of more conventional surveying techniques
Environmental monitoring	Drones with environmental sensors installed can keep an eye on the water and air quality near building projects. This guarantees adherence to environmental laws and eases the detection of possible effects on surrounding communities and ecosystems
Vegetation monitoring	Before, during, and after construction projects, drones can be used to monitor the health of the vegetation and evaluate the ecological impact. This backs conservation and wise land management initiatives
Construction progress monitoring	Drones can collect real-time pictures and videos of building sites, giving stakeholders and project managers the most up-to-date information on the status of the project. This lessens the need for frequent on-site inspections and improves project monitoring
Safety inspections	Drones can be used to carry out safety checks, spot possible risks and make sure safety procedures are being followed. By doing this, the chance of accidents on building sites is reduced and worker safety is improved
Materials management	Construction material tracking and management are made easier with the use of drones and remote monitoring systems. By doing this, resource efficiency is guaranteed, material waste is decreased, and sustainable buying methods are supported
Energy efficiency assessments	Thermal camera-equipped drones can detect energy inefficiencies in construction projects and structures. This enhances insulation, maximises energy use, and lowers total energy consumption
Wildlife monitoring	Drones are useful for assessing biodiversity and keeping an eye on wildlife near development projects. This data aids in reducing the negative effects of construction operations on nearby ecosystems and supports conservation initiatives
Waste management	Drones can be used to keep an eye out for inappropriate waste disposal and illegal dumping at building sites. By doing so, pollution of the environment is avoided, and proper waste management procedures are enforced
Noise monitoring	Drones and other remote monitoring devices, such as sensors, can measure the vibration and noise levels caused by construction activity. The less disturbance this data causes to other populations and ecosystems, the better
Remote project collaboration	Technologies for remote monitoring allow project participants to collaborate virtually. By eliminating the need for actual travel, lowers carbon emissions and promotes environmentally friendly corporate operations
Regulatory compliance	Drones deliver precise and thorough documentation needed for permits. Construction project delays are minimised because of streamlining the approval procedure and guaranteeing compliance with regulatory standards

Table 12.9 Application of augmented reality and virtual reality in supporting sustainable construction

Application	Explanation
Design visualization	Stakeholders may see and interact with construction designs in a virtual setting thanks to AR and VR technologies. Better decision-making, fewer design errors, and effective integration of sustainability features from the outset are all made possible by this immersive experience
Construction planning and simulation	Construction teams may plan and simulate construction activities in a virtual environment thanks to VR technologies. This assists in streamlining processes, cutting down on waste, and determining any environmental effects before construction ever starts
Energy performance analysis	Real-time energy performance data can be superimposed on physical buildings using augmented reality apps. This enables interested parties to evaluate and maximise energy efficiency, pinpoint areas in need of development, and bring about modifications to design that are sustainable
Training and skill development	Construction workers receive virtual reality training in a virtual setting. Without requiring actual on-site experiences, this immersive training improves skills, safety awareness, and knowledge of sustainable construction techniques
Remote collaboration	Stakeholders in a project can collaborate remotely thanks to AR and VR. As a result, there is less need for travel, which lowers transportation-related carbon emissions and promotes sustainable business practices
Occupant engagement	VR can produce virtual tours of buildings for prospective tenants. Using this engagement tool, they can interact with environmentally friendly elements of the building and feel more connected to its green spaces and energy-efficient systems
Waste reduction	Prefabrication procedures can be planned and optimised with the help of VR. Because parts may be accurately planned and produced off-site, faults and unnecessary material utilisation can be minimised, which helps reduce material waste
Construction safety training	Immersion safety training experiences are made possible by AR and VR technologies. By virtually navigating building sites, recognising possible hazards, and following safety procedures, workers can enhance on-site safety and lower the number of accidents
Maintenance and operations	AR applications overlay digital data on physical equipment to help facility managers with maintenance duties. This promotes sustainable facility management, assures effective maintenance procedures, and extends the life of equipment
Green building certification compliance	On-site compliance assessments for green building certifications can be aided by AR tools. This guarantees that the building complies with sustainability guidelines and facilitates certification

(continued)

Table 12.9 (continued)

Application	Explanation
Public engagement	AR has a place in public outreach related to building projects. Transparency and community involvement can be encouraged by enabling stakeholders to observe planned structures in the real world through augmented reality applications
Material selection and visualization	Stakeholders can evaluate and visualise sustainable materials inside the framework of the construction project with the use of augmented reality apps. Making educated decisions helps guarantee that eco-friendly materials are chosen

Sustainable building practices can be improved through better design visualisation, planning, training, cooperation, and overall project management by utilising AR and VR technologies. These technologies support the industry's transition to more socially and environmentally conscious building methods.

Data Analytics and Machine Learning

Construction data can be analysed using data analytics and machine learning to reveal trends, patterns, and possible areas for improvement. This can improve project outcomes, streamline operations, and inform sustainable practices. In sustainable construction, data analytics and machine learning (ML) are being used increasingly to streamline workflows, boost decision-making, and increase overall project efficiency. The use of machine learning and data analytics in sustainable construction in provided in Table 12.10.

Table 12.10 Application of data analytics and machine learning in supporting sustainable construction

Application	Explanation
Energy performance optimization	To forecast and optimise future energy performance, machine learning algorithms can examine historical data on building usage, weather trends, and energy consumption. This aids in optimising the energy efficiency of building design and operation
Smart building management	Real-time building system monitoring, such as occupancy, lighting, and HVAC, is made possible by data analytics. This data can be analysed by machine learning algorithms to find trends and abnormalities, maximising resource utilisation and enhancing building performance
Predictive maintenance	Based on past performance data, machine learning models can forecast when building systems or construction equipment are likely to break down. This minimises the impact of emergency repairs on the environment and enables proactive maintenance, which cuts down on downtime
Waste reduction and recycling	Construction waste streams can be observed and analysed with the use of data analytics. Throughout the construction process, machine learning algorithms can spot chances for recycling, waste reduction, and improved material management

(continued)

Table 12.10 (continued)

Application	Explanation
Supply chain optimization	By predicting demand for building materials, machine learning can improve supply chain logistics and lessen the environmental effect of material production and delivery
Life cycle assessment	Construction materials and processes can have their life cycles evaluated, with the use of data analytics and machine learning. This aids in assessing and reducing a building's environmental impact throughout its whole life cycle
Green building certification assistance	Machine learning algorithms have the potential to help guarantee adherence to green building certification standards, including LEED, BREEAM, GBI, and GBCN. To confirm that construction methods meet sustainability standards, data analysis is needed
Occupant behaviour modeling	Data on occupant behaviour, preferences, and comfort levels can be analysed by machine learning. By using this data, buildings can be designed and operated more efficiently, increasing tenant happiness and using less energy
Renewable energy integration	Machine learning algorithms may allow for the best possible integration of the building's energy system with renewable energy sources, such solar and wind power. This ensures that renewable energy is used effectively and lessens reliance on non-renewable resources
Carbon emission tracking	Carbon emissions related to building operations can be tracked and reported using data analytics and machine learning. This aids in establishing targets for reducing emissions and guaranteeing adherence to sustainability objectives
Safety analytics	By examining past safety data, machine learning models can anticipate and avert possible mishaps on building sites. This lessens the effect of safety accidents on the environment and enhances overall safety performance
Community and stakeholder engagement	Analysing data can be used to evaluate how construction projects affect local communities socially. Potential social risks can be identified by machine learning models, which can also assist in the development of mitigation solutions

Through the utilisation of data analytics and machine learning, sustainable construction methods can yield increased productivity, decreased ecological footprints, and better decision-making in numerous components of the building procedure. These tools support the development of infrastructure and buildings in a more data-driven and sustainable manner.

Digital Platforms for Stakeholder Engagement

Local communities are among the stakeholders with whom digital platforms are utilised to interact and communicate. This encourages openness, diversity in society, and community engagement in environmentally friendly building initiatives. Digital platforms are essential for encouraging communication, cooperation, and transparency among stakeholders involved in sustainable construction. See the applications of digital platforms for stakeholder engagement in sustainable construction in Table 12.11.

Sustainable projects have the potential to foster a more inclusive, transparent, and cooperative stakeholder interaction environment by utilising digital platforms. These platforms help to foster trust, respond to issues, and make sure that construction

Table 12.11 Application of digital platforms in supporting sustainable construction

Application	Explanation
Project information sharing	Digital platforms offer centralised project portals via which community members and other stakeholders can obtain project information. Transparency is promoted by including project periods, milestones, and updates
Virtual public consultations	Stakeholders can engage in virtual public consultations through digital platforms, conveniently participating in discussions regarding the construction project from wherever they are. Participation of the community in decision-making processes is encouraged by this inclusion
Interactive project visualization	Digital platforms use AR and VR technologies to produce interactive construction project visualisations. This enables interested parties to virtually investigate and comprehend the project's effects on the neighbourhood and environment
Forms for online feedback and surveys	Through online surveys and feedback forms, stakeholders can offer input on a range of project-related topics. Project teams can better address issues and make well-informed decisions by using digital technologies to streamline the process of gathering and analysing feedback
Social media engagement	Social media platforms are employed by construction projects to foster communication with stakeholders and the broader community. Frequent updates, photos, and project highlights posted on social media sites would establish a vibrant and easily available communication channel
Collaboration platforms	Project teams, contractors, and stakeholders may communicate and work together seamlessly thanks to digital collaboration platforms. This improves communication of sustainability goals, encourages information exchange, and strengthens coordination
Mobile apps for communication	To inform stakeholders, some building projects use specialised mobile applications. These apps can offer event notifications, real-time updates, and a forum for stakeholders to express concerns or comments
E-learning and training platforms	E-learning courses are available on digital platforms to teach community members and construction personnel about sustainable construction techniques. This encourages knowledge and comprehension of the social and environmental components of the project
Digital community engagement platforms	Stakeholders can collaborate on project-related issues, exchange ideas, and take part in conversations thanks to software specifically made for community involvement. These forums promote diversity and guarantee that a range of viewpoints are considered
Performance reporting	Performance dashboards that highlight the project's environmental effect, sustainability data, and adherence to green building standards can be seen on digital platforms. These dashboards are available to stakeholders so they may monitor results and progress

(continued)

Table 12.11 (continued)

Application	Explanation
Document sharing and version control	The sharing of project papers is made easier by digital platforms that integrate with cloud-based document management systems. This promotes accuracy in communication by guaranteeing that stakeholders have access to the most up-to-date information and updates
Data visualization tools	Data visualisation tools are used by digital platforms to provide complex information understandably. It is simpler for stakeholders to comprehend the social and environmental effects of the construction project when they can engage with visual representations of the data

projects satisfy the requirements and expectations of a wide range of stakeholders while still being in line with sustainability goals.

Blockchain for Sustainable Construction

Many people primarily associate blockchain with Bitcoin and cryptocurrencies. While that was its initial use, blockchain technology has far greater potential. Rather than referring to a single system, "blockchain" encompasses a range of distributed ledger technologies designed to record and monitor various activities and transactions. You might question the need for blockchain when traditional systems already handle data and transaction tracking. However, blockchain provides enhanced security, transparency, and decentralization, making it a valuable innovation beyond digital currencies. Blockchain encourages and promotes strategic collaboration. This collaboration ensures that all decisions are effectively communicated to everyone involved in the contract in real-time. Blockchain use is required where a business or relationship has trust issues. A blockchain can also be used where verification can be done by consensus algorithms. Users agree to create and share data. A wallet is created using a pseudonymous address and stored as a smart contract on the blockchain. Blockchain can facilitate the achievement of sustainable construction practices in many ways. While blockchain will not set sustainability targets, once the targets are set by the clients and agreed upon by stakeholders and documented, they cannot be compromised or changed without everyone's consent. One of the major roadblocks to sustainable construction is the arbitrary shifting of sustainability targets after they are initially set, especially by the design team, perhaps to accommodate conflicting client value systems. This modification occurs due to various reasons, such as cost constraints, project timeline pressures, changing market conditions, and evolving client preferences. As a result, sustainability targets may be deprioritized or modified to align with immediate project demands, thereby compromising long-term environmental and social benefits.

Smart contracts, encouraged by blockchain technology, ensure that contractors and suppliers record their payment terms promptly in the smart contract, and payments and rewards are automated once the contractor and supplier certify that their requirements or conditions have been met. A smart contract is code written on the blockchain. Additionally, blockchain helps combat greenwashing—misleading claims about sustainability—by enabling the tracking of materials from production to on-site use. This transparency ensures that only verified sustainable materials are used while also tracking and minimizing waste. By leveraging blockchain for supply chain traceability, project teams can authenticate the origin and lifecycle impact of construction materials, preventing fraudulent sustainability claims. Moreover, blockchain facilitates the measurement and monitoring of embodied carbon in materials, contributing to more accurate sustainability assessments and reporting. By integrating blockchain with IoT sensors and real-time data tracking, construction projects can measure energy consumption, emissions, and resource utilization more effectively. This data-driven approach enhances decision-making, supports regulatory compliance, and strengthens the credibility of sustainability certifications. Figures 12.1 and 12.2 show how blockchain contribute to sustainable construction targets.

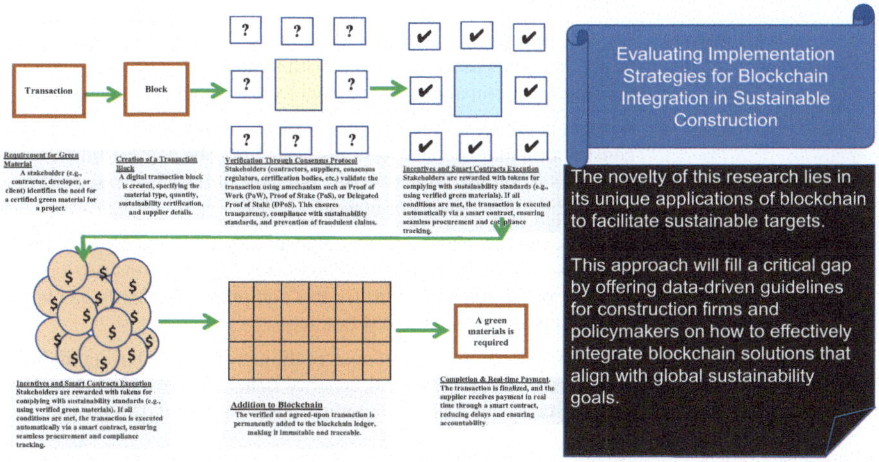

Fig. 12.1 Mechanism for a blockchain for a green material—env. target

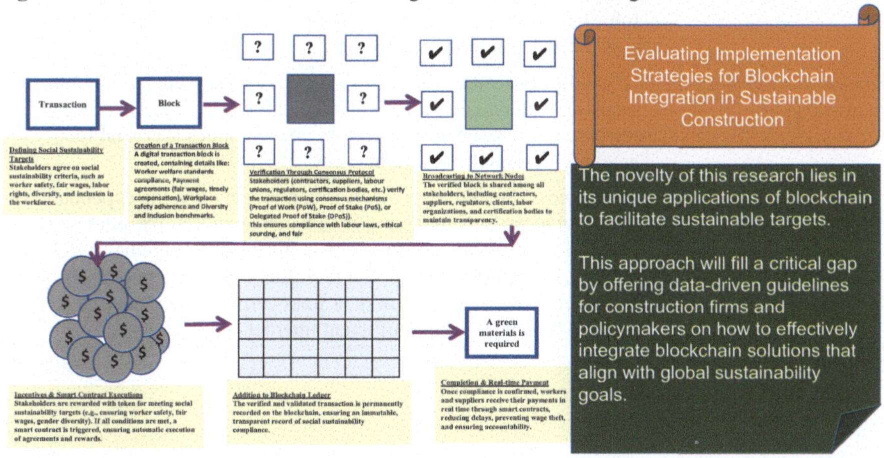

Fig. 12.2 Mechanism for a blockchain for a green material—social target

12.5 Constraints of Digital Construction Tools Applications for Sustainable Construction Management

Table 12.12 contains the main constraints affecting the applications of digital tools in sustainable construction. The use of digital construction technologies in the framework of sustainable construction management is subject to several restrictions. These limitations could change depending on organisational, human, and technological aspects. A systematic approach involving stakeholder participation, continuous training programmes, and a dedication to staying up to date with industry best practices and technological advancements is needed to address these obstacles.

Table 12.12 Constraints for digital tools for sustainable construction management

Constraint	Explanation
Cost of implementation	Digital construction tool implementation and acquisition might come at an excessive cost. For projects with tighter budgets or smaller construction companies, this upfront expense could be a deterrent
Technological barriers	It might be difficult to integrate new digital technologies with current construction management systems. There may be compatibility problems with older systems, necessitating further financial outlays for system replacements or upgrades
Compatibility issues	One potential problem with digital technologies is that they require specialised ability to operate and manage. Teaching construction industry personnel how to utilise this equipment efficiently can take time and possibly need added funding
Data security and privacy concerns	Sensitive data is involved in construction projects, and using digital tools creates privacy and data security issues. It is essential to make sure digital systems abide by industry standards and data protection laws
Limited standardization	Interoperability and collaboration in the construction business may be hampered by the lack of established standards for digital tools. Information sharing might be difficult since different technologies may employ different data formats
Resistance to change	Given the construction industry's reputation for using antiquated methods, implementing new digital technologies may encounter resistance. Using change management techniques is crucial to breaking through this cultural stasis
Reliability and connectivity	The success of building projects is contingent upon the dependability of digital tools, encompassing both hardware and software. Project operations might be interrupted by connectivity problems, system outages, or technical faults. The real-time usage of digital tools in construction projects in remote or impoverished areas may be affected by issues with insufficient internet connectivity

(continued)

Table 12.12 (continued)

Constraint	Explanation
Environmental impact of technology	The creation of electronic waste might result from the quick speed at which technology is developing. To reduce environmental effects, obsolete digital instruments must be disposed of and recycled properly
Regulatory compliance	The rules and specifications of the industry must be followed by digital building tools. It is imperative to guarantee that these instruments comply with regulatory mandates concerning environmental impact, safety, and accessibility
Scalability issues	It is possible that some digital tools are difficult to scale for several types of projects. It is crucial to put into practice solutions that can be adjusted for both small- and large-scale initiatives
Dependency on vendor support	Building projects that depend on outside digital technologies must consider the dependability and durability of their suppliers. Risks can arise from things like vendor bankruptcy or support termination

Strategic steps must be put in place to get beyond the limitations that come with using digital technologies in sustainable construction management. Construction stakeholders can effectively incorporate and utilise digital tools in sustainable construction management by taking proactive actions. This will ultimately lead to increased efficiency and sustainability outcomes. The techniques that be taken to avoid or mitigate the constraints of the applications of digital tools in sustainable construction are provided in Table 12.13.

Table 12.13 Techniques to mitigate the constraints of digital tools for sustainable construction management

Technique	Explanation
Comprehensive needs assessment	Before choosing and putting digital technologies into use, do a thorough assessment of the demands and requirements of the construction project. This guarantees that the selected tools meet the goals and limitations of the project
Pilot programs and testing	Conduct pilot projects to evaluate the viability and interoperability of digital tools in regulated settings. This enables the smaller-scale detection and resolution of potential problems before the large-scale implementation
Cost–Benefit analysis	To assess the financial impact of incorporating digital tools, conduct a detailed cost–benefit analysis. To support the initial expenditure, emphasise the long-term advantages like increased sustainability, decreased waste, and increased efficiency

(continued)

Table 12.13 (continued)

Technique	Explanation
Training and skill development	Invest in educational initiatives that will improve construction industry professionals' digital literacy and competencies. To keep the staff knowledgeable about the newest technologies and best practices, offer continuing education
Interoperability standards	Promote and follow interoperability guidelines in the construction sector. To enable the smooth integration of digital tools, and promote the creation and uptake of common data formats and communication protocols
Data security protocols	Put strong data security procedures and safeguards in place to protect sensitive data. Verify adherence to industry data protection standards and legislation. Update security protocols often to counter new threats
Change management strategies	Create and put into practice change management techniques to deal with reluctance to technological change. Early in the process, involve crucial stakeholders, emphasise the advantages of digital tools, and cultivate an innovative culture
Reliability assessments	Put digital technologies with a record of accomplishment of dependability first. Perform in-depth evaluations of the hardware and software components' dependability and performance. Provide backup plans to ensure that technical difficulties are resolved quickly
Connectivity planning	Consider and prepare for connectivity issues, particularly in isolated locations. Examine alternate options for connectivity, such as mobile networks or satellite internet. Workflows for projects should be designed to handle sporadic connectivity
Environmental considerations	When making your choice, take the effects of digital tools on the environment into account. Select equipment and technology that support sustainability, use less energy, and follow ethical production and disposal procedures
Regulatory compliance	Verify that digital tools adhere to pertinent industry norms and laws. To ensure compliance, keep up with changing regulatory requirements and update your tools appropriately
Scalability planning	Choose digital tools that may be adjusted to various project sizes and are scalable. Make sure the tools can expand to meet the needs of the project and handle scale changes without causing major problems
Diversification of vendors	To mitigate reliance on a solitary vendor, broaden your digital tool procurement options. By doing this, the likelihood of vendor-related problems like insolvency or support termination is decreased
Continuous improvement and feedback	Create systems for ongoing improvement based on user input and changing project requirements. To keep ahead of technological changes and absorb lessons gained, periodically examine and upgrade digital tools and procedures

12.6 Prioritise Construction of Technologies: International Perspective for Digital Twins

This sector reports part of research on the implementation of digital twin technology for construction projects. The respondents are from background and countries. It was an online survey, and the response size small (39 respondents altogether) but the results are meaningful and is an indicate of state of understand of the implementation of digital twin the construction sector. The ranking of was based on correlation analysis. Over of the profile of the respondents are presented in the following table and figures (Table 12.14).

The 'Other' in Table 12.15 are respondents from South Africa, India, Saudi Arabia, and the USA.

Table 12.16 provide insightful results on company's investment on the implementation of digital twin technologies.

Industry Awareness and Adoption: The relatively low percentage of companies investing in digital twins (e.g., 13% investing less than 5%) may indicate that the construction industry is still in the early stages of adopting this technology. This could be due to a lack of awareness about the benefits or uncertainty about the technology's

Table 12.14 Respondent's work experience

Year	Less than 5 years	5–10 years	11–15 years	16–20 years	More than 20 years
Frequency	4	10	8	22.2	8
Percentage	11.1	27.8	22.2	16.7	22.2

Table 12.15 Respondent's location

Year	Malaysia	Singapore	Nigeria	Pakistan	Other
Frequency	9	6	12	3	6
Percentage	25.0	16.7	33.3	8.3	16.7

Table 12.16 Amount respondent's company invested in construction digital twins

Size of investment	Percentage
Less than 5%	13
5–10%	3
10–15%	5
15–20%	3
20–25%	2
25–30%	2
35–40%	1
40–45%	2
50 and above	1

effectiveness. Cost–Benefit Analysis: Companies may be weighing the potential benefits of digital twins against the costs involved in implementing and maintaining them. Factors such as the size of the company, available budget, and perceived return on investment could influence their decision-making process. Technological Maturity: Companies may perceive digital twins as a cutting-edge technology that is still evolving. Some may prefer to wait until the technology matures further or becomes more standardized before making significant investments. Competitive Landscape: Companies operating in highly competitive markets may be more inclined to invest in digital twins as a way to gain a competitive edge. Conversely, companies in less competitive markets or with traditional business models may see less urgency in adopting such technologies. Regulatory Environment: Regulatory requirements and industry standards may also play a role in shaping companies' investment decisions. For example, companies operating in regions with stringent building codes or sustainability mandates may prioritize investments in technologies like digital twins to ensure compliance.

The results may be seen as a positive sign of innovation and forward-thinking in some cases, while in others, it may indicate a need for greater awareness-building or clearer demonstration of the technology's benefits to encourage wider adoption. In particular, the level of investment in construction digital twins can be interpreted in multiple ways depending on the context, ranging from a positive indicator of innovation and strategic foresight to a call for greater awareness-building and evidence-based advocacy to overcome barriers to adoption.

- Positive Sign of Innovation and Forward-Thinking: Some companies may view their investment in construction digital twins as a proactive step towards embracing innovation and staying ahead of industry trends. They may see it as an opportunity to gain a competitive advantage, improve operational efficiency, reduce costs, enhance project outcomes, and differentiate themselves from competitors. In this context, the investment reflects a forward-thinking approach aimed at positioning the company for long-term success in a rapidly evolving industry.
- Need for Greater Awareness-Building: In other cases, the relatively low level of investment in digital twins may signal a lack of awareness or understanding among industry stakeholders about the technology's potential benefits and applications. Companies may be unfamiliar with how digital twins can streamline project planning, design, construction, and maintenance processes, leading to improved productivity, better decision-making, and enhanced project outcomes. Therefore, there is a need for educational initiatives, training programs, industry seminars, and knowledge-sharing platforms to raise awareness about the value proposition of digital twins and demystify misconceptions or uncertainties surrounding their implementation.
- Clearer Demonstration of Technology's Benefits: Additionally, companies may be hesitant to invest in digital twins due to a perceived lack of evidence or case studies demonstrating their tangible benefits and return on investment (ROI). They may require clearer and more compelling demonstrations of how digital twins can address specific pain points, mitigate risks, optimize resource allocation,

accelerate project delivery, improve collaboration among stakeholders, and deliver measurable value throughout the project lifecycle. By showcasing successful use cases, real-world examples, and quantifiable metrics, industry leaders, technology vendors, and research institutions can instill confidence and trust in the efficacy of digital twins (Fig. 12.3 and Table 12.17).

This Table 12.18 presents the distribution of responses regarding the extent to which the adoption and application of construction technology have improved construction productivity. Respondents were asked to indicate the percentage improvement observed in productivity due to the adoption of construction technology. The results indicate a generally positive perception of the impact of construction technology adoption on productivity, with notable variations in the extent of improvement reported by different respondents. These findings underscore the importance of continued investment in technology adoption and ongoing efforts to optimize its integration and utilization within the construction sector to unlock its full potential for productivity enhancement.

Insight to data:

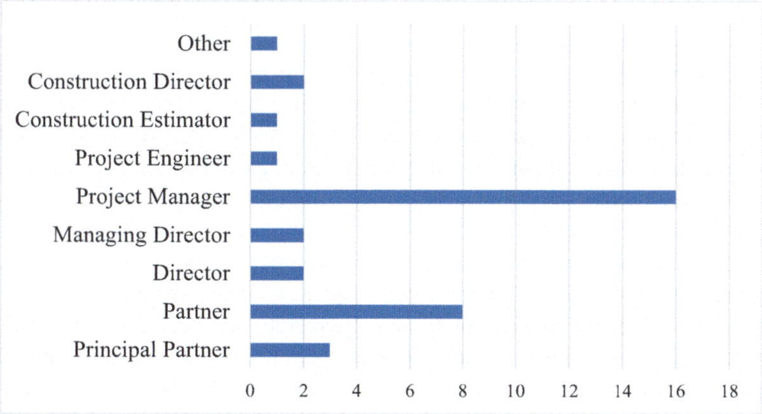

Fig. 12.3 Respondent position in their organisation

Table 12.17 Respondent's level of familiarity with the digital twin technologies

Familiarity	Frequency
Not familiar	11
Less familiar	6
Somewhat familiar	5
Familiar	12
Very familiar	4
Very extremely familiar	1

Table 12.18 Extent to the adoption and application of construction technology has improved construction productivity

Extent (%)	Frequency
Less than 5	3
5–10	3
10–15	3
15–20	4
20–25	4
25–30	6
35–40	5
40–45	1
45–50	2
More than 50	8

- Positive Impact: The majority of respondents (8) reported a significant improvement in productivity, with more than 50% attributing their gains to the adoption of construction technology. This suggests that for a substantial portion of the surveyed population, technology adoption has had a transformative effect on productivity levels.
- Varied Impact: However, there is variability in responses, with some respondents indicating minimal improvements (e.g., less than 5%) and others reporting moderate to substantial gains (e.g., 35–40%). This variation could be due to factors such as the type and maturity of technology adopted, the extent of integration into existing workflows, organizational culture, workforce training, and project-specific dynamics.
- Room for Improvement: While a significant number of respondents reported high levels of improvement, it's notable that a portion of respondents (e.g., 3, 3, and 4) reported relatively modest gains or improvements below 25%. This suggests that there may be untapped potential or barriers hindering the full realization of productivity benefits from construction technology adoption. Addressing these barriers, such as addressing skill gaps, improving technology infrastructure, and fostering a culture of innovation, could further enhance productivity gains across the industry.

12.7 Correlation Analysis of the Rating of Digital Construction

Correlation analysis was used to rank construction technologies. Correlation analysis provides a comprehensive and statistically sound method for identifying the most influential variables within a dataset, facilitating informed decision-making in research or practical applications. In particular, this technique was instead mem of mean techniques or severity index for the following reasons:

Capture Multivariate Relationships

Correlation analysis considers the relationship between each pair of variables simultaneously, capturing multivariate interactions that may not be evident when considering variables individually.

- Direction and Strength of Relationships:

Correlation coefficients provide insights into both the direction (positive or negative) and strength of relationships between variables. This information helps prioritize variables that exhibit stronger associations with each other.

- Statistical Significance:

Correlation coefficients are accompanied by p-values, indicating the statistical significance of the relationships. This ensures that the identified associations are not merely due to chance.

- Robustness to Scale:

Correlation analysis is robust to differences in the scale of measurement across variables, making it suitable for comparing variables measured on different scales or units.

- Objective Ranking:

Correlation analysis provides an objective and data-driven approach to ranking variables based on their relationships with each other, reducing the potential for bias introduced by subjective interpretation.

- Exploratory Analysis:

Correlation analysis allows researchers to explore the complex network of relationships among variables, uncovering potential patterns or clusters that may inform further analysis or hypothesis generation.

In so doing, we can consider the strength and significance of the correlations with other technologies. The details of the results are availed in the appendix. Following are the ranking of the construction technologies based on Pearson correlation coefficients:

1. Augmented Reality
2. Artificial Intelligence
3. Machine Learning
4. Collaboration and Project Management Tools

5. Geographic Information Systems
6. Global Positioning System
7. Smart Sensors
8. Big Data
9. Internet of Everything
10. Mobile and Cloud Technologies
11. Blockchain Technology
12. Customized Cloud Applications
13. Digital Stamping
14. Construction Wearables
15. Pro Crew Schedule
16. Construction Management Cloud Software
17. Document Imaging
18. Automated Workflow
19. Building Information Modeling (BIM)
20. Mobile Applications
21. 3D Printing
22. 5G and Wi-Fi
23. Virtual Reality
24. Robotics
25. Internet of Things
26. Satellite
27. Reality Capture Technologies
28. 3D Laser Scanning
29. Machine Learning
30. Drone.

12.8 Overview of the Ranking

The ranking provides valuable insights into the diverse landscape of digital technologies shaping the future of construction, highlighting areas of opportunity for innovation and improvement in construction processes and project outcomes. At the top of the list, we find Augmented Reality (AR) and Artificial Intelligence (AI) as the most correlated technologies. This suggests that AR and AI play significant roles in enhancing construction processes, perhaps through applications like visualizing project progress, optimizing resource allocation, or automating decision-making tasks. Following closely behind are Machine Learning (ML) and Collaboration/Project Management Tools, indicating the importance of data-driven decision-making and efficient project coordination in modern construction projects. Geographic Information Systems (GIS) and Global Positioning System (GPS) also rank high, underscoring the value of location-based data and precise positioning technologies in construction planning and execution. Smart Sensors and Big Data are also prominently ranked, reflecting the growing trend of leveraging sensor data and analytics to monitor construction activities, predict maintenance needs, and optimize resource utilization. The Internet of Everything (IoE) follows suit, emphasizing the interconnectedness of various devices and systems in construction environments.

Mobile and Cloud Technologies, Blockchain, and Customized Cloud Applications highlight the increasing reliance on mobile devices, cloud platforms, and blockchain technology to streamline communication, data storage, and transaction security in construction operations. Digital Stamping, Construction Wearables, and Pro Crew Schedule rank lower but still demonstrate notable correlations, suggesting their potential to enhance document management, worker safety, and project scheduling efficiency, respectively. It is important to note that while correlation indicates a relationship between variables, it doesn't necessarily imply causation. Therefore, further research and analysis are needed to understand the underlying mechanisms driving these correlations and to determine the most effective strategies for integrating these technologies into construction workflows.

12.9 Theoretical Contribution of the Findings

The theoretical contribution of the findings lies in their ability to advance our understanding of how digital technologies impact the construction industry and contribute to the theoretical frameworks guiding research and practice in this field. Here are some key theoretical contributions:

- Technology Adoption and Integration: The findings provide insights into which digital technologies are most commonly adopted and integrated into construction processes. This contributes to theories of technology adoption, diffusion, and innovation within the construction industry. Understanding which technologies are most correlated with positive outcomes can inform theoretical models of technology adoption by construction firms and project teams.
- Digital Transformation: The ranking sheds light on the extent to which digital transformation is occurring within the construction industry. By identifying the most correlated technologies, the findings contribute to theories of digital transformation, which seek to understand how organizations leverage digital technologies to improve performance, productivity, and competitiveness.
- Resource Allocation and Investment: The findings highlight where construction firms and project teams are investing their resources in terms of technology adoption and implementation. This contributes to theories of resource allocation and investment decision-making, providing insights into how firms prioritize different digital technologies based on their perceived impact on project outcomes and organizational goals.
- Process Optimization and Efficiency: By identifying the most correlated technologies, the findings contribute to theories of process optimization and efficiency within the construction industry. Understanding which technologies are

most closely associated with improved project performance can inform theoretical models of process optimization and help identify best practices for integrating digital technologies into construction workflows.

- Interdisciplinary Collaboration: Many of the highly correlated technologies, such as Building Information Modeling (BIM), Geographic Information Systems (GIS), and Collaboration/Project Management Tools, require interdisciplinary collaboration between construction professionals, engineers, architects, and technology experts. The findings contribute to theories of interdisciplinary collaboration by highlighting the importance of cross-functional teams and integrated project delivery approaches in leveraging digital technologies for construction projects.

12.10 Practical Contribution of the Findings

The practical contribution of the findings is significant for stakeholders in the construction industry, including firms, project teams, technology providers, and policymakers. Here's how the findings can contribute practically:

- Informed Decision-Making: The ranking of technologies based on their correlation with construction processes provides valuable information for decision-makers. Construction firms and project teams can use this information to make informed decisions about which technologies to adopt and integrate into their projects. By focusing on technologies that show high correlation with positive outcomes, stakeholders can allocate resources more effectively and maximize the return on investment in digital tools.
- Optimized Project Performance: Understanding which technologies are most correlated with improved project performance allows stakeholders to optimize their project workflows. By strategically integrating technologies such as Building Information Modeling (BIM), Robotics, and Automated Workflow, construction firms can enhance efficiency, accuracy, and collaboration throughout the project lifecycle. This can lead to cost savings, schedule acceleration, and higher quality deliverables.
- Enhanced Competitiveness: Adopting and leveraging the right digital technologies can enhance the competitiveness of construction firms in the marketplace. By staying abreast of technological advancements and aligning their technology strategies with the findings of correlation analysis, firms can differentiate themselves from competitors, attract clients, and win more projects. Additionally, technology-savvy firms may gain a reputation for innovation and excellence, further strengthening their position in the industry.
- Risk Mitigation: Certain technologies, such as Document Imaging and Blockchain Technology, may be correlated with improved risk management and compliance. By adopting these technologies, construction firms can streamline documentation processes, enhance transparency, and mitigate risks associated with

errors, disputes, and regulatory non-compliance. This can lead to smoother project execution, reduced disputes, and better overall project outcomes.

- Strategic Partnerships and Collaborations: The findings can facilitate strategic partnerships and collaborations between construction firms, technology providers, and other stakeholders. Firms can leverage the expertise of technology providers to implement and customize digital solutions that align with their specific needs and goals. Collaboration between industry players can accelerate innovation, drive technology adoption, and foster ecosystem-wide growth and development.
- Policy and Regulatory Implications: Policymakers and regulatory bodies can use the findings to inform policies and regulations related to technology adoption and digital transformation in the construction industry. By understanding which technologies are most correlated with positive outcomes, policymakers can develop incentives, standards, and guidelines that encourage firms to embrace digital technologies and drive industry-wide modernization and improvement.

12.11 Summary

With an emphasis on sustainability, the chapter examines the revolutionary effects of advanced technologies on the construction sector. The integration of technology innovation and sustainable principles is the comprehensive strategy known as digitalization in sustainable construction. Construction projects can increase productivity, lessen their negative effects on the environment, increase resilience, and contribute to a more resilient and sustainable built environment by strategically utilising digital tools. As the construction industry develops, it is essential to use innovations to tackle environmental issues and improve productivity. Throughout the entire life cycle, the chapter proves how these technologies improve decision-making, reduce waste, and promote energy efficiency. There is also discussion on the use of digital platforms for stakeholder participation and the contribution of technology to green building certifications. In the end, the chapter emphasises how crucial it is for modern technologies to transform construction practices, promote sustainability, and lead the sector into a more technologically and environmentally advanced future.

Bibliography

1. Chen X, Chang-Richards AY, Pelosi A, Jia Y, Shen X, Siddiqui MK, Yang N (2022) Implementation of technologies in the construction industry: a systematic review. Eng Constr Archit Manag 29(8):3181–3209
2. de Ligny EVED, Erkelens P (2008) Construction technology diffusion in developing countries: limitations of prevailing innovation systems. J Constr Dev Countries 13(2)
3. Gharbia M, Chang-Richards A, Lu Y, Zhong RY, Li H (2020) Robotic technologies for on-site building construction: a systematic review. J Build Eng 32:101584
4. Hossain MA, Nadeem A (2019) Towards digitizing the construction industry: state of the art of construction 4.0. In: Proceedings of the ISEC, vol 10, pp 1–6

5. Iacovidou E, Purnell P, Lim MK (2018) The use of smart technologies in enabling construction components reuse: a viable method or a problem creating solution? J Environ Manage 216:214–223
6. Kangari R, Miyatake Y (1997) Developing and managing innovative construction technologies in Japan. J Constr Eng Manag 123(1):72–78
7. Li CZ, Zhong RY, Xue F, Xu G, Chen K, Huang GG, Shen GQ (2017) Integrating RFID and BIM technologies for mitigating risks and improving schedule performance of prefabricated house construction. J Clean Prod 165:1048–1062
8. Li Y, Liu C (2019) Applications of multirotor drone technologies in construction management. Int J Constr Manag 19(5):401–412
9. Lu Y, Li Y, Skibniewski M, Wu Z, Wang R, Le Y (2015) Information and communication technology applications in architecture, engineering, and construction organizations: a 15-year review. J Manag Eng 31(1):A4014010
10. Qi B, Razkenari M, Costin A, Kibert C, Fu M (2021) A systematic review of emerging technologies in industrialized construction. J Build Eng 39:102265
11. Okpala I, Nnaji C, Karakhan AA (2020) Utilizing emerging technologies for construction safety risk mitigation. Pract Period Struct Des Constr 25(2):04020002
12. Sacks R, Girolami M, Brilakis I (2020) Building information modelling, artificial intelligence and construction tech. Dev Built Environ 4:100011
13. Skibniewski MJ, Chao LC (1992) Evaluation of advanced construction technology with AHP method. J Constr Eng Manage 118(3):577–593

Chapter 13
Global Perspectives on Sustainable Construction

Abstract This chapter explores the global landscape of sustainable construction, focusing on the shared goals, varied challenges, and innovative responses shaping the move toward environmentally responsible building practices. Sustainable construction is shaped not only by local conditions but also by international dynamics such as the availability of sustainable materials, technologies, and global certification systems. In many developing countries, reliance on imported sustainable components creates a need for greater cross-border collaboration. Drawing on the Breakthrough Agenda and related frameworks, the chapter outlines the urgent need for coordinated action to reduce carbon emissions and meet net-zero targets in the construction sector. It examines the socio-economic, cultural, and regulatory contexts that inform sustainable practices across different regions, acknowledging both common barriers and context-specific approaches. The chapter also presents emerging technologies including smart building systems, green infrastructure, and sustainable materials that are driving positive change in construction practices worldwide. Additionally, it calls for inclusive policies and stronger international partnerships to create a more balanced and accessible sustainable construction market. By offering a broad overview of global practices and collaborations, this chapter supports a deeper understanding of how sustainable construction can be adopted more widely and effectively in the built environment.

13.1 Learning Outcomes

The stance for sustainable construction is influenced by multiple factors, many of which are outside the purview and power of the local construction market. This is often the case because sustainable materials, sustainable components, and sustainable technologies are not locally available in the local market. Most of the countries still depend on imported sustainable materials and components to meet local demands. As a result, the impact of the global perspective on demand and supply of sustainable construction is high enough to propel investment in the market. Carbon emissions also require global collaboration. Many of the green certifications are used outside their

A. Olanrewaju, *Sustainable Construction Management: Research and Practice Companion*, Management in the Built Environment,
https://doi.org/10.1007/978-981-96-4910-5_13

countries of origin. The report on the Breakthrough Agenda has shown that high-level collaboration is needed to drive faster reductions in global greenhouse gas emissions. The construction sector is not expected to reach net zero emissions by the middle of the century, according to data from the Building Breakthrough Agenda, with emissions rising at a rate of 1% annually on average since 2015. Greater cooperation is required to establish near-zero emissions and climate-resilient buildings as the new standard by 2030, which is six years away. After studying this chapter, you will be:

- Explain the challenges that are faced universally in pursuit of sustainable construction.
- Explain some innovative solutions and technologies that are catalysing positive change in the sustainable construction market.
- Understand collaborative platforms to net zero carbon.

13.2 Introduction

Sustainable construction is more than simply good building practises; it stands for a shared need to create a future that is sustainable. This chapter examines many viewpoints from around the globe, examining the nuances of sustainable construction and exposing the international initiatives, difficulties, and innovations influencing the sustainable development of the construction industry. The pressing need to tackle environmental and social issues has thrust sustainability to the fore of international discourse. Sustainable building practices become increasingly crucial as countries struggle with the effects of resource depletion, urbanisation, and climate change. These practices help to reduce environmental damage and promote resilience. Globally, various locations and nationalities provide distinct difficulties and solutions towards the aims of sustainable construction. In this chapter, the cultural, economic, and regulatory factors that influence sustainable construction practices, methods, and principles are dissected as it moves through these many regional and national approaches. The chapter recognises that, despite the common aim of sustainability, there are universal obstacles to overcome in the quest for sustainable construction. The impact of issues on the global construction landscape is examined, including resource imbalances, balancing economic development with environmental preservation, and the necessity for widespread education on sustainable practices. The chapter also highlights innovative approaches and technological advancements that are spurring favourable changes in the sustainable construction industry. There is hope for a more sustainable built environment thanks to innovations including the use of sustainable materials, the integration of smart building technology, and developments in green infrastructure. Understanding that sustainability transcends national boundaries, the chapter investigates global alliances and partnerships that support efforts in sustainable building. The chapter concludes with an explanation of sustainable construction's worldwide future. It investigates how recent initiatives and discoveries might affect public awareness, industry norms, and legislative agendas, ultimately leading to a paradigm change in how the world creates and supports its built environment.

Regional Variances in Sustainable Practices

The field of sustainable construction is dynamic and heavily influenced by regional differences, making it a need for the entire world. The regional factors listed below impact and direct approaches to sustainable construction methods.

Cultural Dynamics

The geographical variations in sustainable construction techniques are shaped in large part by cultural dynamics. These dynamics are made up of several elements that are the customs, beliefs, and values of a particular cultural setting. The cultural variables behind the regional variations in sustainable construction are listed in Table 13.1.

Community Engagement as a Cultural Tradition: Long-standing community involvement may make people more likely to collaborate in long-term ways. More participation in neighborhood-led initiatives and environmentally friendly construction techniques may arise from this.

Table 13.1 Cultural dynamic for accounting for the regional variance

Factor	Explanation
Cultural attitudes toward nature	Deeply ecocentric regions tend to be more inclined towards sustainable building practices since they see nature as essential to human well. Regions with anthropocentric ideals, on the other hand, can give priority to human-centric development, which would affect how much attention is placed on sustainability
Historical architectural heritage	Rich historical architectural heritage regions must strike a balance between the need for preservation and innovative environmental techniques. The adoption of contemporary eco-friendly technology and materials might be influenced by cultural affinity to old structures
Traditional building practices	Strong attachments to traditional building techniques drive cultures to search for ways to incorporate new sustainability without sacrificing historical significance. Different regions have various levels of willingness to change old behaviours to conform to modern environmentally conscious requirements
Religious and spiritual beliefs	Religious communities that place a high value on environmental stewardship tend to have more of an emphasis on sustainable building practices. This can have an impact on construction methods, material selections, and architectural designs to conform to religious beliefs
Social and community values	Collectivist cultures may place a higher priority on environmental harmony and community well-being, which could result in more collaboratively sustainable projects. Individualistic societies may place more emphasis on personal accountability, which may have an impact on people's adoption of sustainable habits
Education and environmental awareness	Education-focused cultures are often more environmentally conscious. This affects how much community support and involvement there is in sustainable building methods

(continued)

Table 13.1 (continued)

Factor	Explanation
Perception of progress and development	Regions tread carefully when it comes to accepting modernism while yet supporting local customs. Sustainable technology uptake and acceptability may be influenced by how development and progress are viewed in each cultural setting
Cultural Response to Climate challenges	Cultural perceptions of sustainable building are shaped by past reactions to climate-related issues. Areas that have experienced extreme weather in the past might be more proactive when it comes to climate-responsive design and environmentally friendly construction methods
Aesthetic preferences	Deeply ingrained cultural aesthetic preferences might impact building decisions. The choice of materials, colours, and general architectural aesthetics can all be influenced by regional preferences for sustainable designs that complement culturally accepted aesthetics
Government and policy influence	Policy-driven sustainable construction approaches may find more traction in cultures that place more faith in government endeavours. The effectiveness of legal systems that support sustainability may be impacted by cultural perspectives on the function of governance
Generational shifts	New sustainable practices are accepted to varying degrees depending on how well a region's culture adjusts to generational changes. New sustainable technology may be adopted more quickly in areas where change is welcomed
Community engagement tradition	Cultural tradition of community participation: communities that have a long history of community involvement may be more inclined to work together in sustainable ways. This may result in more people being involved in community-led projects and sustainable building methods

Economic Landscapes

The regional differences in sustainable construction practices are largely a result of the economic environments. The resources, priorities, and economic climate of a given area are major determinants of how the construction sector adopts sustainability. Table 13.2 shows the main economic variables that affect how sustainable construction methods vary by area.

Gaining knowledge of these economic environments helps one better understand the incentives, obstacles, and possibilities that influence regional differences in sustainable building methods. A region's commitment to creating a sustainable future is significantly affected by its economic context, regardless of the factors driving it, be they market dynamics, resource restrictions, or economic prosperity.

Table 13.2 Economic factors account for the regional variance

Factor	Explanation
Financial capacities	Various economic prosperity levels show various abilities to invest in environmentally friendly building practices. Richer areas might have more money to spend on eco-friendly products, sustainable building initiatives, and energy-efficient technologies. Conversely, poor areas can give preference to affordable, yet ecologically responsible, options
Resource availability	Resources, such as sustainable materials and renewable energy sources, are not always readily available in every area. Whereas areas with a surplus of renewable resources might concentrate on generating clean energy, areas with a shortage of resources might prioritise effective resource management and construction use
Government policies and incentives	Government incentives and policies affect the economic environment. Eco-friendly building approaches are likely to be adopted more widely in areas with considerable financial help and supportive policies. Conversely, areas lacking such backing could meet financial obstacles while attempting to execute sustainable projects
Market demands and consumer preferences	Consumer-driven sustainability: consumer choices and market needs shape economic environments. Demand for green buildings may be stronger in areas where consumers place a high value on sustainability; this could result in a rise in the use of sustainable building techniques. Economic factors often coincide with consumer preferences and market dynamics
Cost–benefit analysis	One crucial consideration is whether sustainable building initiatives are financially possible. If the long-term advantages—such as energy savings and environmental impact—outweigh the initial expenses, regions that undertake thorough cost–benefit analysis may be more likely to invest in sustainable solutions. Decisions about whether to support or oppose sustainable measures are influenced by this economic viewpoint
Technological innovation	Technological advancements in sustainable construction are more likely to be adopted by economically developed areas that place a high priority on research and development. The rate at which technology is adopted is influenced by economic abilities, which drive investments in green technologies, smart building systems, and novel materials
Cost of energy	Energy-efficient construction techniques may be given priority in areas with higher energy expenditures to save operating costs. The adoption of energy-efficient technology such as solar panels, energy-efficient HVAC systems, and insulating materials that support sustainable construction is influenced by the economic impact of energy prices

(continued)

Table 13.2 (continued)

Factor	Explanation
Local economic priorities	The approach to sustainable construction is influenced by the economic interests of a region, including but not limited to industry diversification and sector specialisation. While specialised economies may have considerations based on their primary industries, regions that are expanding their economy may view sustainability as a potential opportunity
Financial resilience	A region's ability to withstand the upfront expenses of sustainable building is influenced by its resilience and stability economically. Economically sound regions might find it easier to adopt long-term sustainability plans, while economically vulnerable places might find it more difficult to commit to such initiatives
Global economic interactions	International sustainability certifications and standards may have an impact on regions that take part in international trade and economic exchanges. Locally sustainable construction methods may be affected by the acceptance of best practices and standards that are embraced in the global marketplace because of regional economic interdependence
The availability of green financing	Sustainability certifications and standards from around the world may have an impact on regions that trade and interact economically on a global scale. The adoption of standards and best practices by global marketplaces can have an impact on local sustainable construction practices due to the economic interdependence of regions
Economic recovery strategies	Recovery plans implemented after crises have an impact on economic environments. As part of larger strategies for economic recovery, regions appearing from recessions may give priority to sustainable building, using green initiatives as sources of employment and a boost to economic resilience

Policy Dynamics

Policy dynamics have a complex role in influencing regional differences in sustainable construction methods. The legal framework, incentives, and requirements that decide how to approach environmentally friendly construction in each location are shaped by government policies, which act as the driving force behind these initiatives. The policy variables listed in Table 13.3 explain why sustainable construction practices vary by area.

Table 13.3 Policy factors accounting for the regional variance

Factor	Explanation
Regulatory framework	A key role is played by the regulatory framework that governments set forth. Eco-friendly methods are likely to be more widely adopted in areas with strict rules about energy efficiency, environmental standards, and sustainable building norms. In contrast, other methods of sustainability may be taken in areas with more lax rules, depending on market forces and voluntary activities
Incentive programs	Programmes for government incentives have a significant impact on regional differences. Eco-friendly activities often increase in areas that provide financial incentives for sustainable construction projects, such as tax credits, grants, or subsidies. Incentives that aren't monetary, like awards or certificates, influence the environment by motivating developers and builders to adopt sustainable techniques
Mandates and standards	One crucial policy dynamic is the implementation of green building certifications and standards. Places with strict green building standards—like LEED, GBCN, GBI, and BREEAM—tend to have more sustainable construction. Development and construction companies are guided in incorporating eco-friendly aspects into their projects by mandates that demand adherence to these standards
Energy efficiency requirements	Policies that prioritise renewable energy sources and energy efficiency are vital. Construction projects in regions with stringent energy efficiency regulations and a strong emphasis on renewable energy sources are probably going to use more solar panels, wind turbines, and other clean energy solutions
Waste management policies	Variations by region are partly caused by waste management policy. There is a greater likelihood of sustainable construction projects using recycled materials and effective waste management systems in areas with strong legislation supporting a circular economy and recycling methods
Urban planning guidelines	Sustainable building practices are affected by urban planning policies in a long-term way. Construction projects are more likely to use sustainable methods in areas where pedestrian-friendly designs, energy-efficient infrastructure, and green spaces are prioritised in urban planning rules
Public procurement policies	Government rules governing public procurement may have an impact on sustainable building techniques. Governments that prioritise ecologically responsible building in their projects set an example for other private businesses, and these entities are encouraged to follow suit
Research and development funding	Innovation is stimulated by policies that fund R&D in sustainable construction technology. Modern sustainable techniques are often adopted in regions that distribute funding and resources to research and development projects in environmentally friendly construction
Community engagement requirements	Sustainable practices are supported by laws that require or promote community involvement in construction decision-making. Communities that appreciate and include community participation in construction planning may prioritise projects that are in line with community ideals, such as sustainability

(continued)

Table 13.3 (continued)

Factor	Explanation
Adaptation to climate change	Policies aimed at promoting climate resilience may be implemented in climate change-vulnerable areas. In these areas, sustainable building methods could involve strengthening resistance to severe weather, sea level rise, or other climate-related issues
Long-term sustainability goals	Policies that incorporate long-term sustainability aims are likely to encourage an ongoing dedication to environmentally friendly building practices. A roadmap for sustainable development is provided by policies, which help stakeholders coordinate their activities with the main goals of sustainability
Education and training initiatives	Developing a Skilled Workforce: Regional differences are influenced by policies that support programmes for education and training in sustainable construction. Construction projects that incorporate sustainability are more likely to be completed in regions that have taken steps to develop a workforce that is trained in eco-friendly techniques

13.3 Application of Regulation to Influencing the Adoption of Sustainable Construction

The regional variations seen across various geographical areas can be attributed mostly to the strict rules governing sustainable construction techniques. The environmental awareness of the construction sector in each region is greatly influenced by government laws, conventions, and building rules. The regulations that support the implementation, uptake, and effectiveness of sustainable construction methods are listed in Table 13.4.

Table 13.4 Regulation easing the sustainable construction practice

Regulation	Explanation
Environmental commitment	Strongly regulated areas often demonstrate greater environmental commitment. Governments that place a high priority on standards demonstrate their commitment to sustainability and pave the way for a building sector that uses environmentally friendly methods. Sustainable materials and technology are often adopted more widely in these regions
Building code requirements	Tight building rules have a significant impact on how sustainable construction is practised. The incorporation of sustainable techniques into construction projects is eased by regions that have strict building rules that prioritise energy efficiency, resource conservation, and environmental impact reduction
Green building certifications	The acceptance of green building certifications often reflects the degree of stringency in regulations. Strong regulatory environments are typically associated with a greater number of projects pursuing certifications such as LEED, GBCN, GBI, BREEAM, etc., which demonstrate a dedication to meeting strict environmental requirements

(continued)

Table 13.4 (continued)

Regulation	Explanation
Energy efficiency standards	Tight rules for energy efficiency have a significant role in lowering construction projects' carbon footprint. Sustainable construction techniques that give priority to energy-efficient technologies and renewable energy sources are more common in regions with strict regulations for energy performance and integration of renewable energy
Waste management requirements	The implementation of circular economy principles is influenced by the strictness of waste management legislation. The construction sector is encouraged to actively contribute to minimising waste and promoting the reuse of materials in regions where there are stringent regulations about waste reduction, recycling, and sustainable disposal methods
Water conservation measures	Tight water conservation regulations support environmentally friendly building techniques. A construction sector that actively contributes to the preservation of water resources through creative and sustainable ways is seen in regions that have tight regulations for water-efficient technologies, rainwater collection, and water recycling
Resilience and adaptation standards	Regulations about climate change resilience are also stringent. Stricter regulations may apply in areas vulnerable to climate-related problems, such as increasing sea levels or harsh weather. These regulations are intended to promote sustainable building methods that improve resilience and climate change adaptation
Urban planning guidelines	Tight rules found in urban planning standards support the development of sustainable urban areas. Strictly regulated areas promote green areas, pedestrian-friendly layouts, and energy-efficient infrastructure in urban planning, encouraging environmentally beneficial building methods that support the larger goal of developing eco-friendly cities
Incentives for innovation	Strict laws often encourage innovation in the building sector. Environmentally conscious areas may offer incentives for technology development, promoting the use of innovative approaches and environmentally friendly materials that abide by tight guidelines
Public–private collaboration	Tight rules can foster an environment where the public and private sectors work together. Sustainable construction approaches are being developed and implemented with more vigour in areas where legislation encourages collaborations between public and commercial entities as well as research institutions
Community engagement requirements	Tight rules could highlight community involvement in the decision-making process for building. These kinds of regulations guarantee that environmentally friendly building methods complement local values and preferences, promoting inclusivity in making decisions
Long-term environmental goals	The long-term environmental aims of a place are often correlated with the strictness of its legislation. Areas with strict rules have a higher probability of steering towards a sustainable future by implementing policies that ease the attainment of broad sustainability goals

When it comes to sustainable building, the strictness of laws serves as a compass, pointing areas in the direction of environmental responsibility. The building sector adjusts and transforms their practices in response to governments' ongoing efforts to hone and enforce regulations. The strictness of laws becomes a dynamic force that defines the state of sustainable construction today and moves areas closer to a time when environmentally friendly practices are a fundamental part of the built environment.

Environmental Factor

Regional differences in environmentally responsible building methods are deeply ingrained in the landscape of places. The adoption and use of sustainable construction approaches are significantly influenced by the distinct environmental features, opportunities, and problems of a given place. This investigation explores the major environmental variables that contribute to regional differences in sustainable building methods. The environmental component for regional variance is listed in Table 13.5.

Table 13.5 Environmental factor for regional variance

Factor	Explanation
Climate and topography	Sustainable construction methods are strongly affected by the topography and climate of a given area. Extreme climate regions may place a higher priority on materials that can withstand temperature changes, insulation, and energy-efficient building designs. Resilient building techniques may be prioritised in coastal areas because of their susceptibility to sea level rise
Biodiversity and ecosystem sensitivity	Biodiversity-rich regions often incorporate sustainable techniques to reduce their negative effects on the environment. Green roofs, permeable surfaces, and wildlife-friendly landscaping are a few examples of the preservation strategies that construction projects in environmentally sensitive locations may use to protect the native flora and fauna
Water availability and scarcity	Sustainable construction techniques are influenced by the availability of water resources or, on the other hand, their shortage. Water-scarce locations may prioritise technology like rainwater harvesting and water recycling to satisfy conservation goals, while water-rich regions may concentrate on effective water management to protect resources
Natural disaster vulnerability	Sustainable construction techniques are specifically designed to increase resilience in areas that are vulnerable to natural disasters like hurricanes, floods, or earthquakes. To reduce environmental risks, this may entail utilising green infrastructure, elevated building designs, and disaster-resistant materials
Air quality concerns	Sustainable building methods that lessen air pollution may be given priority in areas where air quality problems are present. This could entail installing air purification systems, using low-emission building materials, and creating green building designs that improve ventilation

(continued)

Table 13.5 (continued)

Factor	Explanation
Soil composition and stability	Building techniques are affected by the stability and makeup of the soil in each area. Utilising permeable surfaces, sustainable landscaping techniques, and erosion management strategies can all be used in areas with sensitive soil to reduce soil disturbance and preserve biological balance
Renewable energy potential	Areas with a lot of potential for renewable energy might receive help from sustainable building techniques that emphasise utilising clean energy sources. Integrating solar panels, wind turbines, and other renewable energy technology becomes essential to reducing dependency on conventional energy sources
Urbanization and land use patterns	Urbanised areas often meet difficulties in supporting sustainable construction practices amidst swift development. It becomes imperative to strike a balance between the demand for urban growth and sustainable land use patterns, which will have an impact on decisions about compact development, mixed-use areas, and green urban design
Cultural heritage and conservation	Culturally significant areas might give priority to eco-friendly building techniques that complement conservation initiatives. Reusing ancient structures in new ways, restoring them sustainably, and creating designs that are sensitive to cultural heritage are some ways to strike a balance between modern development and cultural preservation
Resource availability and scarcity	In culturally significant places, green building practices that support conservation efforts may take precedence. Finding a balance between contemporary development and cultural preservation can be achieved, for example, by repurposing historic buildings in novel ways, rehabilitating them sustainably, and developing designs that consider cultural heritage
Noise pollution considerations	High noise pollution areas may use eco-friendly building techniques that improve sound absorption. This could entail reducing the negative effects of noise on occupants using green barriers, soundproofing materials, and well-considered architectural designs
Crisis preparedness and adaptation	Sustainable practices may be proactively integrated in regions that expect future environmental concerns, such as the implications of climate change. In certain areas, adaptive design may be the main emphasis of construction projects, making use of materials and technologies that are resistant to changing climatic circumstances

The complex environmental fabric of each geographic location is the source of regional variations in sustainable construction approaches. The path towards sustainable construction transforms into a dynamic reaction to the opportunities and difficulties given by the local environment as regions negotiate their distinct environmental circumstances. Stakeholders can cooperatively contribute to creating a resilient, environmentally conscious constructed environment for future generations by comprehending and adjusting to the environmental subtleties of a given place.

Technological Adoption: From Tradition to Innovation

Rapid technology breakthroughs are changing the face of sustainable construction, with different technologies being used globally to meet specific opportunities and problems. The geographical variations in sustainable construction methods are mostly attributable to these technological adoptions. Table 13.6 contains the major technological advancements influencing sustainability in construction in various geographical areas.

Table 13.6 Technological factors Influencing

Factor	Explanation
Building information modeling	Building information modelling is often used in areas leading the way in sustainable construction for team-based planning and design. A construction project's various aspects can be visualised and analysed by stakeholders with the use of Building Information Modeling which makes it easier to create detailed 3D models. This technology optimises resource use, reduces errors, and improves communication
Prefabrication and modular construction	Modern technologically advanced areas use modular and prefabricated building techniques. Precision-built off-site prefabricated components save construction waste, speed up project completion, and improve resource efficiency. This strategy is especially helpful in areas where quick and sustainable urban development is the goal
Smart building technologies	Smart building technologies are integrated into sustainable construction techniques to improve energy efficiency and the comfort of occupants. The heating, ventilation, and air conditioning (HVAC) systems, lighting, and general building performance are optimised in regions implementing these technologies using sensors, automation, and data analytics. For operational effectiveness and energy conservation, this is essential
Renewable energy integration	Areas endowed with a wealth of sustainable energy resources, such as wind and solar power, implement energy-efficient technology. To sustainably meet local energy demands, renewable energy sources such as wind turbines and solar panels are strategically installed. The dependability of renewable energy sources is further increased by sophisticated energy storage systems
Advanced water management systems	Modern water management systems are used in technologically advanced areas to ensure sustainable water use. To reduce water waste and guarantee effective water utilisation, smart irrigation, sensor-based leak detection, and water recycling technologies are implemented in line with local water conservation aims
Green roof technologies	The use of green roof technologies is common in metropolitan areas. These green roofs enhance stormwater management, reduce the effects of urban heat islands, and help regulate temperature. Urban planning techniques in technologically sophisticated locations use green infrastructure solutions

(continued)

Table 13.6 (continued)

Factor	Explanation
Digital twins and simulation tools	Regions with superior technology use simulation tools and digital twins to evaluate how construction projects may affect the environment. These solutions minimise unforeseen problems and maximise sustainability outcomes by enabling virtual testing of design concepts, energy performance, and environmental factors before physical implementation
Blockchain for supply chain transparency	The implementation of blockchain technology enhances the traceability and transparency of construction supply networks. Blockchain-adopting areas ensure the sourcing of sustainable resources, reduce the likelihood of unethical behaviour, and foster accountability throughout the supply chain, all of which help to fulfil the growing need for environmentally conscious building
Immersive design and training	Programmes for training in sustainable construction as well as immersive design visualisation make use of augmented reality and virtual reality technology. Construction teams can receive better decision-making, collaboration, and training thanks to these technologies, which enable stakeholders to experience projects virtually
Efficient site surveillance	Drones with cameras and sensors are used for surveillance and site monitoring. Drone technology is being used by regions to track safety compliance, evaluate the status of construction, and gather real-time data on project developments. This guarantees adherence to sustainability targets and improves the efficiency of project management
Innovative material application	The use of 3D printing in construction is being investigated in technologically advanced regions. With the help of this creative method, convoluted constructions made of sustainable materials can be constructed on location. Construction schedules can be accelerated, material waste can be reduced, and innovative design options can be explored with 3D printing
Dynamic energy management	Building systems incorporate AI technologies to optimise energy consumption dynamically. Sensor data is analysed by intelligent algorithms to forecast and optimise patterns of energy consumption. This minimises the impact on the environment and operating expenses by ensuring that buildings run at maximum energy efficiency

Technological improvements are convolutedly intertwined with regional variations in sustainable construction practices. As various regions adopt varied technologies to tackle their problems, the constructed surroundings transform into a fluid platform where sustainability and innovation meet. The future of sustainable construction is being shaped by the continuous advancement of technology, which offers a range of options that suit the distinct qualities and goals of every location.

Community Engagement: The Social Dimension

Because every location has distinct cultural, social, and economic circumstances, community engagement is an essential part of sustainable construction techniques, however how it is implemented differs. The factors that influence community consideration regarding geographical differences in sustainable construction methods are listed in Table 13.7:

Table 13.7 Factors affecting community engagement

Factor	Explanation
Cultural sensitivity and communication	Cultural quirks have a considerable influence on communication preferences and techniques. Understanding regional languages and cultures is necessary for community engagement. While digital platforms could be preferable in some areas, conventional communication routes might be more beneficial in others Strong cultural ties may place a premium on in-person interactions and communal get-togethers. Conversely, areas with advanced technology may depend on online channels for interaction
Power dynamics and decision-making	Communities' power dynamics can have an impact on how much participation they have in decision-making. In certain areas, certain community members' ability to actively take part may be restricted by hierarchical arrangements Consensus-building techniques may be necessary in cultures that prioritise group decision-making, but more centralised decision-making procedures may be helpful in others
Economic disparities and participation	The capacity of community members to engage in active participation might be affected by economic differences. In areas where economic disparities are significant, some groups might not have as many opportunities for engagement While poor regions may struggle to ensure meaningful participation due to resource restrictions, affluent regions may meet issues in ensuring inclusivity
Traditional knowledge and innovation	One area of disagreement may be how to strike a balance between conventional wisdom and creative alternatives. While some communities may welcome innovation, others may oppose modern construction techniques that leave from customary customs. Strong ties to tradition may cause an area to place a higher priority on supporting traditional knowledge, which may have an impact on the adoption of novel but untested sustainable building techniques
Environmental awareness and education	Diverse degrees of environmental consciousness and education influence how well the community understands sustainable practices. It might be difficult to communicate the advantages of sustainable construction in areas with low literacy rates or restricted educational opportunities. Communities that value education and the environment might demand better standards for sustainable building projects, while others might need focused awareness campaigns to foster understanding

(continued)

Table 13.7 (continued)

Factor	Explanation
Social cohesion and collaboration	Various levels of environmental awareness and education affect the community's comprehension of sustainable practices. Promoting the benefits of sustainable construction may be challenging in places with low literacy rates or little access to formal education. Better standards for sustainable building projects may be demanded by communities that place a high value on education and the environment, while other communities may require targeted awareness campaigns to promote understanding
Historical trauma and trust	Conflicts or historical traumas can affect the community's trust in outside endeavours. Negative past experiences with development projects can make people sceptical of and reluctant to take part in sustainable building initiatives Those having a record of accomplishment of well-run community initiatives might be more receptive to cooperation, but those with a history of bad experiences might need more work to earn back people's trust
Regulatory environment and enforcement	Regulations vary by region, which affects how community engagement policies are enforced. While some areas might not have strong legislative frameworks requiring community involvement, others might. While low enforcement may lead to tokenistic involvement, strict regulatory systems may guarantee robust community engagement
Technology accessibility	Inequalities in access to technology can make it more difficult to engage the community. A sizable section of the community may be excluded from online interaction in areas with poor internet access or low levels of digital literacy. Technologically evolved areas might use digital platforms to involve people, while less developed areas might need to use other strategies like printed materials and community gatherings
Customary land tenure and ownership	Different ownership and land tenure regimes have an impact on how communities interact with construction projects. Customary land practices could call for procedures during negotiations While some regions with more standardised land ownership arrangements may adhere to conventional engagement techniques, others with strong customary land practices may require customised approaches to obtain community agreement and involvement

It takes a sophisticated grasp of the unique dynamics and circumstances of each place to address these problems. Incorporating these regional differences into sustainable construction processes that successfully involve communities also fosters teamwork and ensures that projects are in line with the needs and values of the people they affect. Fostering a sustainable way of life is the goal of sustainable construction, not only building structures. Sustainable methods are often backed by the grassroots in areas where environmental issues are deeply ingrained in the culture. It could be necessary for others to actively foster community engagement, stressing the value of including local people in the development of sustainable construction projects. The relationship between cultural sensitivity and community involvement is a dynamic factor that influences project acceptance and success in the field of sustainable construction. A thorough awareness of regional practices, beliefs, and social systems becomes crucial as areas attempt to strike a balance between the demands of modernity and rich cultural tapestries. This investigation explores the complex relationship between cultural sensitivity and community involvement, showing how a sophisticated strategy might reconcile customs with the rapidly changing sustainable construction scenario.

Global Collaboration

Regional variations present several problems for global collaboration in sustainable construction. Navigating the complicated socio-cultural, economic, and regulatory contexts of various areas is essential for successful collaboration. The main concerns with international cooperation about regional differences in sustainable construction methods are listed in Table 13.8.

Handling these problems calls for a careful, flexible strategy that respects and acknowledges the distinctive qualities of every area. The creation of adaptable frameworks that can consider local quirks, capacity-building programmes, and inclusive decision-making procedures should be given top priority in international cooperation in sustainable construction. By addressing these issues, we can make sure that

Table 13.8 Global factors accounting for the regional variances in sustainable construction practices

Factor	
Cultural and linguistic diversity	Language and cultural differences obstruct effective communication and mutual understanding amongst international partners. Misconceptions and misinterpretations can occur, which could make it challenging to match sustainable construction practices and aims
Regulatory disparities	Complementary regulations and requirements in many districts make it difficult to coordinate international sustainable building programmes Disparities in the standards for compliance might lead to disputes and impede the adoption of globally recognised sustainable practices
Technological disparities	The efficacy of cooperative efforts is affected by the digital divide, which is caused by variations in the rates of technological adoption. Technologically advanced regions may surpass others, resulting in unequal participation and impeding the dissemination of best practices
Economic inequalities	Variations in economic growth influence a region's ability to afford sustainable building methods. Certain places may find it difficult to embrace and apply innovative sustainable technology and processes due to a lack of resources

(continued)

Table 13.8 (continued)

Factor	
Educational and skill gaps	Professionals' ability to use sustainable construction procedures is affected by regional differences in education attainment and skill sets. Effective collaboration may be hampered by knowledge gaps, needing further training and capacity-building programmes in some areas
Climate and environmental variability	The choice of sustainable construction technologies is influenced by the distinct climate and environmental difficulties met by various places Effective solutions in one area might not be directly applicable in another, needing changes tailored to that area
Historical context and trauma	Regional viewpoints and approaches to sustainable construction are influenced by historical experiences and tragedies. Areas where bad experiences have occurred in the past might be cautious about collaborating, which could hinder efforts to set up confidence
Social and community dynamics	Regional viewpoints and methods for sustainable building are influenced by past traumas and experiences. Building trust may be hampered in areas where there has been a history of bad experiences with collaboration
Political instability and governance	Historical traumas and experiences shape local perspectives and approaches to sustainable building. In situations where there has been a history of negative collaborative experiences, trust-building may be more difficult
Resource availability	The viability of sustainable building projects is affected by variations in the availability of resources, such as labour and raw materials. It could be challenging in some areas to get sustainable materials or to put resource-intensive sustainable techniques into practice
Public awareness and perception	Different public perceptions and levels of awareness about sustainability have an impact on community participation tactics. For collaboration to be successful, areas with low awareness could need more comprehensive education and awareness initiatives
Time zone and geographical challenges	Real-time collaboration is hampered logistically by different time zones and geographic distances. Delays in coordination and communication could affect how effectively teamwork is done

cooperative efforts successfully contribute to a more resilient and sustainable built environment globally. International cooperation in sustainable building is a dynamic project that depends on the sharing of best practices and knowledge among various geographical areas. As regions of the world confront distinct obstacles and prospects in the realm of sustainable development, the exchange of knowledge turns into a crucial component of advancement. This investigation explores the significance of information sharing and best practices in international cooperation for sustainable building, describing the advantages, difficulties, and tactics to guarantee a cordial flow of ability.

13.4 International Partnerships

The implementation of sustainable construction methods is enhanced by interregional collaborations, which are enabled by global organisations and initiatives. Collaborative efforts, knowledge-sharing forums, and cooperative research projects set up a network where different regions can receive help from each other's experiences, hastening the shift of the world towards sustainable building methods. The importance of global collaborations in the field of sustainable construction cannot be emphasised. Forming global alliances becomes essential as areas pursue sustainable development, which presents a variety of opportunities and problems. This investigation explores the role that international partnerships play in global cooperation for sustainable building, emphasising the advantages, difficulties, and methods for utilising the combined strength of various stakeholders on a global scale.

1. Benefits of International Partnerships

International collaborations bring together a variety of regional specialties. Working together with stakeholders that have different viewpoints enables a global best practice approach to sustainable construction to be taken holistically. Shared resources, such as money, technology, and intellectual property, allow for more extensive and significant projects. Global collaborations enable the sharing of resources, guaranteeing that areas with different capabilities can take part and gain fairly. Innovation is accelerated by the convergence of ideas and experiences. The development and uptake of innovative techniques and technology in sustainable construction are accelerated via international collaborations. Through the establishment of international collaborations, regions can exert a global effect on the discourse and policies around sustainable construction. Working together strengthens the group's voice and promotes sustainable behaviours on a larger scale.

2. Challenges in International Partnerships

International cooperation needs careful consideration of cultural peculiarities. To foster trust and guarantee successful communication, it is essential to understand and appreciate distinct cultural backgrounds. Resolving disparities in power is crucial. Inequalities in political sway, scientific prowess, and economic power might affect how fairly gains are shared in international partnerships. It can be difficult to negotiate legal and regulatory discrepancies. Careful planning and a dedication to bringing international norms into line are necessary for harmonising disparate regulatory systems.

3. Strategies for Successful International Partnerships

- Clear Communication Channels: It's essential to have clear channels of communication. All partners are kept informed, involved, and on the same page in their pursuit of shared aims when there is regular and open communication.

- Working Together to Make Decisions: Encouraging joint decision-making increases each partner's sense of ownership. Every stakeholder can take part in the creation and implementation of sustainable construction projects through inclusive decision-making procedures.
- Initiatives for Building Capacity: By putting capacity-building programmes into action, partners can be sure that everyone has the knowledge and abilities needed to actively take part in joint endeavours. Training initiatives level the playing field and increase the overall effect of global collaborations.
- Mutual Learning Platforms: Creating mutual learning platforms promotes the sharing of best practices and information. Stakeholders have the chance to receive help from one another's experiences through workshops, conferences, and cooperative projects.
- Adaptive Project Management: Using these methods considers how international alliances change over time. Initiatives that are adaptable and receptive to criticism are better able to deal with evolving problems and shifting conditions.
- Dedicated to inclusion: Making sure that inclusion serves as a compass. Global alliances must be founded on the values of fairness, guaranteeing that all parties involved, irrespective of their financial or technological capabilities, play a significant part in cooperative projects.
- Shared Vision and Goals: The success of global collaborations depends on the definition of a shared vision and goals. Setting shared goals guarantees that all partners are working towards the same vision and offers a unifying framework for joint activities.
- Sustainable building is not a homogenous field; rather, it develops in the context of many regional subtleties that are influenced by distinct cultural, economic, and environmental factors. The investigation of the many landscapes that affect sustainable construction methods in various locations is summarised in this synopsis. Each part adds to the complex fabric of sustainable development, ranging from cultural dynamics to economic landscapes, policy dynamics, and technological adoptions.
- Technological Adoptions aim at how innovation and digitalization play a part in technology adoption, stressing the need to strike a balance between innovation and tradition. The investigation ends with insights into the difficulties and possible fixes for managing these regional differences, promoting an international awareness of sustainable building techniques.

13.5 International Collaboration and Standards

Technological Adoptions aim at how innovation and digitalization play a part in technology adoption, stressing the need to strike a balance between innovation and tradition. The advantages of international cooperation and standards in the application of sustainable construction are listed in Table 13.9.

Table 13.9 Benefits of international collaboration and standards in the implementation of sustainable construction

Benefit	Application
Shared knowledge and expertise	The exchange of knowledge and chances for shared learning are produced by international partnership. The varied knowledge, experiences, and innovations that their international counterparts bring to the table can be helpful to stakeholders from other geographical areas
Resource pooling	Pooling resources—financial investments, research abilities, and technological knowledge—are made possible through collaboration. By working together, we can increase economic efficiency and provide stakeholders with greater effect than we could with separate efforts
Uniformity and consistency	The creation of global norms encourages consistency and uniformity in environmentally friendly building techniques. This clarity makes sure that all parties involved follow the same set of rules, which promotes easy communication and comprehension
Market access and trade facilitation	Complying with international standards ease construction products and services access to markets. Businesses that adhere to sustainability standards have an easier time breaking into new markets and taking part in international trade
Incentivizing sustainable practices	The development of incentives and recognition programmes for sustainable construction is often the result of cooperative efforts. This encourages parties involved to embrace eco-friendly activities by providing recognition, prizes, or other material advantages
Global influence and advocacy	Working together internationally gives stakeholders the ability to promote sustainable policies globally. Through concerted efforts, partners may affect global agendas and promote the uptake of eco-friendly practices
Harmonization of practices	International processes are streamlined by standardisation, which harmonises sustainable construction techniques. All parties involved—from contractors and architects to legislators and regulators—gain from this streamlining
Technology transfer	The spread of sustainable construction innovations is eased by collaborations. The adoption of best practices and the rate of innovation can both be accelerated by the global sharing of creative ideas created in one area
Global recognition and branding	Respecting international norms earns recognition on a worldwide scale. Projects and organisations that adhere to accepted standards build a solid reputation and favourable brand image globally
Problem management	Global recognition is earned by adhering to international norms. Projects and enterprises that follow recognised guidelines set up a formidable reputation and positive brand image throughout the world
Capacity building	Building ability and transferring knowledge are supported by collaboration. A more equitable and inclusive global development can be promoted by allowing less experienced regions to gain from the experience of more developed countries in the field of sustainable construction

(continued)

Table 13.9 (continued)

Benefit	Application
Environmental impact	Collaboration eases knowledge transmission and capacity building. Encouraging less experienced regions to receive help from developed countries' sustainable construction experience can contribute to a more equal and inclusive global development
Resilience and adaptability	By enabling the construction industry to adjust to shifting social, economic, and environmental situations, collaborative efforts improve resilience. Stakeholders can predict problems and respond more skillfully when they have shared experiences

The positive global impact of sustainable practices is amplified by international collaboration and standards in sustainable construction. By encouraging a global commitment to ecologically responsible building practices, these initiatives make sure that all relevant parties follow the same standards for a more sustainable built environment.

Constraints of international standards and practices for sustainable construction

International sustainable construction standards and practices have many advantages, but they also have several limitations and difficulties that may reduce their efficacy. Table 13.10 shows the major obstacles related to global norms and guidelines for sustainable building.

Table 13.10 Challenges relating to international standards and recommendations for sustainable building

Constraint	Explanation
Cultural variability	The consistent application of international standards may be hampered by regional and cultural variations in building methods, customs, and values In areas with distinct cultural viewpoints, standards created for one cultural context could not be universally applicable, which could cause resistance or misalignment
Economic disparities	The widespread adoption of sustainable construction practices may be impeded by economic inequities across national boundaries. Financial limitations may prevent developing countries from adopting pricey green technologies. Inequitable participation can result from unequal economic abilities, which restricts the universal implementation of international norms
Regulatory divergence	Harmonising international standards with local legal frameworks is made more difficult by differences in national and regional rules Regulations that conflict with one another might make it difficult for international standards to be integrated smoothly, needing compromise and adaptation to fit different regulatory environments

(continued)

Table 13.10 (continued)

Constraint	Explanation
Technological readiness	Infrastructure and technological readiness differ by area. Certain places may lack the knowledge or resources needed to put specific sustainable building techniques into effect. The adoption of sophisticated sustainable technologies might be uneven, and the worldwide standardisation of these technologies can be hampered by differences in technological maturity
Resource availability	Resources, such as renewable energy sources and sustainable materials, are not equally available in every area. Restricted availability of some resources could make it more difficult to follow certain guidelines, which would affect how practical and affordable sustainable construction methods are
Educational gaps	Diverse levels of acceptance and comprehension of international standards are influenced by variations in educational attainment and awareness of sustainable practices. Lack of information and education may prevent the adoption of sustainable construction practices in areas with lesser awareness
Resistance to change	New and sustainable methods can be adopted more slowly in the construction business if there is resistance to change, particularly among established practices. Even when backed by international standards, the adoption of sustainable construction practices may be hampered by antiquated beliefs and a lack of openness to innovative ideas
Local climate considerations	Local temperature fluctuations and environmental circumstances may not always be taken into consideration by sustainable construction requirements Impact: following standards created for varying climates could lead to less-than-ideal performance, which would lessen the usefulness of sustainable practices in particular areas
Cost implications	Stakeholders' desire to adhere to international standards may be affected by the greater upfront expenses of adopting sustainable construction techniques. Economic factors may cause short-term financial rewards to take precedence over long-term sustainability aims
Enforcement mechanisms	Certain regions or entities may choose not to follow international norms if there are insufficient procedures in place to enforce them. The impact and influence of international standards may be lessened in the absence of strong enforcement, which could limit their ability to bring about meaningful change

Given the various worldwide contexts in which sustainable construction approaches are applied, addressing these limitations calls for a flexible and nuanced strategy. Overcoming these obstacles and promoting a more widely recognised framework for sustainable construction require cooperation, education, and customised solutions.

13.6 The Breakthrough Agenda

The goal of the Breakthrough Agenda is to increase global cooperation so that by 2030, sustainable solutions and clean technologies will be the most accessible, economical, and appealing choice across all areas and crucial industries.

At COP26, the Breakthrough Agenda was introduced to aid in bridging the "collaboration gap" and quickening global climate action to fulfil the Paris Agreement's global decarbonisation goals. Governments and businesses worldwide can collaborate more effectively to take immediate, decisive action on the world's largest challenge thanks to the Breakthrough Agenda, which provides a framework for prioritising, coordinating, and improving collaborative action across seven high-emitting sectors.

13.7 The Buildings Breakthrough

The Buildings Breakthrough is a component of the Breakthrough Agenda, an alliance of leading international public–private, corporate, and governmental initiatives that provides a framework for governments, businesses, and civil society organisations to band together and step up their annual action in significant emitting sectors.

By 2030, the new initiative seeks to increase global cooperation in the decarbonisation of the building industry and to position clean technologies and sustainable solutions as the most feasible, accessible, and appealing options across all areas.

13.8 Some Initiatives for the Construction Sector

UN Environment Programme (UNEP)

- The UNEP is the world's leading environmental voice. It provides leadership and encourages collaboration in environmental care by motivating, educating, and empowering nations and peoples to enhance their standard of living without compromising that of future generations.

Global Alliance for Buildings and Construction (GlobalABC)

- Established during COP21 and hosted by UNEP, the GlobalABC boasts 290 members from 41 countries, making it the premier worldwide platform for all building stakeholders united by a shared vision: a construction industry focused on durable, efficient, and emission-free structures.

13.9 Summary

There is an urgent need for enhanced collaboration at both local and international levels to bolster demand and supply for sustainable construction. Many sustainable construction materials are sourced from developing countries, yet the standards and practices for sustainable construction are predominantly established by developed nations. This disparity highlights the need for a more integrated approach to construction management that fosters synergies across borders. To address this, countries, regardless of their economic status, should develop and implement policies and regulations that guide the construction sector and regulate the market for sustainable materials and practices. These policies should be crafted with input from a diverse range of stakeholders, including professionals, suppliers, and manufacturers within the construction industry. By incorporating the perspectives of all relevant parties, these regulations can better align with the realities of both producing and using sustainable construction materials, thereby supporting a more equitable and effective transition to sustainable practices globally.

Bibliography

1. Aras G (ed) (2015) Sustainable markets for sustainable business: a global perspective for business and financial markets. Gower Publishing, Ltd.
2. Arowoiya VA, Oke AE, Aigbavboa CO, Aliu J (2020) An appraisal of the adoption internet of things (IoT) elements for sustainable construction. J Eng Des Technol 18(5):1193–1208
3. Arroyo P, Tommelein ID, Ballard G (2016) Selecting globally sustainable materials: a case study using choosing by advantages. J Constr Eng Manag 142(2):05015015
4. Bhatt R, Macwan JEM (2012) Global weights of parameters for sustainable buildings from consultants' perspectives in Indian context. J Archit Eng 18(3):233–241
5. Kibert CJ (2003) Sustainable construction at the start of the 21st century. Int. E-Journal Constr 2003:1–7
6. Kibert CJ (2007) The next generation of sustainable construction. Build Res Inf 35(6):595–601
7. Liu B, Gao Q, Liang L, Sun J, Liu C, Xu Y (2021) Ecological relationships of global construction industries in sustainable economic and energy development. Energy 234:121249
8. Mehta HS, Porwal V (2013) Green building construction for sustainable future. Civ Environ Res 3(6):7–13
9. Mistri A, Patel CG, Pitroda JR (2019) Analysis of causes, effects and impacts of skills shortage for sustainable construction through analytic hierarchy process. Int J Tech Innov Mod Eng Sci 5(5):168–176
10. Murtagh N, Scott L, Fan J (2020) Sustainable and resilient construction: current status and future challenges. J Clean Prod 268:122264
11. Olawumi TO, Chan DW (2018) A scientometric review of global research on sustainability and sustainable development. J Clean Prod 183:231–250
12. Rafindadi ADU, Mikić M, Kovačić I, Cekić Z (2014) Global perception of sustainable construction project risks. Procedia Soc Behav Sci 119:456–465
13. Shan M, Hwang BG, Zhu L (2017) A global review of sustainable construction project financing: policies, practices, and research efforts. Sustainability 9(12):2347
14. Shan M, Hwang BG (2018) Green building rating systems: global reviews of practices and research efforts. Sustain Cities Soc 39:172–180
15. Sfakianaki E (2019) Critical success factors for sustainable construction: a literature review. Manage Environ Qual Int J 30(1):176–196

Chapter 14
Sustainable Construction Management Organisation and Practice

Abstract The chapter details the processes involved in establishing and developing a sustainable construction company. The chapter highlights the importance of team dynamics, organisational behaviour, and market forces in driving sustainable outcomes. Readers are introduced to the structure and registration of professional construction management organisations committed to sustainability. The chapter also addresses the behavioural considerations within construction teams and companies, recognising their influence on decision-making and long-term project performance. Key themes include project benchmarking, regulatory compliance, stakeholder engagement, and the use of innovative technologies and certifications to promote sustainability. By providing a comprehensive framework of principles, methods, and management strategies, the chapter offers a guide for professionals aiming to reduce environmental impact, support social equity, and enhance economic value across construction projects. Ultimately, it equips readers with the tools and understanding needed to contribute effectively to the creation of a sustainable built environment.

Keywords Construction organisation · Project performance · Project benchmark

14.1 Learning Outcomes

This chapter lays the foundation for sustainable construction management practices and procedures. Conventional construction practices and procedures, though, have some similarities with sustainable construction management practices and procedures. Sustainable construction management practice and procedures encompass the principles, methods, and strategies employed to ensure that construction projects are designed, executed, and managed in a manner that minimises environmental impact, promotes social equity, and enhances economic viability. This approach involves integrating sustainability considerations into all stages of the project lifecycle, from initial planning and design to construction, operation, and end-of-life phases. Sustainable construction management practice and procedures aim to balance environmental,

A. Olanrewaju, *Sustainable Construction Management: Research and Practice Companion*, Management in the Built Environment,
https://doi.org/10.1007/978-981-96-4910-5_14

social, and economic considerations to create construction and infrastructure that meet the needs of the present without compromising the ability of future generations to meet their own needs. After studying this chapter, you will be able to:

- Explain the establishment and development of a sustainable construction company
- Explain behavioural matters in a construction company and its projects
- Explain sustainable construction market
- Explain sustainable construction teams
- Understand the registration of professional sustainable construction management organization.

14.2 Introduction

As sustainability concerns have grown in importance, the global construction sector has seen a significant transition in recent years. This change is a result of a greater understanding of how critical it is to address problems related to social fairness, resource depletion, and environmental deterioration in the built environment. Because of this, the idea of sustainable construction management has become essential for directing the sector towards more accountable and robust methods. This chapter examine the professional practices and procedures that define the domain of sustainable construction management. From pre-construction to post-contract cost control, sustainable construction managers employ a range of methodologies and tools to optimize sustainable outcomes while adhering to stringent practices and regulatory standards. Through careful management, and negotiation, sustainable construction managers navigate complex contractual frameworks to safeguard the interests of all project stakeholders in shaping a sustainable built environment. From stakeholder engagement and green building certifications to innovation in construction technologies, this chapter explores the multifaceted dimensions of sustainable construction management practices.

14.3 Establishment and Development of a Sustainable Construction Company

Creating and fostering a firm that minimises adverse environmental effects, encourages social responsibility, and guarantees long-term economic sustainability is essential to starting and growing a sustainable construction company. A sustainable construction company that incorporates these concepts into its operations not only helps to conserve the environment and promote social well-being, but it also sets itself up for long-term success in a continuously evolving sector. The elements of a sustainable construction company are as follows:

1. Vision and Mission: A company that practices sustainable construction starts with a mission and vision that are clearly based on sustainability. This include pledges to lower carbon emissions, preserve natural resources, advance social justice, and produce top-notch building projects. The sustainable construction firm establishes a solid foundation for its operations and conveys its commitment to generating beneficial environmental, social, and economic effects in the built environment by clearly defining its vision and mission statements, which incorporate these sustainability concepts. The assignments entail going into detail about every facet of sustainability and how it fits into the organization's larger objectives, and which would include:

   a. Reducing Carbon Emissions: Through the use of energy-efficient techniques, the company is committed to reducing its carbon footprint during the building process. This could entail implementing low-emission construction practices, maximising the use of renewable energy sources, and streamlining transportation logistics to cut emissions.
   b. Natural resource conservation is a major goal. These resources include raw materials, land, and water. This can be accomplished by employing tactics to reduce resource depletion and ecosystem degradation, such as locating sustainable construction materials, putting water-saving measures into place, and embracing sustainable waste management techniques.
   c. Promoting Social Equity: By encouraging inclusive practices that support diversity, equity, and fair labour standards, the company places a high priority on social responsibilities, duties and privileges. These could be offering fair job opportunities, making sure that everyone has safe working conditions, and actively interacting with the community to address its needs and concerns.
   d. Delivering High-Quality Construction Projects: Delivering construction projects of the highest calibre is essential to the mission. This entails using state-of-the-art technologies, hiring qualified personnel, and upholding strict quality standards to guarantee that projects meet or beyond customer expectations while upholding sustainability goals.

2. Green Construction Practices: In order to lessen its influence on the environment, the company uses green construction practices, such as employing sustainable construction techniques, energy-efficient designs, and the use of renewable resources. The organisation exhibits its dedication to energy conservation, sustainable development, and environmental stewardship by actively incorporating these green building principles into its projects. This helps to create built environments that are healthier and more robust for both present and future generations. The company's adoption of sustainable construction techniques includes describing particular tactics and programmes meant to lessen their negative effects on the environment:

   a. Utilizing Renewable Materials: The company places a high priority on using sustainable materials that come from recycled materials, fast renewable resources, and sustainably managed forests. Materials like bamboo, salvaged wood, steel that has been repurposed, and low-impact concrete substitutes are examples of this.

b. Implementing Energy-Efficient Designs: In order to minimise energy usage and lower greenhouse gas emissions, the company incorporates energy-efficient design ideas into its construction projects. This could entail adding high-performance insulation, utilising energy-efficient HVAC systems and appliances, and orienting the structure to maximise natural light and ventilation.
c. Incorporating Sustainable Construction Methods: The company prioritises resource efficiency and reduces waste output by using sustainable construction methods. This covers techniques like prefabrication to cut down on waste generated on site, recycling and reusing building debris, and water-saving strategies like rainwater collection and low-flow fixtures.
d. Embracing Sustainable Technologies: The company adopts cutting-edge sustainable solutions to improve the construction environmental performance. The incorporation of renewable energy systems, such wind turbines and solar panels, smart building technologies for energy management and optimisation, and the installation of green roofs to enhance thermal insulation and lessen the effects of urban heat islands are a few examples of this.

3. Stakeholder Engagement: It is crucial to interact with stakeholders, including staff members, customers, suppliers, and communities. Establishing trust and promoting cooperation, entails promoting open communication, paying attention to criticism, and resolving issues. The company's sustainability strategy relies heavily on effective stakeholder involvement to make sure that its projects meet the needs and expectations of all parties involved while being socially, environmentally, and commercially viable. Building trusting relationships with a variety of stakeholders is a top priority for a sustainable construction firm to promote cooperation and guarantee the success of its projects:

   a. Transparent Communication: In order to ensure that information is communicated honestly and transparently, the company encourages open and transparent lines of contact with all stakeholders. To inform stakeholders about the status, obstacles, and milestones of the project, frequent newsletters, community meetings, and updates should be included.
   b. Active Listening and Feedback: The company actively hears what its stakeholders-employees, customers, suppliers, and communities-have to say and takes their wants and concerns into consideration. This entails gathering feedback from stakeholders via focus groups, surveys, and one-on-one conversations in order to comprehend their viewpoints and take their suggestions into account when making decisions.
   c. Addressing Concerns and Resolving Problems: In order to promptly and effectively address stakeholder concerns, the company takes proactive measures. This could entail carrying out remedial measures, updating project

schedules, or offering more assistance and resources in order to satisfy the demands and expectations of stakeholders.

d. Building Trust and Collaboration: Stakeholder engagement is given top priority by the company, which fosters cooperation and confidence among all parties participating in its projects. This cooperative strategy encourages shared accountability and a sense of ownership, which improves project outcomes and builds longer-lasting connections.

4. Compliance and Certification: Respecting pertinent environmental laws and earning certifications shows a company's dedication to sustainability and builds its credibility. An essential part of the business's sustainability plan is adhering to environmental laws and obtaining reliable certifications. The organisation shows its dedication to environmental responsibility, improves its reputation, and advances sustainable construction methods by placing a high priority on compliance and certification. Acquiring credible certifications and meeting and surpassing environmental norms are top priorities for the sustainable building company:

    a. Environmental Regulations Compliance: Construction-related municipal, national, and international environmental requirements are scrupulously followed by the company. This covers specifications for waste management, energy efficiency, sustainable resource use, and the quality of the air and water. The business reduces its environmental impact and stays out of trouble by making sure certain standards are followed.
    b. Certification Pursuit: The organisation proactively strives for certifications, that are internationally acknowledged standards for the performance of sustainable buildings. Strict adherence to requirements pertaining to energy efficiency, indoor environmental quality, material selection, and innovation is necessary to obtain these certifications. By earning these certifications, the business stands out in the industry, wins over clients that care about the environment, and shows its dedication to sustainability.
    c. After receiving its initial certification, the company is dedicated to continuing to analyse and improve its sustainability performance. To find opportunities for improvement and make sure that sustainability practices are incorporated into every facet of company operation, this entails conducting routine assessments, audits, and reviews. The company keeps its competitive advantage and stays at the forefront of sustainable construction methods by consistently aiming for improvement.
    d. Educating Stakeholders: The business informs all of its stakeholders—clients, staff members, subcontractors, and the general public—about the significance of sustainability certifications and how they relate to environmental stewardship. By increasing the industry's adoption of sustainable building techniques and highlighting the significance of these certifications, the company promotes positive environmental effects on a broader scale.

5. Innovation and Technology: It is essential to embrace innovation and make use of technology in order to create sustainable construction solutions. This could entail using cutting-edge building materials, clever construction methods, and digital technologies for monitoring and project management. The sustainable construction company advances sustainable building methods, improves project outcomes, and promotes continuous development by embracing innovation and technology. The company stays flexible, competitive, and resilient in a construction industry that is changing quickly thanks to the strategic use of new materials, smart building techniques, digital tools, and data-driven optimisation tactics. The environmentally conscious construction company is at the forefront of utilising state-of-the-art developments to promote efficiency and sustainability in all of its projects:

    a. Advanced Building Materials: The company investigates and incorporates cutting-edge construction materials that provide exceptional durability and environmental performance. Materials with a high percentage of recycled content, a low embodied carbon, and improved energy efficiency are examples of this. The organisation ensures the sustainability of its constructions while mitigating environmental impact through the integration of these materials into its projects.
    b. Smart Construction Techniques: The organisation may maximise resource utilisation, reduce waste generation, and expedite project timeframes by using smart construction practices including prefabrication, modular building, and off-site manufacturing. The company expedites project delivery, improves quality control, lowers construction waste, and increases cost-effectiveness by prefabricating construction components in regulated manufacturing conditions.
    c. Digital Tools for Project Management: The organisation leverages digital technologies to improve decision-making, teamwork, and project management. Stakeholders can use digital tools to simulate and visualise designs, spot any conflicts or inefficiencies, and maximise building performance over the course of a project. In order to further foster efficiency, accountability, and transparency among project teams, the company also makes use technologies for document sharing, real-time communication, and progress tracking.
    d. Monitoring and Optimization: The company uses technologies to develop real-time optimisation and monitoring tactics that improve building performance and occupant comfort. The organisation finds potential for energy savings, operational improvements, and preventative maintenance by gathering and evaluating data on temperature, occupancy patterns, indoor air quality, and energy use. Over time, this proactive strategy improves building resilience and operational efficiency while also lessening its impact on the environment.

6. Continuous Improvement: A sustainable construction company should identify ways to enhance its operations, assesses its methods on a regular basis, and adjusts to new sustainability trends and best practices. This calls for continuous efforts in development, research, and teaching. A sustainable construction firm maintains its leadership position in sustainability innovation, promotes positive industry change, and provides enduring value to its stakeholders and the environment by placing a high priority on continual development. By adopting a comprehensive strategy that includes training, research, certification, performance monitoring, and stakeholder involvement, the organisation manages to be flexible, strong, and sensitive to the changing needs of the sustainable building industry. A sustainable construction company must promote a culture of creativity, education, and flexibility:

   a. Education and Training: The business funds staff education and training initiatives that emphasise green building techniques, upcoming technology, and sustainability concepts. Regular training sessions, workshops, and certifications enable staff members to remain up to date on industry innovations and best practices, enabling them to successfully implement sustainable solutions.
   b. Research and Development: The company devotes funds to R&D projects with the goal of identifying novel building methods, materials, and technologies that have improved sustainability features. Through partnerships with industry players, academia, and technology companies, the company investigates novel approaches to mitigate environmental effects, optimise energy efficiency, and elevate construction performance throughout its portfolio of projects.
   c. Performance Monitoring and Feedback: By putting strong monitoring systems and performance feedback mechanisms in place, the company can keep tabs on crucial sustainability measures, identify areas that need work, and track its advancement over time. By gathering information on waste production, carbon emissions, energy usage, and tenant satisfaction, the company may better understand how it is performing and make targeted improvements to correct inefficiencies or flaws.
   d. Benchmarking and Certification: The company evaluates its performance in relation to sustainability certifications, benchmarks, and industry norms. By obtaining and upholding certifications, the company strengthens its standing and competitiveness in the market by displaying its dedication to constant improvement and adherence to strict sustainability standards.
   e. Stakeholder Engagement: Interacting with stakeholders, such as customers, suppliers, trade groups, and regional communities, promotes cooperation, information exchange, and creativity. By asking a variety of stakeholders for their opinions, thoughts, and suggestions, the company is able to see things from new angles, spot new prospects, and fortify its bonds with them-all of which promote innovation and constant development throughout its operations and initiatives.

7. Social Responsibility: The corporation places a high priority on social responsibility by encouraging diversity, worker safety, fair labour standards, and community involvement in addition to environmental concerns. A sustainable construction firm should place a high priority on social responsibility to produces excellent work, and to makes a beneficial impact on the welfare of its workforce, local communities, and society as a whole. The company demonstrates its commitment to ethical business conduct and sustainable development by taking a comprehensive approach that includes worker safety, fair labour practices, diversity, inclusion, community engagement, and supply chain responsibility. This approach has a significant positive impact on the lives of people and the communities the company serves. A sustainable construction company incorporates social justice, ethical behaviour, and community involvement into its basic operations:

    a. Worker Safety and Well-being: The health, safety, and welfare of employees must always come first. The company should establish a culture of safety awareness on building sites, offers thorough training programmes, and puts strict safety regulations into practice. The company fosters a positive work environment that increases productivity and lowers the chance of accidents or injuries by placing a high priority on worker protection and reducing occupational hazards.
    b. Fair Labor Practices: Maintaining fair labour policies shows the company's dedication to treating workers with respect and dignity. These practices include equal pay, manageable working hours, and chances for professional progression. The organisation encourages job happiness, staff retention, and long-term loyalty among its workers by abiding by labour laws and industry norms.
    c. Diversity and Inclusion: Accepting diversity and cultivating an inclusive work environment broadens the company's horizons, encourages innovation, and fortifies its capacity for creativity. In addition to offering equitable chances for career advancement and a supportive work environment free from harassment or discrimination, the company actively supports diversity in recruiting. The organisation fosters a dynamic and inclusive workforce that mirrors the larger community it serves by valuing differences and embracing diversity.
    d. Community Engagement: The company's dedication to being a responsible corporate entity is demonstrated through its interactions with stakeholders, the local community, and nonprofit organisations. The company makes a beneficial impact on the social fabric of the communities in which it operates by means of volunteer programmes, charity projects, and community outreach efforts. The company cultivates a favourable reputation and expands its social licence to operate by investing in community development projects, supporting local businesses, and attending to community needs. These actions also help to deepen connections and acquire trust.
    e. Supply Chain Responsibility: By choosing vendors and subcontractors who uphold its moral principles and values, the company extends its commitment to social responsibility throughout its supply chain. The organisation

reduces the possibility of unethical behaviour or human rights violations within its supply chain by performing due diligence, monitoring supplier performance, and encouraging responsible sourcing practices. This guarantees that its operations uphold the highest standards of social responsibility from start to finish.

8. Economic Viability: Even if sustainability is the company's top priority, economic viability is nonetheless given careful consideration. This entails providing sustainable construction projects while controlling expenses, allocating resources as efficiently as possible, and guaranteeing profitability. A sustainable construction company guarantees stability, profitability, and long-term success by giving equal weight to sustainability goals and economic viability. The company attains a balance between environmental responsibility and commercial profitability by means of strategic cost management, resource optimisation, risk management, market positioning, and ROI analysis. This leads to favourable outcomes for the company's stakeholders as well as the wider community. To guarantee long-term profitability and commercial success, a sustainable construction company carefully strikes a balance between social responsibility and environmental stewardship:

    a. Cost Management: The organisation employs economical tactics to oversee project expenditures while upholding environmental objectives. Using sustainable construction techniques, maximising material utilisation, and reducing waste at every stage of the project's lifecycle are all part of this. Through process simplification, cost-saving measures, and the utilisation of economies of scale, the organisation optimises productivity and minimises project expenses, hence augmenting its competitive advantage in the industry.
    b. Resource Optimization: In order to achieve economic viability in sustainable construction projects, efficient resource allocation is essential. The company believes in the circular economy, uses renewable resources first, and uses as little resources as feasible. Using cutting-edge design techniques, sustainable construction materials, and energy-saving technologies, the company gradually minimises its environmental effect, maximises resource use, and cuts operating expenses.
    c. Risk Management: To protect the company's finances and reduce any project hazards, proactive risk management is crucial. To safeguard against unanticipated events or interruptions, the organisation carries out comprehensive risk assessments, recognises possible dangers, and creates strong problem mitigation plans. Through the implementation of efficient risk management practices, the organisation reduces financial uncertainty, improves project resilience, and maintains profitability over the course of the project.
    d. Market Positioning: By highlighting its experience in completing high-value, ecologically conscious projects and differentiating its sustainable construction products, the company strategically positions itself in the market. The company's emphasis on sustainability helps it draw in environmentally

concerned clients, land valuable contracts, and stay ahead of the competition in the sector.

e. Return on Investment: To show stakeholders the value proposition of sustainable construction techniques, the company assesses the short- and long-term financial gains associated with them. The company emphasises the positive ROI of sustainable investments by quantifying social and environmental benefits in addition to traditional financial metrics. This helps it establish itself as a preferred partner for clients searching for sustainable solutions that provide measurable economic value throughout the project lifecycle.

14.4 Introduction to the Establishment and Development of a Construction Markets

The creation and expansion of construction markets are essential to the development of infrastructure, the economy, and society advancement. Growing urbanisation, population expansion, and technology breakthroughs are changing the face of the world, and the construction sector is essential to addressing a range of societal demands and promoting sustainable development.

14.5 Sustainable Construction Market Dynamics

The different forces and causes that impact the development, patterns, and behaviour of this industry are referred to as the sustainable building market's dynamics. In order to recognise possibilities, overcome obstacles, and spur innovation, stakeholders in the sustainable construction market must have a thorough understanding of these dynamics. Through monitoring of the current market trends and utilising strategic insights, companies can profit from the increasing need for environmentally friendly construction solutions. The primary drivers and factors influencing the market's expansion for sustainable construction are listed below:

a. Environmental Concerns: Growing public awareness of environmental deterioration and climate change has increased demand for sustainable construction methods. Market expansion is fuelled by laws intended to lower carbon emissions and increase energy efficiency.
b. Regulatory Landscape: The market is shaped by laws and rules requiring sustainable construction certifications and standards. Sustainable construction practices are frequently used as a result of compliance with these criteria.
c. Technological Innovations: Innovation in the sustainable construction sector is fuelled by developments in smart construction technologies, renewable energy systems, and building materials. These developments lower waste, increase energy efficiency, and improve building performance.

d. Market Demand: Demand for sustainable construction projects is driven by rising consumer knowledge of environmental issues and preferences for sustainable projects. The industry is growing because developers and investors see the long-term worth and marketability of sustainable structures.
e. Investment and Financing: Investment in sustainable construction projects is encouraged by the availability of green financing options, subsidies, and incentives. Strong sustainability credentials are becoming increasingly crucial to investors and financial institutions.
f. Industry Collaboration: To advance sustainable construction techniques, cooperation between stakeholders is crucial, including legislators, developers, contractors, and architects. Innovation and the adoption of best practices are fuelled by collaborations and knowledge-sharing programmes.
g. Global Megatrends: Global megatrends such as population expansion, urbanisation, and resource scarcity are what are driving demand for sustainable construction solutions. By determining infrastructure requirements and development priorities, these developments have an impact on market dynamics.

14.6 The Regulatory Frameworks for the Sustainable Construction Market

Governments and trade associations provide laws, regulations, standards, and certifications that support sustainable construction techniques, which together form the regulatory frameworks for the sustainable construction market. Stakeholders in the sustainable construction market must comprehend and abide by these regulatory frameworks in order to maintain legal compliance, satisfy industry standards, and obtain incentives and certifications that improve their reputation and competitiveness in the market. The various features of the regulatory frameworks for the market for sustainable construction are summarised in the following:

a. Building Codes and Standards: Governments create building codes that outline the minimal specifications needed for remodelling and new projects. These rules are becoming more and more specific about features like indoor air quality, water conservation, energy efficiency, and using sustainable materials.
b. Green Building Certification Programmes: Programmes for voluntary certification set high requirements for sustainable construction methods. Projects that receive certification prove that they adhere to certain requirements, which frequently results in rewards on the market and financial recognition.
c. Energy Efficiency Regulations: Regulations requiring energy performance criteria for constructions are available many countries. Minimum energy efficiency ratings for building envelopes, lighting, HVAC, and lighting systems may be mandated by these laws. Performance testing and energy audits are common components of compliance.

d. Renewable Energy Policies: Policies to encourage the use of renewable energy technologies in buildings may be put in place by governments. Subsidies, tax breaks, net metering agreements, and laws requiring a certain percentage of a building's energy requirements to come from renewable sources are a few examples of this.
e. Waste Management Regulations: Construction projects are impacted by recycling and waste management regulations. Permits, reporting duties, and fines for non-compliance can all be used to enforce waste reduction, recycling, and disposal regulations.
f. Water Efficiency Standards: Regulations designed to save water in buildings are motivated by worries about a shortage of water. Standards might cover graywater reuse, rainwater harvesting systems, and energy-efficient plumbing fixtures.
g. Occupational Health and Safety Regulations: Rules protecting the health and safety of construction workers must be followed. Hazard assessments, safety training, PPE, and emergency response plans are examples of compliance-related practices.

14.7 The Economic Trends for Sustainable Construction Market

The market dynamics that are shaped by factors including technology breakthroughs, regulatory interventions, market demand, and societal preferences are reflected in the economic patterns of the sustainable construction sector. For those involved in the sustainable construction sector to take advantage of opportunities, overcome obstacles, and promote innovation in building techniques and technologies, they must have a thorough understanding of these economic trends. These are the main economic developments influencing the market for sustainable building:

a. Rising Demand: There is a growing need for sustainable construction due to worries about energy prices, the environment, and human health. Buildings that provide energy efficiency, resource conservation, and healthy interior environments are in high demand from consumers, corporations, and governments.
b. Government Incentives: To encourage the use of sustainable construction techniques, numerous governments provide grants, subsidies, tax exemptions, and other financial incentives. Developers and construction users are encouraged to invest in green infrastructure, renewable energy systems, and energy-efficient buildings via these incentives.
c. Cost Reduction: Sustainable construction materials, technologies, and practices are becoming more affordable due to economies of scale and technological developments. The economic argument for sustainable construction is strengthening as these solutions become more affordable, making it more competitive with traditional construction techniques.

d. Market Integration: The use of sustainable construction techniques is spreading throughout the construction sector. In response to market demand and legal requirements, developers, contractors, engineers, construction managers, quantity surveyors, maintenance managers, architects, and suppliers are integrating sustainability principles into their business models and product offerings.
e. Green Finance: In order to finance sustainable construction projects, financial institutions are progressively providing green financing options, such as green bonds, loans, and mortgages. Due to the recognition of sustainable constructions' durability and long-term economic benefits by lenders and investors, the sector has seen an increase in investment.
f. Job Creation: There are more job prospects in green construction design, construction, retrofitting, and operation as a result of the change to sustainable construction. There is a great need for skilled personnel in fields including smart technologies, green materials, renewable energy, and energy efficiency.
g. Global Market Growth: Globally, the market for sustainable construction is expanding rapidly due to factors such population increase, urbanisation, initiatives to mitigate climate change, and regulatory requirements. As they grow, emerging economies are embracing sustainable building techniques more and more, which is growing the market.

14.8 The Forces Driving Market Growth for Sustainable Construction

The market for sustainable construction is propelled by a number of factors that support its development and growth. These factors include growing public and corporate demand for environmentally conscious infrastructure, strict laws and policies encouraging green building practices, growing awareness of and concern for environmental sustainability, and technological advancements enabling more effective and sustainable construction methods. The growing expense of conventional resources and energy sources has also increased consumer interest in sustainable alternatives, which is driving the market's expansion. Furthermore, stakeholders from a range of industries are seeing the long-term advantages of sustainable construction, including lower operating costs, increased property value, and better tenant health and well-being. Because of these diverse variables, the market for sustainable construction keeps growing and changing.

a. Increasing Environmental Awareness: Growing concerns about environmental deterioration and climate change have made people, companies, and governments more conscious of the necessity of sustainable construction methods.
b. Regulatory Initiatives: Governments all throughout the world are making laws and policies into place to encourage sustainability in construction sector. Examples of these include building codes that require energy efficiency standards and prerequisites for green certifications.

c. Technological Advancements: Sustainable construction practices are becoming more practical and economical due to advancements in construction materials, processes, and technologies. Examples include the creation of renewable energy technology, energy-efficient building systems, and digital tools for construction management and design.
d. Consumer Demand: Due to concern about comfort, health, and the environment, there is a growing demand from tenants and buyers for sustainable structures. The adoption of sustainable construction approaches by developers, contractoes, and property owners is being influenced by this demand.
e. Cost Savings: Long-term cost savings can be achieved through sustainable construction by lowering energy usage, increasing property value, and requiring less maintenance. Businesses and investors are more likely to fund sustainable construction projects as they become aware of these financial advantages.
f. Resource Scarcity: The construction sector is seeking alternative, renewable materials and energy sources due to the depletion of natural resources and the rising costs of standard building materials. This is pushing innovation in sustainable construction techniques.
g. Health and Well-being: The adoption of sustainable construction features, such as natural lighting, enhanced indoor air quality, and biophilic design elements, is being driven by the focus on occupant health and well-being. These characteristics enhance overall enjoyment and productivity.

These variables interact in convoluted ways to @@@mould the market for sustainable construction, propelling its expansion and development throughout time.

14.9 Emerging Trends for Sustainable Construction Market

As the market for sustainable construction moves towards more resilient, sustainable, and human-centered built environments, these new trends reflect the shifting goals, technology, and practices that are influencing it, which may include:

a. Net Zero Buildings: The idea of creating net zero energy constructions-those that generate as much renewable energy as they use in a year-is becoming more and more popular. Innovations in building design techniques, renewable energy systems, and energy-efficient technology are what are driving this trend.
b. Circular Economy Practices: The circular economy's emphasis on material reduction, reuse, and recycling as a means of reducing waste and resource consumption is finding its way into sustainable construction. This covers tactics like using recycled or upcycled materials, deconstructing and recovering resources, and building modular structures.
c. Construction technologies: Real-time monitoring, control, and optimisation of energy consumption, indoor environmental quality, and building operations are made possible by the increasing integration construction technologies in the

projects. This pattern enhances operating efficiency, occupant comfort, and building performance.

d. Biophilic Design: Sustainable construction is seeing a rise in the use of biophilic design principles, which integrate natural patterns and features into the built environment to improve human well-being. A connection to nature is encouraged and tenant health and productivity are enhanced by features including green roofs, living walls, natural lighting, and access to outside spaces.
e. Resilient Design: As climate-related catastrophes become more frequent and severe, resilient design strategies are increasingly crucial for sustainable building. This entails using materials that are fire- and flood-resistant, constructing structures to endure harsh weather, and putting policies in place to improve adaptability and disaster preparedness.
f. Prefabrication and Modular Construction: Prefabrication and modular construction techniques are being used to increase quality control, decrease waste, and boost construction efficiency. Building components can be fabricated off-site to expedite construction, save costs, and have a smaller environmental impact.
g. Green Building Certifications: Green building certifications have a lasting impact on sustainable construction practices because they offer frameworks and standards for evaluating and honouring sustainability, occupant health, and environmental performance.

14.10 The Behavioural Matters in a Sustainable Construction Company and Its Projects

The difficulties relating to human conduct and interactions that may impede the effective execution of sustainability measures are referred to as behavioural problems in sustainable construction projects. These problems can occur among staff members, managers, clients, and other stakeholders, as well as at different levels within the project team. A sustainable construction company's performance as a whole and its ability to meet sustainability objectives may be negatively impacted by behavioural problems, which can result from a variety of causes. A diverse strategy is needed to address the behavioural difficulties, one that involves communication techniques, education and training, employee involvement, leadership commitment, and the promotion of an innovative and sustainable culture. Several prevalent behavioural problems and their effects:

a. Resistance to Change: Because of inertia, a fear of the unknown, or a lack of confidence in the advantages of sustainability, employees may be reluctant to adopt new sustainable practices. This may make it more difficult for the company to revolutionize and implement sustainable initiatives.
b. Lack of Awareness: Apathy or disinterest towards sustainability initiatives may result from certain employees' ignorance of or lack of comprehension of the principles of sustainable construction. Promoting a culture of sustainability within the workforce requires education and awareness-building.

c. Silos and Communication Barriers: Teams or departments within the organisation might function independently, with little interaction and cooperation on environmental projects. Integrated and comprehensive approaches to sustainability depend on dismantling organisational silos and fostering cross-functional cooperation.
d. Short-Term Focus: Prioritising short-term gains above long-term sustainability goals may result from pressure to satisfy short-term deadlines or financial ambitions. It is crucial to promote a well-rounded strategy that considers the implications for sustainability over the long and short terms.
e. Lack of Leadership Support: Employees may get disengaged or demotivated from sustainability projects if there is a lack of strong leadership support and a commitment to sustainability. To promote long-lasting change, leadership should set an example of clarity, provide resources, and give staff members authority.
f. Inadequate Training and Skills: It is possible that staff members lack the knowledge or experience needed to successfully adopt sustainable practices. By funding skill-development and training programmes, employers can close knowledge gaps and provide staff members the tools they need to support sustainable projects.
g. Resistance from External Stakeholders: Adopting sustainable practices may be resisted by customers, suppliers, manufacturers, or subcontractors because of financial worries, perceived hazards, or a lack of knowledge. Overcoming opposition from outside stakeholders can be facilitated by establishing connections based on openness, trust, and similar ideals.
h. Complacency: To prevent complacency, it is imperative to engage in ongoing monitoring, review, and the establishment of high yet attainable sustainability targets.

14.11 The Sustainable Construction Teams

Sustainable construction teams are the teams of specialists with diverse backgrounds who work together to design, develop, and oversee infrastructure and structures with an emphasis on social, economic, and environmental sustainability. Various professionals work together in these teams to implement sustainable concepts throughout the project lifetime. Sustainable construction teams work together to develop sustainable construction projects by bringing together a variety of perspectives and experience. Some crucial professionals in sustainable construction teams are listed below:

a. Architects: In order to imagine and plan sustainable buildings, architects are essential. They concentrate on incorporating biophilic components, natural lighting, passive heating and cooling techniques, and energy-efficient designs into the structure's architecture.

b. Engineers: In teams working on sustainable construction projects, structural, mechanical, electrical, and civil engineers are crucial components. They use sustainable materials and technologies to optimise construction systems for resilience, energy efficiency, and environmental performance.
c. Sustainable construction managers: Environmental impact analysis, life cycle assessment, energy modelling, stakeholder management, economic analysis and green building certifications are among the areas in which sustainability consultants specialise. They support project teams in adhering to legal requirements, establishing sustainable objectives, and putting best practices into action. They are in charge of making sure that sustainability goals are fulfilled in accordance with project schedules, financial constraints, and quality requirements.
d. Construction Managers: Construction managers oversee construction operations on the job site, guaranteeing the safe and efficient use of sustainable construction methods. During construction, they manage subcontractors, monitors work, and resolve any problems that may come up.
e. Building Scientists: Thermal comfort, indoor air quality, energy consumption, and building performance analysis are the areas of expertise for building scientists. To improve sustainability and occupant well-being, they carry out energy audits, make recommendations for energy-saving strategies, and optimise building envelope design.
f. Green Building Advocates: Green building proponents encourage sustainable construction methods in businesses, local communities, and government agencies. They promote legislative changes, encourage stakeholder engagement, and increase public knowledge of the advantages of green building.
g. Quantity surveyors: Quantity surveyors identify opportunities to optimize project costs without compromising sustainability goals through value engineering. They manage the procurement process for sustainable construction materials and services and negotiate contracts, and ensure compliance with sustainability criteria. They advise project teams on sustainability best practices and help navigate regulatory hurdles related to sustainable construction. Quantity surveyors monitor project progress and performance, tracking key sustainability metrics such as energy consumption, water usage, and waste generation. They develop risk mitigation strategies to address sustainability-related risks and ensure project resilience in the face of environmental and social challenges.

14.12 Sustainable Construction Team Formation

Several crucial actions must be taken in order to ensure that varied viewpoints and areas of expertise are integrated when forming a sustainable construction team. As sustainable construction teams, project teams may effectively create and work to deliver projects that prioritise social responsibility, environmental stewardship, and economic sustainability. The essential guidelines for assembling a sustainable construction team are provided below:

a. Identify Project Goals and Objectives: Describe the project's sustainability aims and objectives, including resource conservation, energy efficiency, and occupant health and well-being. Make the project's timetable, budget, and scope clear.
b. Assess Stakeholder Needs: Determine the crucial parties involved, such as end users, investors, clients, and regulatory bodies. Recognise their sustainability-related needs, preferences, and priorities, and take their input into account while developing the project.
c. Assemble a Multidisciplinary Team: Assemble a diverse team of experts in project management. Identify experts that have a strong commitment to sustainability and have worked on sustainable construction projects.
d. Define Roles and Responsibilities: Assign each team member specific duties and tasks according to their training and experience. Assign project leadership positions to supervise team operations and guarantee consistency with project goals, such as sustainability coordinator and project manager.
e. Promote Collaboration and Communication: Encourage a culture of cooperation and communication among team members so they may exchange concepts, knowledge, and best practices. To promote teamwork and problem-solving, schedule frequent brainstorming sessions, workshops, and team meetings.
f. Provide Training and Education: Provide team members with chances for education and training to improve their knowledge of green building certifications, sustainable construction techniques, and emerging technology. Promote lifelong learning and career advancement.
g. Integrate Sustainability into Design and Planning: When designing and planning a project, incorporate sustainable principles by considering several aspects like material selection, energy efficiency, water conservation, site selection, and indoor environmental quality. Make informed design decisions by utilising methods like daylighting analysis, life cycle evaluation, and energy modelling.
h. Establish Performance Metrics: To monitor the project's sustainability performance during the design, building, and operation stages, set quantifiable performance indicators and goals. Track your progress towards sustainability objectives and make necessary adjustments to maximise your results.
i. Engage with External Partners and Stakeholders: Work together with outside partners to take advantage of their resources and experience, including sustainability experts, green building associations, suppliers, contractors, and community organisations. Interact with regulatory bodies to guarantee adherence to green building guidelines and requirements.
j. Promote a Culture of Sustainability: Encourage a sustainable culture throughout the group and company by highlighting accomplishments, exchanging success tales, and honouring contributions to environmentally friendly building techniques. Promote originality, ingenuity, and ongoing development in sustainability projects.

14.13 Leadership in Sustainable Construction Management

Leadership in sustainable construction is the proactive roles and duties that people, groups, and communities play in encouraging the construction sector to embrace sustainable methods. These are crucial characteristics to know are:

a. Vision and Commitment: A clear vision and a dedication to sustainability principles are the first steps in becoming a leader in sustainable construction. In construction projects, leaders stress the significance of social responsibility, environmental stewardship, and economic viability.
b. Strategic Planning: In order to incorporate sustainability into every facet of construction project-from design and planning to execution and operation-leaders create strategic strategies and objectives. The leaders establish goals for waste minimization, resource conservation, energy efficiency, and tenant health and well-being.
c. Innovation and Collaboration: Leaders in sustainable construction cultivate an environment that values creativity and cooperation, promoting the investigation of novel materials, technologies, and methods that reduce ecological footprints while optimising efficiency.
d. Regulatory Compliance: To guarantee adherence and encourage the finest practices in sustainable construction, leaders keep up with industry standards, environmental laws, and green building rules.
e. Stakeholder Engagement: Leaders interact with stakeholders, such as clients, staff, suppliers, regulators, and members of the community, in order to obtain feedback, forge agreements, and handle issues pertaining to sustainability.
f. Education and Training: In order to give staff members, the information and abilities they need to successfully apply sustainable construction techniques, leaders fund education and training initiatives. They encourage certification in green building standards and professional growth.
g. Monitoring and Reporting: Leaders set up systems to track and report on sustainability performance, monitors crucial indicators including water and energy use, waste production, indoor air quality, and energy use. They use data to track their progress towards sustainability targets and pinpoint areas that need improvement.
h. Continuous Improvement: Leaders in sustainable construction are dedicated to ongoing development, feedback gathering, post-project assessments, and using lessons learned to better projects in the future.
i. Public Advocacy: At the municipal, national, and international levels, leaders promote laws and programmes that encourage sustainable building methods. To exchange expertise and advance sustainability, they take part in forums, conferences, and industry groups.
j. Inspiration and Influence: Above all, people are inspired by sustainable construction leaders because of their deeds, honesty, and enthusiasm for sustainability. They set an exemplary example, proving that sustainable building is crucial for

the long-term prosperity of the construction sector as well as for the environment and society.

14.14 Sustainable Construction Team Motivation

Motivating team members to actively participate in and contribute to sustainable construction projects entails motivating and energising them. When team members are encouraged in their professional growth, feel that their work has a purpose, and are acknowledged for their efforts towards creating a more sustainable future, they are more motivated to work towards sustainable construction projects. Below are some pointers to inspire your sustainable building team:

a. Purpose and Meaning: Teams working on sustainable construction projects are frequently motivated by a common sense of meaning and purpose. Team members are motivated by the desire to use their work to positively benefit society, the environment, and future generations.
b. Environmental and Social Values: sustainable team members are frequently very driven in green construction initiatives. They are dedicated to lowering carbon emissions, protecting the environment, advancing social justice, and raising standard of living.
c. Intrinsic Rewards: Self-satisfaction, self-worth, and contentment are examples of intrinsic rewards that are crucial in inspiring sustainable construction teams. Members of the team find fulfilment in knowing that their efforts help create a more sustainable future.
d. Professional Development: Team members who are driven place a high importance on chances for growth and advancement in the field of sustainable construction. They have a strong desire to expand their skill set, get insight into green building methodologies, and remain current with developments and trends in the sector.
e. Recognition and Appreciation: Team motivation can be increased by expressing gratitude and acknowledging their contributions. Fostering a good and supportive work environment is achieved through publicising triumphs in sustainable building projects, recognising milestones, and acknowledging individual and collective efforts.
f. Team Collaboration: In sustainable construction projects, motivation requires collaborative effort. Team members are more likely to be motivated to work towards shared objectives when they feel appreciated, respected, and included in decision-making processes.
g. Goal Alignment: Motivation requires that team members' professional and personal aspirations line up with the goals of sustainable construction projects. People are more engaged and motivated when they understand how their efforts help to accomplish project goals and organisational purposes.

h. Continuous Improvement: Teams that are driven to construct sustainably embrace a culture of constant improvement. They actively seek input, pinpoint opportunities for development and innovation, and work to produce projects with excellent, long-lasting results.

14.15 Sustainable Construction Managerial Roles

Positions in management that supervise and assist the integration of sustainability concepts into building projects fall under the broad category of sustainable construction duties. Sustainable construction managers must possess a variety of skills, including strong leadership, strategic thinking, engaging stakeholders, and a dedication to developing sustainability in the built environment. In sustainable construction teams, the primary managerial positions are as follows:

a. Strategic Planning: The creation of strategic plans that support sustainability goals is under the purview of managers working in sustainable construction. In order to achieve environmental, social, and economic sustainability throughout the project lifespan, goals, KPIs, and strategies must be established.
b. Policy Development: When it comes to creating and executing policies and processes that promote sustainable construction methods, managers are essential. This include making certain that environmental laws are followed, encouraging green construction certifications, and including sustainability specifications into project agreements.
c. Stakeholder Engagement: Engaging stakeholders in a productive manner is necessary for sustainable construction projects. In order to ensure alignment with sustainability goals, address issues, and promote collaboration, managers interact with a variety of stakeholders, including clients, suppliers, contractors, government agencies, and community people.
d. Resource Management: Sustainable construction managers effectively allocating resources, such as waste, water, energy, and materials. They advocate the use of sustainable materials and renewable resources, spot opportunities for resource conservation, and put policies in place to lessen their negative effects on the environment.
e. Problem Management: Problems related to supply chain disruptions, stakeholder conflicts, regulatory compliance, and the environment are all evaluated and reduced by managers when working on sustainable construction projects. To deal with possible obstacles and guarantee project success, they create risk management techniques.
f. Performance Monitoring: Managers monitor project performance in relation to sustainability standards and targets. For the purpose of monitoring advancement, identifying areas in need of development, and reporting project results, this entails gathering and evaluating data on energy usage, carbon emissions, waste production, and other sustainability indicators.

g. Team Leadership: In addition to offering project teams leadership and direction, sustainable construction managers foster an environment of accountability and sustainability. They encourage teamwork, provide team members the freedom to think creatively and solve problems, and make sure that everyone is aware of their roles and responsibilities in accomplishing sustainability goals.
h. Continuous Improvement: To improve the sustainability performance of building projects, managers spearhead initiatives for continual improvement. In order to develop sustainable construction methods, this entails carrying out post-project evaluations, disseminating lessons gained, putting best practices into action, and promoting innovation and technology uptake.

14.16 Stakeholder Engagement in Sustainable Construction

In sustainable construction, stakeholder engagement is the process of involving and communicating with all stakeholders interested in or impacted by a construction project. The objective is to promote accountability, transparency, and teamwork throughout the project's lifecycle. In addition to the clients, architects, quantity surveyors, construction managers, engineers, and contractors, stakeholders in sustainable construction might also include communities, NGOs, government organisations, suppliers, and end users. In sustainable construction, effective stakeholder engagement can reduce conflicts, increase project legitimacy, foster trust, and ultimately produce more successful and long-lasting results for all parties involved. Stakeholder involvement in sustainable construction mostly consists of the following:

a. Identification: Finding every relevant stakeholder for the project, including those who might be influenced by the construction activities directly or indirectly.
b. Communication: Establishing open lines of communication to request input and comment from stakeholders and to keep them updated on project plans, status, and possible effects.
c. Collaboration: Promoting cooperation between stakeholders to take advantage of their knowledge, assets, and viewpoints in creating long-term fixes and reducing hazards.
d. Participation: Giving interested parties the chance to take part in decision-making procedures through advisory committees, design charrettes, and public consultations.
e. Transparency: Preserving openness in the governance and decision-making processes of projects, including sharing pertinent data regarding their objectives, performance, and results.
f. Accountability: Encouraging all parties involved to accept responsibility for their respective roles and duties in promoting sustainable building goals and realising intended project results.

14.17 Client Engagement in Sustainable Construction Practices

Client involvement in sustainable construction practices is concern with engaging clients in the adoption, application, and promotion of sustainability principles across the course of a building project Incorporating sustainable construction techniques with clients cultivates alliances founded on mutual accountability, trust, and shared values. Construction professionals may create projects that fulfil client needs and advance social responsibility and environmental stewardship by actively including clients in sustainability decision-making and implementation. The following are the essential conditions for client participation in sustainable building:

a. Needs Assessment: Involving clients in the project's early phases to learn about their needs, priorities, and sustainability goals. This is evaluating their goals in terms of the environment, society, and economy in order to create solutions that are sustainable and satisfy their demands.
b. Education and Awareness: Teaching clients the advantages of sustainable construction techniques, such as lower costs, better building performance, and environmental preservation. Giving customers access to knowledge on standards, best practices, and certifications for green buildings can assist them in making well-informed decisions.
c. Collaborative Decision-Making: Including clients in the choices made about energy-efficient technology, sustainable design options, and other aspects of green buildings. Working together with clients makes it possible to include sustainability concerns into project schedules and requirements.
d. Performance Monitoring: Defining precise performance measures and sustainability KPIs to monitor project progress and results. Transparency and accountability are maintained when project sustainability performance, energy use, trash reduction, and other pertinent indicators are routinely shared with clients.
e. Stakeholder Engagement: Encouraging communication and cooperation between different project stakeholders, such as contractors, architects, construction managers, engineers, and regulatory bodies, in order to guarantee that sustainability objectives are aligned and that everyone is committed to sustainable results.
f. Continuous Improvement: Encouraging customers to adopt a culture of continuous improvement by getting input, assessing output, and spotting chances to improve sustainability practices before, during, and after projects.

14.18 Determination of the Fees Structure in Sustainable Construction Projects

The scale, complexity, and sustainability objectives of the project are among the many elements that must be considered when determining the price structure for sustainable building projects. The primary actions listed below will assist a sustainable building company in figuring out the charge schedule:

a. Understand Project Requirements: Start with having a complete understanding of the project's requirements, including its goals for sustainability, scalability, and scope. Consider elements like social responsibility programmes, energy efficiency targets, trash reduction goals, and green building certifications.
b. Assess Project Complexity: Examine the project's complexity, considering the site's characteristics, architectural style, engineering specifications, stakeholder expectations, and regulatory compliance. Complex projects might need more knowledge and resources, which would affect the cost structure.
c. Define Services and Deliverables: Clearly state the services and products your company will offer all the way through the project's lifecycle. This could involve commissioning, design development, construction management, sustainability consulting, and post-occupancy assessments.
d. Consider Sustainable Practices: Include sustainable techniques in the pricing structure, such as life cycle assessments, energy modelling, waste management plans, material procurement, and green building certifications. These extra services could call for specific knowledge and resources.
e. Evaluate Market Rates: Examine local market prices and industry norms for sustainable construction services. When choosing your price structure, consider elements including labour costs, overhead, project location, and competitive pricing.
f. Customize Fee Proposal: Provide a tailored fee proposal that considers the services rendered, project needs, and level of complexity. Provide a clear description of the project's deliverables, schedule, terms of payment, and cost structure (fixed fee, hourly rate, percentage of construction cost, etc.).
g. Provide Value Proposition: Stress the value proposition of your company's sustainable construction services, emphasising your knowledge, experience, creativity, and extra advantages like less of an impact on the environment, long-term financial savings, and improved building performance.
h. Negotiate and Finalize: Collaborate with the client to discuss the cost schedule and respond to any concerns or inquiries they might have. Before completing the price proposal, be open and honest about the costs, conditions, and expectations to ensure that everyone is on the same page.
i. Document Agreement: Once a pricing schedule has been decided upon, put the terms and conditions in writing by creating a legal contract or agreement. To safeguard the interests of both parties, precisely define the tasks involved, who is responsible for what, when payments are due, and any other pertinent information.

j. Monitor and Adjust: monitor performance and advancement of the project to make sure the cost structure is still reasonable. Fee adjustments should be made as needed to preserve profitability and client satisfaction while accommodating modifications to the project's scope, schedule, or needs.

14.19 Formation of Sustainable Construction Management Office

Establishing a sustainable construction management office is charge of supervising and carrying out sustainability efforts throughout the project lifecycle is the first step in meeting the sustainability goals of the built environment. The main actions needed to establish such an office are as follows:

a. Leadership Commitment: The creation of a dedicated office to spearhead sustainability initiatives must be supported by top management, who must also show a strong commitment to sustainability.
b. Define Objectives: Clearly state the aims, purposes, and boundaries of the office of sustainable construction management. This could entail lessening the influence on the environment, fostering social responsibility, increasing resource efficiency, and encouraging innovation.
c. Team Formation: Assemble a diverse team with members that are experts in engineering, architecture, sustainability, construction management, and other related subjects. The group should consist of committed experts in charge of organising, carrying out, and overseeing sustainability projects.
d. Develop Policies and Procedures: Provide thorough regulations, guidelines, and processes that control the use of sustainable building techniques. Throughout the project lifecycle, these documents should specify the standards, procedures, and best practices that must be adhered to them.
e. Integration into Project Planning: From the launching of the projects, incorporate sustainability concerns into the planning and decision-making stages of the project. Make sure that sustainability objectives are integrated into the phases of design, procurement, construction, and operation, and that they are in line with project objectives.
f. Stakeholder Engagement: Including clients, members of the project team, suppliers, subcontractors, law enforcement, and local communities in sustainability activities is crucial. To accomplish common sustainability goals, promote cooperation, communication, and interest alignment.
g. Capacity Building: In order to improve the skills and understanding of project team members in sustainable construction methods, provide training, education, certification, accreditation, and capacity-building efforts. Provide training sessions, conferences, and certifications to advance knowledge and proficiency in sustainability.

h. Performance Monitoring and Reporting: Provide mechanisms for tracking, evaluating, and reporting sustainability performance indicators. Use technology and tools to monitor crucial performance metrics, assess development, and pinpoint areas in need of improvement.
i. Continuous Improvement: Encourage a culture of continuous improvement by asking for input, carrying out evaluations on a regular basis, and putting the knowledge gained from previous projects into practice. Promote experimentation, creativity, and adaptability to new sustainability trends and best practices.
j. Stakeholder Communication: Inform internal and external stakeholders about the sustainable construction management office's goals, accomplishments, and mission. Communicate openly and proactively to establish credibility, trust, and support for sustainability projects.

14.20 Membership Composition in Sustainable Construction Management Office

A sustainable construction management office's membership is usually made up of people who are committed to promoting sustainability objectives in the construction sector and have a variety of backgrounds, specialties, and jobs. The following are some crucial positions and duties that could be part of the membership makeup:

a. Sustainability Manager/Director: Oversees the creation and execution of sustainability plans, policies, and programmes. Leads the sustainable construction management office. In charge of establishing objectives, organising tasks, and guaranteeing compliance with company goals.
b. Project Managers: In charge of incorporating sustainability concerns into the design, implementation, and completion of projects. Achieve sustainability objectives within the project's limitations, spending limit, and timeframe.
c. Environmental Specialists: Offer your knowledge on environmental laws, compliance, and effect evaluation. Create plans to support resource efficiency, cut waste, and lessen environmental impact.
d. Social Responsibility Coordinators: Pay attention to social dimensions of sustainability, such as participation in the community, labour rights, diversity, and inclusion. Strive to advance fair labour practices, improve social fairness, and cultivate a good rapport with stakeholders.
e. Energy Efficiency Experts: Focus on energy-efficient building design, renewable energy systems, and energy management. Create plans for cutting carbon emissions, increasing energy efficiency, and integrating green building technologies.
f. Green Building Consultants: Offer technical assistance for certification programmes for green buildings. Help with the planning, development, and certification of sustainable buildings.

g. Supply Chain Managers: Oversee the procurement procedure and supplier interactions to guarantee the utilisation of sustainable resources, goods, and services. Encourage ethical procurement methods, supply chain transparency, and responsible sourcing.
h. Community Engagement Specialists: Involve stakeholders, NGOs, and local people in addressing social and environmental issues pertaining to construction projects. Encourage communication, cooperation, and joint ventures to advance sustainable development.
i. Quantity surveyors: They provide accurate cost estimates for incorporating sustainable features into the project, helping stakeholders understand the financial implications of sustainability initiatives. They consider factors such as initial costs, maintenance expenses, energy efficiency, and end-of-life disposal to make informed decisions that minimize life cycle costs while maximizing sustainability benefits. They prepare reports and analysis to communicate project sustainability outcomes to stakeholders, including clients, investors, and regulatory bodies. They advise project teams on sustainability best practices and help navigate regulatory hurdles related to sustainable construction.
j. Data Analysts: Compile, evaluate, and publish key performance indicators and sustainable performance metrics. Using data-driven research, offer insights and suggestions for enhancing sustainability outcomes.
k. Communications and Outreach Coordinators: Create communication tactics, both internal and external, to increase awareness, encourage participation, and highlight sustainable accomplishments. Oversee marketing initiatives, stakeholder communications, and public relations.

The sustainable construction management office may effectively promote positive change, innovation, and continuous improvement in the construction sector by putting together a broad workforce with competence in a range of sustainability-related areas. In construction projects, team members must work together and coordinate well to get sustainable results.

14.21 Registrations for Professional Sustainable Construction Management Organisation

Joining a reputable organisation for sustainable construction management has many advantages, such as resource access, networking possibilities, and industry recognition. The company can showcase its dedication to excellence, expertise, and ongoing education in the field of sustainable construction by registering with a reputable organisation for sustainable construction management. Below is the standard registration procedure:

a. Research: Investigate reputable associations for sustainable building management first. Seek out companies that share your interests, skills, and professional objectives.

b. Membership Criteria: Check if your company meet the qualifications for membership in each organisation by reviewing their membership criteria. Qualifications could include training, work history, professional credentials, and moral behaviour.
c. Application Process: Once you have located organisations that fit your needs, submit your application as required. Usually, this entails filling out an online application and sending in any necessary files, such a CV, certification of qualifications, and references from the professional bodies' ad individual.
d. Membership Fees: Be aware of any costs related to joining the organisation as a member. The cost of membership may differ based on the size of the organisation, the services provided, and the type of membership (e.g., corporate, professional, or student).
e. Benefits: Consider about the advantages of membership, like job boards, conferences, publications, networking events, professional development opportunities, and educational materials. Consider how these advantages correspond with your professional goals.
f. Engagement: After joining, take an active part in the organisation by volunteering for leadership positions, attending events, joining committees or special interest groups, and contributing to publications. Making connections with other members through networking can yield insightful information, cooperative opportunities, and career support.
g. Continued Professional Development: Participate in the organization's continuing education programmes, webinars, workshops, and conferences to stay up to date on industry trends, best practices, and emerging technology.
h. Maintain Membership: Renewing your dues each year and continuing to participate in the organization's events will keep your membership active. Maintaining your involvement can help you grow professionally, build a larger network, and promote sustainable construction management techniques.

14.22 Sustainable Construction Management Organisation

Registering associations for sustainable construction management is a great way to get recognised in the field, acquire resources, and network with other professionals. Professionals in sustainable construction management can network with colleagues, stay up to date on industry developments, and show their dedication to both professional and environmental development by enrolling with these professional associations. Some noteworthy organisations and the advantages of registering are listed below:

a. U.S. Green Building Council (USGBC): Membership provides access to LEED (Leadership in Energy and Environmental Design) resources, training programs, and networking events. Members of USGBC get access to lobbying opportunities, the latest information on green building trends, and a platform to demonstrate their dedication to sustainability.

b. Green Building Councils (GBCs) Worldwide: GBCs operate in many countries, and promoting green building practices and providing support to professionals in the construction sector all over the world. Membership of the GBC provides opportunities for education, collaboration, and recognition through green building certification programmes.
c. International Living Future Institute (ILFI): Membership of ILFI offers membership to professionals committed to creating regenerative and sustainable built environments. Members gain access to resources, events, and the Living Building Challenge, a rigorous green building certification programme.
d. Building Research Establishment (BRE): BRE membership provides access to research, standards, and certification schemes related to sustainable construction. Members can contribute to sector knowledge, collaborate on research projects, and access training and development opportunities.
e. The Dubai Green Building Evaluation System (DGBES) is a comprehensive framework established by the Dubai Municipality to promote sustainable building practices in the United Arab Emirates. The system aims to enhance the environmental performance and energy efficiency of buildings, reduce resource consumption, and mitigate the environmental impact of urban development.
f. BEAM Plus is a green building assessment scheme developed by the Hong Kong Green Building Council (HKGBC) to promote sustainable building practices in Hong Kong. The HKGBC offers a thorough framework for assessing the sustainability and environmental performance of buildings in a range of contexts, including institutional, commercial, residential, and industrial.
g. Green Mark is developed by the Building and Construction Authority of Singapore to promote environmental sustainability in the built environment. It evaluates buildings based on their environmental impact and performance across various categories, encouraging the adoption of sustainable design, construction, and operation practices.
h. DGNB (Deutsche Gesellschaft für Nachhaltiges Bauen_German Sustainable Building Council). DGNB Certification plays a significant role in promoting sustainable architecture and construction practices in Germany by providing a robust framework for assessing and certifying the environmental performance of buildings. DGNB supports Germany's endeavours to accomplish its sustainability objectives and shift towards a constructed environment that is more resource-efficient and ecologically friendly.
i. The Green Star rating system is a comprehensive framework for assessing and certifying the sustainability performance of buildings and communities in Australia. The Australian construction industry benefits greatly from Green Star accreditation, which sets standards for best practices, fosters innovation, and accelerates positive social and environmental change. It aids in the transition of Australia to a built environment that is more sustainable and helps international initiatives to combat resource depletion and climate change.
j. Haute Qualité Environnementale (HQE)/ High Environmental Quality serves as a valuable tool for promoting sustainable building practices in France and beyond, fostering innovation, and contributing to the creation of healthier, more

environmentally responsible built environments. HQE is in favour of France's dedication to sustainability and its shift to a resource-efficient, low-carbon economy.

k. Comprehensive Assessment System for Built Environment Efficiency (CASBEE) serves as a valuable tool for promoting sustainability in the built environment, supporting Japan's commitment to environmental conservation, energy efficiency, and quality of life in urban areas. In order to promote a more resilient and sustainable built environment, CASBEE offers a framework for evaluating and enhancing the environmental performance of structures and urban developments.
l. GRIHA plays a crucial part in advancing sustainability in the Indian built environment by setting standards, fostering innovation, and driving the adoption of green building practices across the country. It is an effective instrument for accomplishing India's objectives for sustainable development and for tackling environmental issues in the construction sector.
m. BREEAM SE is a sustainability assessment method and certification scheme designed specifically for buildings and developments in the Southeast Asian region. BREEAM SE is recognized globally as a leading sustainability assessment method and certification scheme. It is widely used by developers, investors, governments, and industry stakeholders to assess and benchmark the sustainability of buildings and developments in Southeast Asia. BREEAM SE-certified projects contribute to regional sustainability goals and showcase leadership in green building practices.
n. Miljöbyggnad (Environmental Building) is used to assesses buildings based on their environmental impact, energy efficiency, indoor environmental quality, and resource use. Miljöbyggnad offers different certification levels, including Silver, Gold, and Platinum, based on the building's performance.
o. The Chinese Green Building Label (CGBL) or Three Star certification, is a national green building certification system developed by the Chinese government to promote sustainable construction practices and reduce the environmental impact of buildings. The CGBL evaluates buildings using standards for indoor environmental quality, site sustainability, innovation, materials selection, energy efficiency, and water conservation.
p. The Green Building Council of Nigeria (GBCN) is dedicated to promoting sustainable building practices and green building certification in Nigeria. GBCN facilitates the adoption of internationally recognized certification systems and promotes their implementation in Nigeria. By means of its endeavours and programmes, GBCN advances sustainability in the built environment and propels the expansion of green building techniques throughout the nation.
q. Guidelines and tools are made available by the New Zealand Energy Efficiency and Conservation Authority (EECA) to encourage energy conservation and efficiency in a range of fields, including the built environment. These recommendations are meant to encourage environmentally friendly behaviour and assist businesses and individuals in lowering their energy usage and environmental

effect. In order to guarantee uniformity and adherence to energy efficiency standards, EECA guidelines are frequently in line with New Zealand's building norms and standards. Building owners, developers, and tenants can make well-informed decisions to optimise energy consumption, save operating costs, and create more sustainable and healthy built environments by adhering to EECA rules.

r. BOMA BEST is a national environmental certification programme administered by the Building Owners and Managers Association Canada. It assesses the environmental performance of existing buildings based on energy efficiency, waste management, indoor air quality, and other sustainability criteria. ENERGY STAR for Buildings: The ENERGY STAR programme, administered by Natural Resources Canada (NRCan), helps building owners and operators improve energy efficiency and reduce greenhouse gas emissions. ENERGY STAR certification is available for commercial and institutional buildings that meet specific energy performance requirements.

14.23 Key Deliverables for Sustainable Construction Management

Sustainable construction management deliverables are distinct because they prioritize integrating sustainability principles directly into the management processes that guide a construction project from inception to completion. Unlike conventional deliverables, which often focus primarily on time, cost, and quality, sustainable construction management deliverables are designed to align with environmental, social, and economic sustainability objectives at every stage of the project. These deliverables represent specific outputs that ensure the project is managed in a way that reduces environmental impact, enhances resource efficiency, and fosters social responsibility. This includes not only adhering to environmental compliance practices but also setting frameworks that incorporate sustainable resource management, waste reduction strategies, energy-efficient practices, and compliance with sustainability regulations and certifications. Sustainable construction management deliverables also address ongoing performance metrics, stakeholder engagement, and training, equipping the project team with the knowledge and processes needed to sustain these practices throughout the project lifecycle. Ultimately, these deliverables serve as guiding tools that embed sustainability into the core of construction management, ensuring that the construction process itself actively contributes to sustainability goals and standards, not just the final outcome. Following are main deliverables specific to sustainable construction management that ensure effective sustainability-focused management and project oversight:

Project Charter: The project charter outlines the purpose, objectives, and scope of the sustainable construction project, with a focus on aligning all stakeholders on sustainability goals. This foundational document includes sustainability-focused project milestones, key resources, budgetary guidelines, and primary stakeholders.

It serves as an authorization for the project and sets the groundwork for sustainable practices across all phases.

Business Case Documentation: This document provides a comprehensive justification for the project from a sustainability and financial perspective, analyzing the long-term benefits of sustainable construction management. It details expected returns on sustainable investments, such as energy savings, improved occupant health, and regulatory compliance. By highlighting the economic and environmental impacts, the business case demonstrates the value of prioritizing sustainability to secure stakeholder commitment.

Sustainable Project Planning Documentation: Development of comprehensive planning documents that integrate sustainable objectives, specifying the use of sustainable materials, methods, and timelines for each phase of the project.

Resource and Waste Management Plans: Detailed plans for efficient resource usage, including waste reduction, recycling, and reuse strategies throughout the project lifecycle. This includes documentation for reducing energy and water consumption on-site.

Risk and Compliance Reports: Ongoing reports assessing the project's adherence to sustainable practices, environmental regulations, and health and safety standards. These reports identify risks related to sustainability and propose mitigation strategies.

Stakeholder Communication and Engagement Plans: Deliverables that establish protocols for engaging with stakeholders, including local communities, environmental agencies, and contractors, to ensure that the project maintains social and environmental responsibility.

Sustainability Performance Metrics and Monitoring Systems: Implementation of systems to monitor, track, and report on sustainability metrics such as carbon emissions, energy use, and waste reduction. These deliverables provide data to assess the project's sustainability progress.

Training Programs and Guidelines: Deliverables include training sessions for workers on sustainable practices, along with handbooks or guides on following sustainable construction protocols, safety measures, and compliance.

Lifecycle Cost Analysis (LCA) Reports: Detailed reports analyzing the total environmental and economic costs associated with materials and energy consumption throughout the project's lifecycle. This includes long-term sustainability costs and benefits, helping make financially sound yet sustainable decisions.

Green Procurement and Supplier Audits: Documentation of sustainable procurement practices and periodic audits of suppliers to ensure materials and services meet sustainability standards. This includes prequalification requirements for green suppliers and detailed assessments of their sustainability practices.

Adaptive Reuse and Circular Economy Plans: Deliverables that outline methods for reusing existing structures, materials, or equipment and strategies for applying circular economy principles to reduce waste and promote material lifecycle extensions.

Environmental Impact Assessments (EIA): Comprehensive reports that evaluate potential environmental impacts of the project, outlining mitigation strategies to minimize harm to ecosystems and surrounding communities. These assessments are crucial for regulatory compliance and sustainable management.

Energy and Water Efficiency Audits: Conducted at project milestones, these audits measure energy and water use efficiency, ensuring alignment with sustainability goals. Reports from these audits inform adjustments to reduce consumption and improve conservation measures.

Social and Community Impact Reports: Deliverables that assess how the project impacts local communities, considering social equity, accessibility, and potential improvements to local quality of life. This includes periodic updates to ensure ongoing community engagement and address emerging concerns.

Post-Construction Sustainability Review: A comprehensive review conducted after project completion to assess sustainability outcomes and identify lessons learned. This final analysis provides insights for future projects, allowing continuous improvement in sustainable construction management practices.

Health and Wellness Assessments: Reports assessing indoor air quality, lighting, noise levels, and other factors impacting occupant health and well-being. These assessments ensure that sustainability efforts also support human health in constructed spaces.

Resilience and Climate Adaptation Plans: Documents that outline strategies for making structures resilient to climate change impacts, such as extreme weather events. These plans detail adaptive measures in construction and materials that can handle future climate risks effectively.

Final Sustainability Compliance and Certification Documentation: Documentation required to achieve certifications for sustainable practices (e.g., LEED or BREEAM), ensuring that the project's management processes meet recognized standards in sustainability.

Bibliography

1. Bakhtiar KA, Shen LY, Misnan SHB (2008) A framework for comparison study on the major methods in promoting sustainable construction practice. Jurnal Alam Bina 12(3):55–69
2. Chinowsky P, Songer A (eds) (2011) Organization management in construction. Spon Press, p. 204

3. Clough RH, Sears GA, Sears SK, Segner RO, Rounds JL (2015) Construction contracting: a practical guide to company management. Wiley
4. Frein J (2012) Handbook of construction management and organization. Springer Science & Business Media
5. Gunhan S (2019) Analyzing sustainable building construction project delivery practices: builders' perspective. Pract Period Struct Des Constr 24(1):05018003
6. Hardin B, McCool D (2015) BIM and construction management: proven tools, methods, and workflows. Wiley
7. Harris F, McCaffer R, Baldwin A, Edum-Fotwe F (2021) Modern construction management. Wiley
8. Hwang BG, Leong LP, Huh YK (2013) Sustainable green construction management: schedule performance and improvement. Technol Econ Dev Econ 19(sup1):S43–S57
9. London KA, Kenley R (2001) An industrial organization economic supply chain approach for the construction industry: a review. Constr Manag Econ 19(8):777–788
10. Matar MM, Georgy ME, Ibrahim ME (2008) Sustainable construction management: introduction of the operational context space (OCS). Constr Manag Econ 26(3):261–275
11. McGeorge D, Zou PX (2012) Construction management: new directions. Wiley
12. Osman WNB, Udin ZM, Salleh D (2012) Adoption level of sustainable construction practices: a study on Malaysia's construction stakeholders. J Southeast Asian Res 2012:1
13. Paranagamage P, Carrillo P, Ruikar K, Fuller P (2012) Lessons learned practices in the UK construction sector: current practice and proposed improvements. Eng Proj Organ J 2(4):216–230
14. Pesotskaya E, Selyutina L, Egorova L (2019) Actual aspects of modeling method application in organization of construction management. In: IOP conference series: materials science and engineering. IOP Publishing, vol 687, no 4, p 044005
15. Ruparathna R, Hewage K (2015) Review of contemporary construction procurement practices. J Manag Eng 31(3):04014038
16. Sears SK, Sears GA, Clough RH (2010) Construction project management: a practical guide to field construction management. Wiley
17. Serpell A, Ferrada X, Rubio L, Arauzo S (2015) Evaluating risk management practices in construction organizations. Procedia Soc Behav Sci 194:201–210
18. Seeley IH (1997) Quantity surveying practice. Macmillan, London
19. Shelbourn MA, Bouchlaghem DM, Anumba CJ, Carrillo PM, Khalfan MM, Glass J (2006) Managing knowledge in the context of sustainable construction. J Inf Technol Constr 11:57–71
20. Shi L, Ye K, Lu W, Hu X (2014) Improving the competence of construction management consultants to underpin sustainable construction in China. Habitat Int 41:236–242
21. Shirazi B, Langford DA, Rowlinson SM (1996) Organizational structures in the construction industry. Constr Manag Econ 14(3):199–212
22. Shurrab J, Hussain M, Khan M (2019) Green and sustainable practices in the construction industry: a confirmatory factor analysis approach. Eng Constr Archit Manag 26(6):1063–1086
23. Tabassi AA, Roufechaei KM, Ramli M, Bakar AHA, Ismail R, Pakir AHK (2016) Leadership competences of sustainable construction project managers. J Clean Prod 124:339–349
24. Walker A (2015) Project management in construction. Wiley
25. Wang W (2021) The concept of sustainable construction project management in international practice. Environ Dev Sustain 23(11):16358–16380
26. Wu P, Low SP (2010) Project management and green buildings: lessons from the rating systems. J Prof Issues Eng Educ Pract 136(2):64–70

Appendix

Project: XXXXXXXX
Issued for Tender
Sustainability Specifications

Part 1: General

1.1 Introduction

A. This specification has been prepared by XXX Construction Company (Hereinafter called "the YYY Consultant") to guide the specification that relates to Sustainability and YYY, to be included in the tender package of the Project by the Lead Consultant.
B. Project: xxxxx aims to achieve certification at YYY **Certified** level under "YYY. It is the Contractor's responsibility to ensure adherence to this sustainability specification and to obtain the YYY Certified certification at construction completion ultimately.
C. This specification has been prepared at the design stage of the project. During the construction process, changes may occur to the project design, which may require changes to the YYY strategy and targeted credit points. In all instances, the Contractor is required to ensure that the mandatory prerequisites and minimum credits are achieved, such that the project achieves YYY certification at the 'Certified' level.
D. Where discrepancies exist between this specification and other specification documents, the Contractor shall seek clarification from the Client, Lead Consultant and the YYY Consultant on the requirements to be followed. In such cases, the more stringent requirement shall always be followed, provided that the YYY Certified certification is achieved.

A. Olanrewaju, *Sustainable Construction Management: Research and Practice Companion*, Management in the Built Environment,
https://doi.org/10.1007/978-981-96-4910-5

1.2 Related Documents

A. Drawings, Specifications, YYY Action Plan, and general provisions of the Contract, including General and Supplementary Conditions, and other Division 01 Specification Sections, apply to this Section.

1.3 Scope of Work Summary

A. The section includes general and specific requirements and procedures for compliance with required prerequisites and credits needed for the Project to obtain certification at the Certified level based on "YYY's requirements".

   1. It is the responsibility of the YYY Consultant to provide the necessary guidance to achieve YYY certification. However, it is the responsibility of the Contractor to fully cooperate with the YYY Consultant to achieve YYY certification. This includes:
      a. Coordinate with the Client, Design Consultants, the YYY Consultant, and other relevant parties, to fulfill all YYY Design and Construction requirements.
      b. Constructing the building in full compliance with the project targeted credits.
   2. Other Requirements

Requirements for YYY are also included in other Sections. YYY prerequisites and credits required for achieving YYY certification depend on "Material" selections and may not be specifically identified as YYY requirements. Certification requirements needed to obtain YYY prerequisites and credits may be used as one criterion for evaluating substitution and comparable product requests.

1.4 Definitions

A. YYY is a widely used green building rating system in many countries. Available for virtually all building, community, and home project types, YYY provides a framework to create healthy, highly efficient, and cost-saving green buildings.

1.5 YYY Coordinator

The Contractor shall appoint a YYY Coordinator before the commencement of works, who shall be an integral and full-time member of the contractor's team. He/She shall be an integral part of the contractor's QA/QC process and shall work on a full-time basis to ensure that all YYY requirements are achieved, from the start of the work through YYY certification. He/She shall monitor the implementation of YYY, track YYY progress, collate YYY documents, develop YYY monthly reports, and liaise with the Client, Design Consultants, and YYY Consultant on YYY matters.

This entity shall be a qualified construction professional with a minimum of 5 years' experience in the construction industry and preferably with previous experience in green building construction.

He/She shall be the primary and single point of contact for all communications with the Client, Design Consultants, and YYY Consultant on the construction execution and submissions related to sustainability.

The YYY coordinator may also serve as the qualified person responsible for implementing the Construction Activity Pollution Prevention Plan, Construction Indoor Air Quality Management Plan, Construction and Demolition Waste Management Plan, and similar YYY-related tasks on this project.

The YYY Coordinator shall be interviewed and approved by the YYY Consultant before his/her appointment on the project.

Also, the YYY Coordinator shall be fluent in English.

1.6 Third-Party Consultant(S) to Support the Contractor

The contractor may appoint a third-party consultant(s) to support the YYY scope of work, though it is not mandatory.

However, if the Contractor decides to appoint such a consultant, the same party should be approved by the YYY Consultant.

1.7 YYY Meetings/Workshops

A. Preconstruction YYY Kick-off Meeting: After award of contract and before the commencement of work, the Contractor shall organize and direct a Preconstruction Meeting/YYY Kickoff Meeting with the participation of the Client, the Lead Consultants, YYY Consultant, Commissioning Authority, and all the other required parties, to review the YYY requirements necessary for YYY Certified certification and detail out an Action Plan for YYY certification. Participation of the senior members of the construction team is mandatory for this meeting. This meeting can be a teleconferencing meeting.
B. YYY Construction Workshop: Before the commencement of work on-site and once the Contractor's project team is fully mobilized on-site, the Contractor shall organize and direct a YYY Construction Workshop. This workshop shall be held in xxxxxxx and will be participated by the YYY Consultant in person, who will educate the contractor's team on the YYY requirements and certification process. The workshop venue & all the other necessary facilities including audio-visual requirements shall be provided by the Contractor.
C. Progress Meetings: The Contractor shall organize and direct these meetings with the Client, Design Consultants, and YYY Consultant, to discuss the progress of YYY. These meetings shall be held monthly, at a minimum. The YYY Consultant will join these meetings via teleconferencing.

1.8 Administrative Requirements

A. Respond to questions and requests from Design Consultants, the YYY Consultant, and the USGBC-GBCI regarding YYY credits that are the responsibility of the Contractor, that depend on product selection or product qualities, or that depend on the Contractor's procedures until the USGBC-GBCI has made its determination on the Project's YYY certification application. Document responses as informational submittals.

B. Provide necessary documentation to USGBC-GBCI submissions and assist YYY Consultant in responding to questions and requests from USGBC-GBCI regarding YYY prerequisites and credits that are the responsibility of the Contractor, that depend on calculations, analysis, product selection or product qualities, or that depend on Contractor's procedures until the USGBC-GBCI has made its determination on the Project's YYY certification application. Document responses and actions are taken as informational submittals.

1.9 Action Submittals

The Contractor shall perform the following to achieve YYY certification at the Certified level.

A. General: Submit additional sustainable design submittals required by other Specification Sections.
B. YYY Calculations

Construction and Demolition Waste Calculation, demonstrating compliance with the "Materials and Resources Prerequisite—Construction and demolition waste management planning" requirements.

A. YYY submittals are in addition to other submittals.
   1. If the submitted item is identical to that submitted to comply with other requirements, include an additional copy with other submittals as a record copy of compliance with indicated YYY requirements instead of separate sustainable design submittal. Mark additional copy "Sustainable design submittal."
   2. It is essential that "Materials" comply with all the Work requirements, and not just the YYY requirements.
B. YYY Documentation Submittals:
   1. Documentation for Signage requirements
      a. Signage communicating designated preferred parking spaces for carpooling as per LT Credit: Reduced Parking Footprint. Provide photos after installation.
      b. Signage communicating designated preferred parking spaces for green vehicles as per LT Credit: Green Vehicles. Provide photos after installation.
      c. Signage communicating reserved parking for plug-in electric vehicles as per LT Credit: Green Vehicles. Provide photos after installation.
      d. Signage communicating the project 'no smoking policy' as per EQ Prerequisite: Environmental Tobacco Smoke Control. Provide photos after installation.
   2. Product data highlighting charge capability, compliance with IEC 62196 or relevant regional or local standard, and internet addressability or network capability for the Electrical Vehicle Supply Equipment (EVSE), as described in LT Credit: Green Vehicles.

3. Documentation for plant material species used for landscaping, demonstrating native/adapted plant attributes.
4. Product data for roof materials for all buildings targeting YYY certification shall have a three-year aged SRI of at least 64 OR if not available, have an initial SRI of at least 82.
5. Product data for hardscape paving materials that have a three-year SR value of at least 0.28 OR, if three-year aged data is not available, have an initial SR of at least 0.33 at installation.
6. Product data for exterior lighting fixtures complying with YYY Backlight, Uplight, and Glare (BUG) requirements as per SS Credit: Light Pollution Reduction.
7. Product data for pressure compensating drip irrigation system.
8. Product data for water fixtures and fittings, highlighting WaterSense labelled, including the flush and flow rates that meet the efficiency requirement used in WE Prerequisite and WE Credit: Indoor Water Use Reduction.
9. Product data for water meters/submeters that comply with the requirements described in WE Prerequisite and WE Credit: Water Metering, including as-built layout drawing indicating the type, and locations of all metering. Photographic evidence confirming the installation of specified meters and BMS system.
10. Product data for building envelope, electrical systems, process loads, mechanical and plumbing systems that meet the efficiency requirement by EA Prerequisite and EA Credit: Energy Performance. This will include demonstrating compliance with the targeted U-values before procurement of the envelop components, through the provision of test certificates and updated area-weighted average U-value calculations and drawings.
11. Product data for energy meters/sub-meters that comply with the advanced metering characteristics described in EA Credit: Advanced Energy Metering, including as-built single-line diagram drawings indicating the type and location of all metering. Photographic evidence confirming the installation of specified meters and BMS system.
12. Product data of refrigerants that comply with the requirements described in EA Prerequisite: Fundamental Refrigerant Management.
13. Product data for occupancy sensors and daylight sensors.
14. Product data for monitoring devices (outdoor airflow monitoring devices, airflow switches or similar monitors) that comply with the YYY requirements described in EQ Prerequisite: Minimum Indoor Air Quality Performance.
15. Product data of filters with a minimum efficiency reporting value (MERV) of 13 or higher OR Class F7 or higher, to demonstrate compliance with filtration media requirements described in EQ Credit: Enhanced Indoor Air Quality Strategies.
16. Product data of CO_2 sensors as per EQ Credit: Enhanced Indoor Air Quality Strategies.
17. Product data of task lights with multi-level controls that comply with the YYY requirements described in EQ Credit: Interior Lighting.
18. Construction Indoor-Air-Quality Management Plan (CIAQMP):

a. Indoor Air Quality (IAQ) Management Plan for the construction and preoccupancy phases of the building. The Contractor must ensure site audits are conducted regularly, a minimum of once a week, to confirm adherence to the plan.
b. Product data for temporary filtration media, if used
c. Product data for filtration media used during occupancy.
d. Construction Documentation: Date-stamped photographs taken every week during the construction period, along with a brief description of the SMACNA approach employed, documenting the implementation of the IAQ management measures, such as protection of ducts and on-site stored or installed absorptive materials.
e. A draft Construction IAQ Management Plan has been provided as part of this specification document. The Contractor shall develop the final project-specific Construction IAQ plan and submit it to the YYY Consultant's approval within fourteen (14) working days of the date for the start of work and prior to any groundwork on site.

19. Indoor Air Quality Assessment:

a. A signed statement describing the building air flush-out procedures, including the dates when flush-out was begun and completed, and a statement that filtration media was replaced after flush-out.
b. Product data for filtration media used during flush-out and occupancy.

20. Construction Activity Pollution Prevention

a. Erosion and Sedimentation Control (ESC) Plan that conforms to the erosion and sedimentation requirements of the 2017 U.S. Environmental Protection Agency (EPA) Construction General Permit (CGP)
b. The Contractor must ensure site audits are conducted regularly, a minimum of once a week, to confirm adherence to the plan.
c. Construction Documentation: Date-stamped photos taken weekly, inspection logs or reports, and descriptions of corrective action in response to problems.
d. A draft ESC plan has been provided as part of this specification document. The Contractor shall develop the final project-specific ESC plan and submit it to the YYY Consultant's approval within fourteen (14) working days of the date for the start of work and before any groundwork on site.
e. Also, the ESC plan should be implemented for the whole development, including the Ambassador Residence building, and not limiting to YYY boundary.

21. Construction and Demolition Waste Management

a. Construction and Demolition Waste Management Plan (CDWMP)
b. The Contractor must ensure site audits are conducted regularly, a minimum of once a week, to confirm adherence to the plan.

c. Construction Documentation: Receipts of all the waste that has been generated, landfilled, and sent off-site for recycling—date-stamped photos taken weekly, demonstrating compliance with the plan.
d. A draft CDWMP has been provided as part of this specification document. The Contractor shall develop the final project-specific CDWMP and submit it to the YYY Consultant's approval within fourteen (14) working days of the date for the start of work.

22. Commissioning

a. Commissioning (Cx) process activities for mechanical, electrical, plumbing, and renewable energy systems and assemblies shall be conducted per ASHRAE Guideline 0–2005 and ASHRAE Guideline 1.1–2007 for HVAC&R Systems, as they relate to energy, water, indoor environmental quality, and durability.
b. The contractor is to execute the testing, adjusting, balancing, and various measurements in coordination with the Commissioning Authority (C × A) designated by the Client who shall lead, review, and oversee the completion of the commissioning process.
c. The contractor is to submit As-Built drawings and pre-functional and Functional Checklists for all the commissioned systems, including Technical Data Sheets/Specification, O&M Manuals, Warranty Certificates, and Sequence of Operation.
d. The C × A is to complete the documentation demonstrating compliance with the YYY requirements described in EA Prerequisite and EA Credit: Fundamental/Enhanced Commissioning.

1.2 Informational Submittals

A. YYY Progress Reports: Submit monthly YYY progress report prepared and reviewed by the Project YYY Coordinator and approved by the Contractor's Construction Manager/Project Manager, to document compliance with the project YYY requirements, for the work completed. This pertains to the following:
   1. Construction Activity Pollution Prevention. Demonstrate compliance with the Erosion and Sedimentation Control Plan through Date-stamped photos taken at regular intervals, inspection logs or reports, and descriptions of corrective action in response to problems. Refer to the attached draft ESC Plan.
   2. Construction and Demolition Waste Management Plan: Demonstrate compliance with the plan, including waste progress reports and logs showing the weight of material going to landfills and the weight of material going to recycling. Hauling records and receipts for all diverted and landfilled to be maintained. Refer to the attached draft Construction and Demolition Waste Management Plan.

3. Construction IAQ Management Plan: Demonstrate compliance with the plan with date-stamped photos for each of the IAQ measures implemented for the project. Refer to the attached draft Construction IAQ Management Plan.

B. Status of YYY Submittals for all YYY Credits being pursued by the Project.
C. All YYY related documents shall be submitted along with the monthly progress report at the end of every month. Non-submission of YYY monthly report will be considered as 'Default' on the Contractor's part by the Client during the invoice approval process.
D. All YYY submittals must include copies of the manufacturer's information, HIGHLIGHTED to verify information related to YYY.

1.3 Quality Assurance

A. The Contractor is advised of the stringent Quality Control (QC) requirements required of the USGBC-GBCI YYY certification process. The QC requirements shall be implemented from concept to completion.
B. The Contractor shall submit Method Statements on Quality Assurance (QA) and QC procedure to ensure that all YYY requirements to achieve YYY Certified certification are implemented from the inception of the project within Thirty (30) working days of the date of the award of the Contract.
C. The YYY Submission Documents (Drawings, Plans, Records, and Datasheets pertinent to the applicable credits and prerequisites) are to be furnished.
D. The detailed Management and Supervision of the Work throughout the Contract shall be by the Contractor.
E. The Contractor will be responsible for appointing a YYY Coordinator during the contract execution stage as described in Section 1.5 of this document.

Part 2 Execution

2.1 Nonsmoking Building

A. Smoking is not permitted within the building or 25 ft (8 m) of entrances, operable windows, or outdoor air intakes. This applies during the construction phase, on-site as well.

2.2 Construction Activity Pollution Prevention

A. Develop & implement Erosion and Sedimentation Control Plan
B. Demonstrate compliance through Date-stamped photos taken weekly, inspection logs or reports, and descriptions of corrective action in response to problems.

2.3 Commissioning

A. Perform Commissioning (Cx) process activities per ASHRAE Guideline 0-2005 and ASHRAE Guideline 1.1–2007 for HVAC&R Systems, as they relate to energy, water, indoor environmental quality, and durability.
B. Execute the testing, adjusting, balancing, and various measurements along with the Commissioning Authority (CxA) designated by the Client who shall lead, review, and oversee the completion of the commissioning process.
C. The CxA is to complete the documentation demonstrating compliance with the YYY requirements described in EA Prerequisite and EA Credit: Fundamental/Enhanced Commissioning

2.4 Construction and Demolition Waste Management

A. Develop & Implement Construction and Demolition Waste Management Plan
B. Demonstrate compliance with the plan through date-stamped photos, including waste progress reports and logs showing the weight of material going to landfills and weight of material going to recycling. Hauling records and receipts for all diverted and landfilled to be maintained.

2.5 Construction IAQ Management

A. Comply with SMACNA's "SMACNA IAQ Guideline for Occupied Buildings under Construction."
   1. Develop and implement Construction IAQ Management Plan.
   2. If the Owner authorizes the use of permanent heating, cooling, and ventilating systems during the construction period, install MERV 8 filter media at each return-air inlet for the air-handling system used during construction.
   3. Replace air filters immediately with MERV 13 before occupancy.
   4. Must ensure that all adsorptive materials stored on-site and/or installed are protected from moisture damage.
   5. Demonstrate compliance with the IAQ Management Plan with date-stamped photos taken weekly for each of the IAQ measures implemented for the project.

2.6 IAQ Assessment

A. Flush-Out:

After construction ends, before occupancy and with all interior finishes installed, perform a building flush-out by supplying a total volume of 14,000 cu. ft. (4,300,000 L) of outdoor air per sq. ft. (sq.m) of floor area while maintaining an internal temperature of at least 60 °F (16 °C) and relative humidity no higher than 60%.

a. Contractor to define and submit operating requirements.

–End–

The manufacturer's authorised representative in the EU is Springer Nature Customer Service Centre GmbH, Europaplatz 3, 69115 Heidelberg, Germany. If you have any concerns regarding our products, please contact ProductSafety@springernature.com

Printed and bound by CPI Group (UK) Ltd, Croydon, CR0 4YY
01/10/2026
02225930-0017